Mastering Differential Equations: The Visual Method

Robert L. Devaney, Ph.D.

PUBLISHED BY:

THE GREAT COURSES
Corporate Headquarters
4840 Westfields Boulevard, Suite 500
Chantilly, Virginia 20151-2299
Phone: 1-800-832-2412
Fax: 703-378-3819
www.thegreatcourses.com

Copyright © The Teaching Company, 2011

Printed in the United States of America

This book is in copyright. All rights reserved.

Without limiting the rights under copyright reserved above,
no part of this publication may be reproduced, stored in
or introduced into a retrieval system, or transmitted,
in any form, or by any means
(electronic, mechanical, photocopying, recording, or otherwise),
without the prior written permission of
The Teaching Company.

Robert L. Devaney, Ph.D.
Professor of Mathematics
Boston University

Professor Robert L. Devaney is Professor of Mathematics at Boston University. He received his undergraduate degree from the College of the Holy Cross in 1969 and his Ph.D. from the University of California, Berkeley, in 1973. He taught at Northwestern University and Tufts University before moving to Boston University in 1980.

Professor Devaney's main area of research is dynamical systems, primarily complex analytic dynamics, but also more general ideas about chaos in dynamical systems. Lately, he has become intrigued with the incredibly rich topological aspects of dynamics, including such things as indecomposable continua, Sierpinski curves, and Cantor bouquets.

The author of more than 100 research papers and a dozen pedagogical papers in the field of dynamical systems, Professor Devaney is also the coauthor or editor of 13 books. These include *An Introduction to Chaotic Dynamical Systems*, a text for advanced undergraduate and graduate students in mathematics and researchers in other fields; *A First Course in Chaotic Dynamical Systems*, written for undergraduate college students who have taken calculus; and the series of 4 books collectively called *A Tool Kit of Dynamics Activities*, aimed at high school students and teachers as well as college faculty teaching introductory dynamics courses to science and nonscience majors. For the last 18 years, Professor Devaney has been the principal organizer and speaker at the Boston University Math Field Days. These events bring more than 1000 high school students and their teachers from all around New England to the campus for a day of activities aimed at acquainting them with what's new and exciting in mathematics.

Professor Devaney has delivered more than 1400 invited lectures on dynamical systems and related topics in every state in the United States and in more than 30 countries on 6 continents. He has designed lectures for research-level audiences, undergraduate students and faculty, high school students and faculty, and the general public. He has also been the chaos consultant for several theaters' presentations of Tom Stoppard's play *Arcadia*. In 2007 he was the mathematical consultant for the movie *21*, starring Kevin Spacey.

Since 1989 Professor Devaney has been director of the National Science Foundation's Dynamical Systems and Technology Project. The goal of this project is to show students and teachers how ideas from modern mathematics such as chaos, fractals, and dynamics, together with modern technology, can be used effectively in the high school and college curriculum. As part of this project, Professor Devaney and his students and colleagues have developed numerous computer programs for exploring dynamical systems. Another long-standing National Science Foundation project has been an attempt to revitalize the study of ordinary differential equations by thoroughly incorporating material from dynamical systems theory and taking a more visual approach. Professor Devaney's coathored *Differential Equations* textbook resulting from that project is now in its fourth edition. He has also produced the Mandelbrot Set Explorer, an online, interactive series of explorations designed to teach students at all levels about the mathematics behind the images known as the Mandelbrot and Julia sets.

In 1994 Professor Devaney received the Award for Distinguished University Teaching from the Northeastern Section of the Mathematical Association of America. In 1995 he was the recipient of the Deborah and Franklin Tepper Haimo Award for Distinguished University Teaching. Professor Devaney received the Boston University Scholar/Teacher of the Year Award in 1996. In 2002 he received a National Science Foundation Director's Award for Distinguished Teaching Scholars as well as the International Conference on Technology in Collegiate Mathematics Award for Excellence and Innovation with the Use of Technology in Collegiate Mathematics. He received Boston University's Metcalf Award for Teaching Excellence in 2003, and in 2004 he was named the Carnegie/CASE Massachusetts Professor of the Year. In 2005 he received the Trevor Evans Award from the Mathematical Association of America for an article entitled "Chaos Rules," published in *Math Horizons*. In 2009 Professor Devaney was inducted into the Massachusetts Mathematics Educators Hall of Fame, and in 2010 he was named the Feld Family Professor of Teaching Excellence at Boston University. ∎

Table of Contents

INTRODUCTION

Professor Biography i
Course Scope 1

LECTURE GUIDES

LECTURE 1
What Is a Differential Equation? 4

LECTURE 2
A Limited-Growth Population Model 22

LECTURE 3
Classification of Equilibrium Points 44

LECTURE 4
Bifurcations—Drastic Changes in Solutions 63

LECTURE 5
Methods for Finding Explicit Solutions 82

LECTURE 6
How Computers Solve Differential Equations 101

LECTURE 7
Systems of Equations—A Predator-Prey System 118

LECTURE 8
Second-Order Equations—The Mass-Spring System 139

LECTURE 9
Damped and Undamped Harmonic Oscillators 157

LECTURE 10
Beating Modes and Resonance of Oscillators 174

Table of Contents

LECTURE 11
Linear Systems of Differential Equations 194

LECTURE 12
An Excursion into Linear Algebra 216

LECTURE 13
Visualizing Complex and Zero Eigenvalues 237

LECTURE 14
Summarizing All Possible Linear Solutions 256

LECTURE 15
Nonlinear Systems Viewed Globally—Nullclines 273

LECTURE 16
Nonlinear Systems near Equilibria—Linearization 293

LECTURE 17
Bifurcations in a Competing Species Model 311

LECTURE 18
Limit Cycles and Oscillations in Chemistry 329

LECTURE 19
All Sorts of Nonlinear Pendulums 346

LECTURE 20
Periodic Forcing and How Chaos Occurs 366

LECTURE 21
Understanding Chaos with Iterated Functions 385

LECTURE 22
Periods and Ordering of Iterated Functions 405

LECTURE 23
Chaotic Itineraries in a Space of All Sequences 422

Table of Contents

LECTURE 24
Conquering Chaos—Mandelbrot and Julia Sets 438

SUPPLEMENTAL MATERIAL

Solutions..455
Types of Differential Equations Cited 507
Using a Spreadsheet to Solve Differential Equations....................509
Timeline 516
Glossary 520
Bibliography..528

Mastering Differential Equations: The Visual Method

Scope:

The field of differential equations goes back to the time of Newton and Leibniz, who invented calculus because they realized that many of the laws of nature are governed by what we now call differential equations.

A differential equation is an equation involving velocities or rates of change. More precisely, it's an equation for a missing mathematical expression or expressions in terms of the derivatives (i.e., the rates of change) of these expressions. Differential equations arise in all areas of science, engineering, and even the social sciences. The motion of the planets in astronomy, the growth and decline of various populations in biology, the prediction of weather in meteorology, the back-and-forth swings of a pendulum from physics, and the evolution of chemical reactions are all examples of processes governed by differential equations.

In the old days, when we were confronted with a differential equation, the only technique available to us to solve the equation was to find a specific formula for the solution of the equation. Unfortunately, that can rarely be done—most differential equations have no solutions that can be explicitly written down. So in the past, scientists would often come up with a simpler model for the process they were trying to study—a model that they could then solve explicitly. Of course, this model would not be a completely accurate description of the actual physical process, so the solution would only be valid in a limited setting.

Nowadays, with computers (and, more importantly, computer graphics) readily available, everything has changed. While computers still cannot explicitly solve most differential equations, they can often produce excellent approximations to the exact solution. More importantly, computers can display these solutions in a variety of different ways that allow scientists or engineers to get a good handle on what is happening in the corresponding system.

In this course, we take this more modern approach to understanding differential equations. Yes, we grind out the explicit solution when possible, but we also look at these solutions geometrically, plotting all sorts of different graphs that explain the ultimate behavior of the solutions.

In each of the 4 sections of this course, we begin with a simple model and then use that model to introduce various associated topics and applications. No special knowledge of the field from which the model arises is necessary to understand the material. Two broad subthemes appear over and over again in the course: One is the concept of bifurcation—how the solutions of the equations sometimes change dramatically when certain parameters are tweaked just a little bit. The second subtheme is chaos—we will see how extremely unpredictable behavior of solutions occasionally arises in the setting of what would seem to be a relatively simple system of differential equations.

In the first part of the course, we discuss the simplest types of differential equations—first-order differential equations—using several population models from biology as our examples. The simplest model is the unlimited population growth differential equation. Later examples, such as the logistic population growth model (a limited growth model), take into account the possibility of overcrowding and harvesting. Using elementary techniques from calculus, we solve these equations analytically when possible and plot the corresponding qualitative pictures of solutions, such as the slope field, the phase line, and the bifurcation diagram. We also describe a simple algorithm that the computer uses to generate these solutions and show how this technique can sometimes fail (often because of chaos). Finally, we introduce a parameter into our model and see how very interesting bifurcations arise.

In the second part of the course, we turn our attention to second-order differential equations. Our model here is the mass-spring system from physics. We see relatively straightforward behavior of solutions as long as the mass-spring system is not forced, but when we introduce periodic forcing into the picture, the much more complicated (and sometimes disastrous) behaviors known as beating modes and resonance occur.

The next part of the course deals with systems of differential equations. Our model here is the predator-prey model from ecology. At first we concentrate on linear systems.

We see how various ideas from linear algebra allow us to solve and analyze these types of differential equations. Bifurcations recur as we investigate the trace determinant plane. We then move on to a topic of considerable interest nowadays, nonlinear systems. Almost all models that arise in nature are nonlinear, but these are the differential equations that can rarely be solved explicitly. We investigate several different models here, including competitive systems from biology and oscillating chemical reactions. Another model will be the Lorenz system from meteorology; back in the 1960s, this was the first example of a system of differential equations that was shown to exhibit chaotic behavior.

In the final part of the course, we turn to the concept of iterated functions (also called difference equations) to investigate the chaos we observed in the Lorenz system. Our model here is an iterated function for the logistic population model from biology, a very different kind of model than our earlier logistic differential equation. In this case, we see that lots of chaos emerges, even in the simple iterated function, and we begin to understand how we can analyze and comprehend this chaotic behavior.

While calculus is a central notion in differential equations, we will not delve into many of the specialized techniques from calculus that can be used to solve certain differential equations. The only topics from calculus that we presume familiarity with are the derivative, the integral, and the notion of a vector field. Any other relevant concepts from calculus or linear algebra are introduced before being used in the course.

By the end of this course, you will come to see how all the concepts from algebra, trigonometry, and calculus come together to provide a beautiful and comprehensive tool for investigating systems that arise in all areas of science and engineering. You will also see how the field of differential equations is an area of mathematics that, unlike algebra, trigonometry, and calculus, is still developing, with many new and exciting ideas sparking interest across all disciplines.

What Is a Differential Equation?
Lecture 1

An **ordinary differential equation** (ODE) is an equation for a missing function (or functions) in terms of the derivatives of those functions. Recall that the derivative of a function $y(t)$ is denoted either by $y'(t)$ or by dy/dt. Suppose $y(t_0) = y_0$. Then the derivative $y'(t_0)$ gives the slope of the tangent line to the graph of the function $y(t)$ at the point (t_0, y_0).

In the old days, the only tools we had to solve ODEs were analytical methods—a variety of different tricks from calculus that sometimes enabled us to write down an explicit formula for the solution of the differential equation. Unfortunately, most ODEs cannot be solved in this fashion. But times have changed: Now we can use the computer to approximate solutions of ODEs. And we can use computer graphics and other geometric methods to display solutions graphically. So this gives us 2 new ways to solve differential equations, and these are the methods that we will emphasize in this course.

Probably the best-known differential equation (and essentially the first example of a differential equation) is Newton's second law of motion. Drop an object from your rooftop. If y measures the position of the center of mass of the object, then we would like to know its position at time t, that is, $y(t)$. Newton's law tells us that mass times acceleration is equal to the force on the object. So, if m is the mass, then we have $my'' = F(y)$, where F is the force acting on the object when it is in position $y(t)$. So we have a differential equation for $y(t)$.

Another example of a differential equation is the mass-spring system (or harmonic oscillator). Here $y(t)$ is the missing function that measures the position of a mass attached to a spring attached to a ceiling. When we pull the mass down and let it go, $y(t)$ gives us the resulting motion of the mass over time. The function $y(t)$ is determined by the differential equation $y'' + by' + ky = G(t)$.

In this case, we have parameters like b and k as well as a forcing term $G(t)$ that depends only on t. We will see very different behaviors for the mass-spring system depending on these parameters and the forcing term. Sometimes a small change in one of these parameters creates a major change in the behavior of solutions. Such phenomena are called **bifurcations**, one of the subthemes we will encounter often in this course.

Most of our course will be spent considering systems of differential equations. Systems of ODEs involve more than one missing function in the equation. We will consider numerous such examples, but the most famous is undoubtedly the Lorenz system from meteorology, below.

$$x' = -10x + 10y$$
$$y' = -xz + Rx - y$$
$$z' = xy - 8/3\, z$$

The Lorenz system of equations was the first example of a system of ODEs that exhibits chaotic behavior, another subtheme that we will encounter throughout the course.

Let's look at our first differential equation: the unlimited population growth model from biology, which is perhaps the simplest nontrivial differential equation. Suppose we have a species living in isolation (with no predators, no overcrowding, and no emigration) and we want to predict its population as a function of time. Call the population $y(t)$.

Our assumption is that the rate of growth of the population is directly proportional to the current population. This translates to the ODE $y' = ky$. Here k is a constant (a parameter) that depends on which species we are considering. This is an example of a first-order ODE, since the equation depends on at most the first derivative of $y(t)$. Usually we wish to find the solution of an **initial value problem**, that is, a specific solution of the ODE that satisfies $y(0) = y_0$ where y_0 is the given initial population.

For simplicity, suppose $k = 1$, so our differential equation is $y' = y$. One solution is the exponential function $y(t) = e^t$, since the derivative of the function e^t is e^t (so that y' does indeed equal y). Another solution is $y(t) = Ce^t$, where C is some constant. Note that when $C = 0$, the solution is the constant function $y(t)$ identically equal to zero. This is an **equilibrium solution**, one of the most important types of solutions of differential equations. The **general solution** of the equation is $y(t) = Ce^t$, since we can solve any initial value problem using this formula. That is, given $y(0) = y_0$, the solution satisfying this initial condition is given by setting $C = y_0$.

More generally, we can consider the equation $y' = ky$, where k is some constant, say $k = 2$. As before, other solutions are of the form Ce^{2t}, where C is an arbitrary constant. Again we see that the solution with $C = 0$ is an equilibrium solution. We can solve any initial value problem by choosing C as our initial value $y(0)$, so Ce^{2t} is the general solution of this differential equation.

Here are 2 of the simplest methods for visualizing solutions of differential equations.

1. The right-hand side of the differential equation tells us the slope of the solution of the ODE at any time t and population y. So in the t-y plane, we draw a tiny straight line with slope equal to the value on the right-hand side. A collection of such slopes is the slope field below.

Figure 1.1

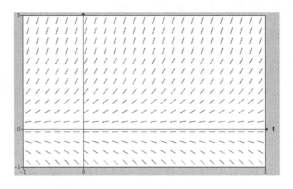

2. Then a solution must be everywhere tangent to the slope field, so we can sketch in the graphs of our solutions.

Figure 1.2

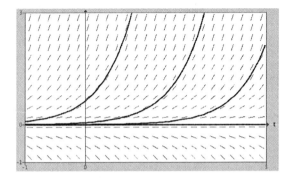

Note the constant solution, $y(t) = 0$. This is our equilibrium solution.

Important Terms

bifurcation: A major change in the behavior of a solution of a differential equation caused by a small change in the equation itself or in the parameters that control the equation. Just tweaking the system a little bit causes a major change in what occurs.

equilibrium solution: A constant solution of a differential equation.

general solution: A collection of solutions of a differential equation from which one can then generate a solution to any given initial condition.

initial value problem: A differential equation with a collection of special values for the missing function such as its initial position or initial velocity.

ordinary differential equation (ODE): A differential equation that depends on the derivatives of the missing functions. If the equation depends on the partial derivatives of the missing functions, then that is a partial differential equation (PDE).

Suggested Reading

Blanchard, Devaney, and Hall, *Differential Equations*, chap 1.1.

Guckenheimer and Holmes, *Nonlinear Oscillations*.

Hirsch, Smale, and Devaney, *Differential Equations*, chap 1.1.

Roberts, *Ordinary Differential Equations*, chap 1.1–1.3.

Strogatz, *Nonlinear Dynamics and Chaos*, chap 1.1.

Relevant Software

Blanchard, Devaney, and Hall, *DE Tools*, HPG Solver.

Problems

1. Let's review some ideas from calculus that we used in this lecture.

   a. Compute the derivative of $y(t) = t^3 + e^t$.

   b. Compute the second derivative of $y(t) = t^3 + e^t$.

   c. Find a function $f(t)$ whose derivative is $t^3 + e^t$.

d. Sketch the graph of the exponential function $y(t) = e^t$.

   e. Find the solution of the equation $e^{(2t)} = 1$.

2. Sketch the slope field and solution graphs for the differential equation $y' = 1$.

3. What are some solutions of the differential equation in problem 2?

4. Repeat problems 2 and 3 for $y' = t$.

5. In the unlimited population growth model $y' = ky$, what happens to solutions if k is negative?

6. Find the general solution of the differential equation $y' = t$.

7. Find all of the equilibrium solutions of the differential equation $y' = y^2 - 1$. Also plot the slope field. What do you think will happen to solutions that are not equilibria?

8. What would be the general solution of the simple differential equation $y' = 0$? What is the behavior of all of these solutions?

9. Consider the differential equation given by Newton's second law, $my'' = g$, where we assume that m and g are constants. Can you find some solutions of this equation?

Exploration

Think about the fact that back in the 1600s, Isaac Newton was able to come up with not only the second law of motion in physics but also the basics of calculus and differential equations (plus everything else he discovered in the sciences). He was quite an intellectual! There is plenty to read about his life and discoveries on the web. For an introduction, go to http://www.newton.ac.uk/newtlife.html.

What Is a Differential Equation?
Lecture 1—Transcript

Hello. Welcome to this course on differential equations. My name is Bob Devaney. I look forward to sharing with you over the next 24 lectures some of the incredibly interesting and exciting things that are happening now in this field of mathematics.

The field of differential equations goes back to the time of Isaac Newton in late 1600s who basically invented calculus because he realized that many of the laws of nature are governed by differential equations. Think force is mass times acceleration. That is a differential equation. Then in the ensuing centuries, differential equations were approached more or less the same way as Newton did, just using the techniques from calculus.

What I hope you'll see in these lectures is that first, the field of math known as differential equations has changed dramatically in the past 20 or 30 years. Secondly, it now has applications in all areas of science and engineering and even beyond. During this course, we'll see how differential equations arise in such diverse areas as physics and biology, chemistry and meteorology. I won't go into a lot of details of these specific applications. You don't have to be an expert in these areas. It's just nice to see how many different areas differential equations come up.

Why have differential equations changed recently? Well, in the old days, the only way you had to understand differential equations was to solve them by hand, to find an explicit formula for a solution. For most differential equations, it turns out that's impossible. You cannot find explicit formulas for the solutions.

So what would people do? Well, one thing they'd do is simplify the differential equation. For example for the differential equations for the pendulum, I remember when I was taking differential equations way back when, the differential equation for a pendulum was simplified so that we can solve it, but then the solution said that the pendulum would only swing back and forth. It could never swing around 360 degrees. We had simplified the differential equation to get an approximate solution that worked in some cases.

Secondly what we'll do is develop very special mathematical tricks to solve differential equations. I remember when I first started teaching differential equations we had I don't know maybe 36 lectures during the semester, and there were 36 tricks. Each lecture was a different trick to solve a different kind of differential equation. By the time we were in lecture 35 I could hardly remember trick number 7.

In any event, nowadays we have easy access to computers. Computers like humans can't solve differential equations. On the other hand, using numerical techniques that we'll talk about later, they can approximate solutions very well. They can get closer and closer to the actual solution most of the time. So this makes this an ideal time to plunge into differential equations. We've got computers to help us out.

Also, not only does a computer approximate the solution, but it displays these solutions graphically. More importantly, we've developed over the years a number of other qualitative or geometric ideas about solutions of differential equations. This means together with computer graphics and these qualitative methods, we'll be taking a very visual approach to differential equations.

You'll see during this course that I often use software to display solutions of differential equations. This software comes from a book that I coauthored with Glen Hall and Paul Blanchard. Many of the different programs in there come from that book. Later on we'll actually use a number of different java applets that are freely available on my Web site that's posted in the course outline. Also, you'll see that one of the greatest tools to understand differential equations is spreadsheets. So we'll use spreadsheets a number of times to understand the behavior of differential equations.

Then finally one thing that's really blossomed in the last 25 years or so is we now know that many differential equations behave chaotically. So in fact, there's no way even to approximate these solutions exactly. Chaos comes up and really changes things. We'll see that toward the end of this course.

Let me begin by first explaining what a differential equation is. Then I'll give you a couple of examples that we'll actually use later on in this course. You'll see then how the visual and computational approach arises.

So what is a differential equation? Well, roughly speaking it's an equation involving the velocities or rates of change. For example I'm traveling at this speed now, and I'm traveling a little faster a little later, and I know when I get on the highway I'm going to be traveling much fast. So my question is when will I get to my destination? Here's another one. The stock market's going up at this rate this week. It went up this rate last year. I think it's going to go up this rate next year. When can I retire? Or I'm throwing a cocktail party and this person is drinking at this rate at my cocktail party. When will he or she fall over?

Those are all equations involving velocities or rates of change and we want to predict something, when I can retire, when I can get to my destination, when the person will fall over. Those are roughly speaking differential equations. More precisely a differential equation is an equation for a missing function or collection of functions in terms of the derivatives of those functions. Remember derivatives in calculus measure, velocity and rate of change.

So right there you see calculus. Almost all the techniques in calculus were developed with an eye toward solving differential equations. Calculus comes up throughout this course. I won't spend time at the beginning reviewing all of the calculus. Rather what I'll do is each time a different idea from calculus comes up, I'll remind you what's happening there.

When I teach this course at the college level, usually the students have all had three semesters of calculus before entering the differential equations class. Three semesters of calculus, wow. Some of the techniques from way back when has really tired them out. But what often happens is students come up to me in the differential equations course and say ah-hah, now I see why all those techniques that I learned in calculus are important. I learned them three semesters ago. I've forgotten them since then, but now they all come back in differential equations.

In fact, a while back one of my students actually came up at the end of the semester and gave me a present. It was a bumper sticker. It's now actually on my car. The bumper sticker says calculus, the agony and dx/dt. Hopefully by the end of this course you too will see that the ecstasy comes up in differential equations.

Actually what you'll mostly see here is dy/dt. Most of our functions here will be a function $y(t)$. The reason is first of all, most differential equations depend on time. You want to see what's happening when time is changing, so time is our independent variable. So then, I plot the function y vertically.

So let me remind you what the derivative is. We call a derivative of $y(t)$ sometimes it's $y'(t)$. Sometimes it's dy/dt. What we do is we have some expression $y(t)$, we plot its graph t horizontally, y vertically. Then we're given some point. Let's say the point t_0. Then $f(t_0)$ we compute $y(t_0)$. Let me call that y_0. What is the derivative of this function at time (t_0)? From calculous, the derivative is just the slope of the tangent line to the graph of that function at the point t_0, y_0.

Another way to think of it is y' or dy/dt is also the instantaneous rate of change of $y(t)$. How $y(t)$ varies, what the rate of change is at that specific instance. Also, we can think of that as the velocity. Similarly in calculus we often do the second derivative. The second derivative measures the concavity of the graph of $y(t)$ whether it is concave down or concave up. Or in terms of time, it's the acceleration y'', the second derivative is the acceleration.

So let me just remind you of a couple of derivatives that'll come up all the time. Suppose you have the function $y(t) = t^2$. Remember its derivative $y'(t) = 2t$. More generally, if you have the function t^n then its derivative is nt^{n-1}. A specific example, take the function $1/t$, that's t^{-1}, so its derivative is $-1t^{-2}$. Or if for example you had $3t^5$ as your function, then the derivative of that function is 3 times the derivative of t^5. That's 3 times 5 times t^4, $15t^4$.

And also, the exponential function. That's going to come up over and over again in this course. The exponential function e^t, what's its derivative. Its derivative is very special, it's $e(t)$ itself. The derivative of the exponential is the exponential.

One of the tools that arises all of the time in differential calculus is the chain rule. The chain rule tells you how to compute the derivative of what we call a composition of two functions. If you first do $g(t)$ and then apply f to it, that's the composition of f and g. F applied to $g(t)$. How do you differentiate $f(g(t))$? The chain rule tells you how to do it. The chain rule says that derivative is first take the derivative of f, but evaluate it at $g(t)$, and then multiply that by the derivative of $g(t)$. So it's $f'(g(t))$ times $g'(t)$.

As an example, take for example the exponential of $4(t)$. We're composing the exponential with function $4(t)$. By the chain rule, the derivative is the exponential of $4(t)$ times the derivative of $4(t)$ which is 4. Another example, let $y(t)$ be the exponential of $t^2 + 1$. Then by the chain rule, the derivative of that function is the derivative of the exponential, that's itself, evaluate it at $t^2 + 1$, so $e(t^2 + 1)$ times the derivative of $t^2 + 1$, that's $2t$.

So, now as I said, let me just give you a couple of examples of differential equations. Most of these will come up much later in the course. I just want to show you how we use the visual method to understand some of them. First though, the most basic differential equation based to the first differential equation, Newton's law of motion, second law of motion actually. That's mass times acceleration equals the force on whatever is moving. Mass times acceleration equals the force. Acceleration is y''. We're looking for a function $y(t)$. We know something involving its second derivative. That's an equation for a missing function in terms of its derivative. That's a differential equation.

So think for example go up on your roof, bring a ball up there, drop the ball. What happens? Well, Newton's law takes over. You let $y(t)$ be the position of the ball, then $m(y'')$ equals whatever force is acting on the ball. For example, obviously the force of gravity is pulling it down. That's one said force. Or there may be a lot of wind, and that's a second force that would make the ball move in different directions. Or you could live in a house under water, and then when you let the ball go, it goes in the opposite direction. A different force is acting on the ball. So that's the first example of a differential equations historically speaking in the sense.

Here's another one. Here's one that'll come up a lot in the middle of the course. It's the mass-spring system. What basically you do is take a spring, attach it to a wall someplace, attach a mass to it, a ball to it, pull the spring down or push it up, let the ball go, and we ask the question, what happens to that mass. Well, the differential equation as we'll see later is y'', y here is the position of the mass as times goes on. y'' plus some constant b times y' plus another constant $ky = 0$. That's an equation for the missing function $y(t)$, the position of the mass in terms of both the first and second derivative of that function.

As a quick aside, you saw some constants in there, b is what we call the damping constant, and k is called the spring constant. Those are parameters. Each one of them gives you a different differential equation, and as we'll see throughout this course, often we'll be changing these parameters and seeing different things happen. When a major different thing happens, what we'll call that is a bifurcation. That'll be one of the subthemes that come up.

So let me turn to the computer and show you exactly what we're going to look at when we look at the mass-spring system. So what we have over here is on the left our mass spring. We see the spring is attached to the ceiling. There's the mass in red. What I'm going to plot in the middle is what's called the phase plane, I'll explain that later on in the course. Then over here what we'll plot is the graph of $y(t)$, the position of our spring.

Notice that well in this region, we've got the y-axis going vertically, and we've got another axis going horizontally. If I let the spring go, you see what happens. You see the motion of the spring as it gradually oscillates down to rest. You see in the phase plane a certain curve that indicates what's happening to that solution. Over here on the right, you see the graph of $y(t)$, the height of that mass-spring system.

Now let me change one of those parameters, the damping constant. Let me let b be large. The damping is large. What happens to the spring? Well, instead of oscillating down to rest, it goes directly to rest. The spring does not oscillate with a large amount of damping. It just goes directly down to its rest position; a change. Something has happened here when we change the damping constant. That's a bifurcation.

Here's another example, let me let the damping constant be 0. What happens to the mass spring? We'll discuss this later. What you see is again the spring changes. It does not go down to its rest position, now it keep running around the circle in the phase plane, and it behaves periodically if you look at the graph of $y(t)$. Again, there's a little different change when we vary a parameter. That's a bifurcation. That's one of the subthemes we'll see in this course.

Here's a third example that will come up toward the end of the course, the famous Lorenz differential equations. You look at those equations, there are three equations now, one for x, one for y, one for z. So we've got solutions $x(t)$, $y(t)$ and $z(t)$. Don't worry about the details now, I'll explain them later. You see a constant in there called R. That's again a parameter much like in the mass-spring system. What this equation is, it's an equation for meteorology. It's the simplest possible meteorological model. It's a model of the weather in a planet that has only one molecule of air. Lorenz, a famous meteorologist/mathematician decided to see if he could understand the simplest possible planet, the weather on a planet with just on molecule of air. These are the differential equations that he came up with. Watch what happens.

Here's the Lorenz model. The differential equation depends on x, y, and z. Over here in the phase space, I'll just plot what's happening for y and z. Again, I'll explain what's going on there. Here's what the Lorenz model actually represents. We've got a planet with one molecule of air that's heated from below and cooled from above. So this molecule tends to run around on what I call convection rolls. Lorenz wanted to know if he could predict what would happen. With this differential equation what he saw was, well when that constant R was relatively low, then we just had solutions running around, our molecule running around in a convection roll. If we went over to this side, the solutions would just sort of glide down to rest in the phase space. Meanwhile over in the planet, you just see the particle rotating maybe clockwise, maybe to counterclockwise motion.

Then the constant R can be changed. What happens if we raise R quite a bit? What Lorenz found was this, now solutions don't tend to those special points in the phase space. Rather sometimes, they run around clockwise, and sometimes they run around counterclockwise. These solutions never do something that is terminal, they just keep going, sometimes around the left, sometimes around the right. This is one of the first equations that exhibited what's known as chaotic behavior. A very simple differential equation in three dimensions illustrates this crazy behavior called chaos, sensitive dependence on initial conditions as we'll call it. Things are going crazy. What's happening here? It's a very difficult question to answer. In any event, this chaos is something that will come up over and over again later in the course. It's one of the things that has really changed the subject of differential equations.

Now let me turn to differential equations. Let me give you an actual example of a differential equation and show you some of the qualitative things that when used, understand it. Here's the simplest possible differential equation. From biology, it's the unlimited population growth model. Let's assume we have some species living in isolation, and there are no predators or anything else to effect the population growth, it just grows.

As with all models that we'll introduce throughout this course, I'll begin by specifying an assumption. I'll give you an assumption specified in English that then will translate into a differential equation. So here's the assumption to the unlimited population growth model. I'm going to assume that the rate of growth of the population is directly proportional to the current population. That's an assumption. The rate of population growth is directly proportional to the current population.

The way I explain that to my students is think of it this way. Take two people, lock them in a room, and come back in a year. How many people will there be there? There'll be 3, assuming certain initial conditions. But now take 200 people and lock them in a room, and come back a year later. How many will there be? There'll be 300, again with certain initial conditions. What you see is 2 in one year goes to 3, but 200 in one year goes to 300. The rate of growth is proportional to the current population.

Now let's translate that to a differential equation. What's the differential equation going to be? Well let me let $y(t)$ be the population at time t. Our assumption is that the rate of growth, the rate of growth is something. Well the rate of growth is just the derivative of $y(t)$, it's $y'(t)$. We're assuming it's directly proportional to something. That means it's equal to a constant times something. We're assuming it's directly proportional to the current population. So the rate of growth is directly proportional to y. So that says that our differential equation is rate of growth y' is directly proportional to equals k times current population, $y(k)$.

This is what we call a first order differential equation because there's only a first derivative there. Secondly, when we introduce differential equations, they usually come with an initial condition, an initial value problem. The initial value will be the population here at time 0. So we're given an initial population, we'll call it say y_0. Our question is can you predict what will happen to the population as time goes on. That differential equation will allow you to do that.

Incidentally, I should mention that this is the same differential equation as you get for the savings model where the interest is compounded continuously. So we have the differential equation $y' = ky$. Let me assume for a moment that k is 1, so we have the differential equation $y' = y$. Can you solve that differential equation? Do you know a function whose derivative equals itself? Of course you do, the exponential. Solve that equation $y(t)$ is e^t is a solution to that differential equation.

Do you know another solution, another function whose derivative equals itself? Yes, how about any constant times the exponential function. The derivative of that is just itself. So we've got a whole bunch of solutions to this differential equation. Let me just notice that there's one very special case here. That's when the constant is equal to 0. Let me say our solution is Ce^t, if C is equal to 0, then our solution is just the constant function 0. Those constant solutions are very special. They're what we call equilibrium solutions. We'll see them over and over again in the future.

Secondly Ce^t is actually the general solution to this differential equation. I call it the general solution because with that family of solutions, depending on C, you can solve any initial value problem. You want to find a solution to the unlimited population growth model that starts out at population 100, just let C be 100. $100e^t$ is a solution, and at time 0, the population is 100. You want to find the solution that starts out at time 0 with a population of 1,000, let C be 1,000. We can solve any initial value problem with that family of solutions.

Now let me turn to the computer and see how we can interpret those actual explicit solutions in terms of graphical qualitative means. So here's what we'll do. Here we have a picture of the plane. The t-axis runs horizontally, so y-axis runs vertically. Our differential equation is $y' = y$. What is that saying to you? Remember y' is the derivative. It's the slope of the tangent line to the graph of $y(t)$ so in fact this differential equation is giving you the slopes of the solutions to the differential equation.

So if you go at any time t and y value, and compute the right hand side of the differential equation, you are just getting the slope of the solution. So as we move our point along the $y(t)$ plane, each point corresponds to a different little slope y. For example at that time there, we could put down, we could calculate that $y' = y$ means our slope looks like around 1. If y goes larger, the right hand side of the differential equation says that the slope is larger; y goes larger still, the slope is larger.

On the other hand, if y is small, the slope is smaller. If $y = 0$, the slope is 0. So the right hand side of this differential equation gives us what we call a slope field. Everywhere in the $y(t)$ plane we can put a bunch of little slope lines. That's dictated just by the formula on the right of the differential equation. So what we do is take a representative collection of points, plot the slope field, and we know that our solutions must be tangent to that slope field. That gives us a first visual approach to what our solutions look like. Just looking at this slope field, we know our solutions must go up hill. Of course we already know that they're exponential, so we know that, but this is saying that graphically.

Here's the graph of one solution. Notice that the graph is everywhere tangent to the slope field. Here's another solution, a third solution, all of them are exponentials. All of them are tangent to the slope field. Way down here at 0, we see our constant solution. What we call the equilibrium solution, y identically equal to 0. You never have negative solutions for population growth models, but for $y' = y$ we don't know whether it's positive or negative.

You could have negative solutions to that general differential equation. They would look like decaying exponentials.

So there are the first two of many pictures that we'll see that involve differential equations. In this course we're going to take a visual approach to differential equations as we go along. Yes, when we have a differential equation whose explicit formula we can find, we'll certainly go ahead and find that. But most often, we cannot do that.

Next time what we'll do is I'll first of all introduce a more realistic population model than our unlimited growth model that says that the population just goes off to infinity. That will be what we call the logistic or limited growth model. Secondly we'll bring in a lot of other qualitative picture that help us understand that new population model. See you then.

A Limited-Growth Population Model
Lecture 2

We begin this lecture by investigating a more complicated (but more realistic) population model, the limited-growth population model (also known as the logistic population growth model). The corresponding slope field gives us an idea of how solutions will behave, at least from a qualitative point of view.

This does not always happen, however. Look at the slope field for the differential equation $y' = y^2 - t$, and put a little target at (2, 2). Can you see where a solution in the lower left quadrant should begin so that it hits the given target?

Figure 2.1

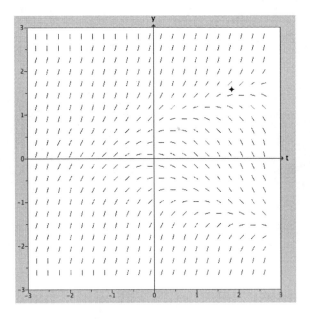

Well, maybe not. Nearby solutions tend to veer off from one another as y and t increase, as can be seen from the slope field, so the slope field does not tell us everything in this case.

Figure 2.2

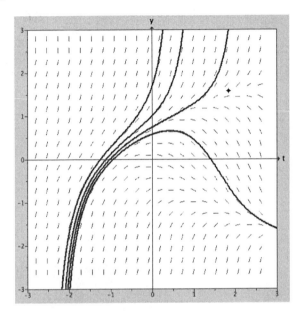

For the limited population growth model, we will assume that overcrowding may occur, which will hinder population growth or lead to population decline (if the population is too large). We make 2 assumptions about our population, $y(t)$:

- If $y(t)$ is small, the rate of population growth is proportional to the current population (as in our unlimited growth model).

- There is a **carrying capacity** N (an "ideal" population size) such that

  $y(t) > N$ means that $y(t)$ decreases ($y'(t) < 0$), and

  $y(t) < N$ means that $y(t)$ increases ($y'(t) > 0$).

The first assumption tells us that $y'(t) = ky$ if $y(t)$ is small. So we let our differential equation assume the form

$$y' = ky(?),$$

where the expression (?)

1. is approximately equal to 1 if y is small;
2. is negative if $y > N$;
3. is positive if $y > N$; and
4. is 0 if $y = N$.

Then one possibility (the simplest) for the term (?) is $(1 - y/N)$.

This yields the limited population growth ODE

$$\frac{dy}{dt} = ky(1 - y/N).$$

For simplicity, let's consider the case where $k = N = 1$:

$$\frac{dy}{dt} = y(1 - y).$$

This doesn't mean our carrying capacity population is just one individual (that would not be a very interesting environment). Rather, think of $y(t)$ as measuring the percentage of the carrying capacity. We will solve this differential equation analytically in a later lecture, but for now we will use qualitative methods to view the solutions.

Here is the slope field for the limited population growth model equation. Note the horizontal slopes when $y = 0$ or $y = 1$; these are our equilibrium points.

Figure 2.3

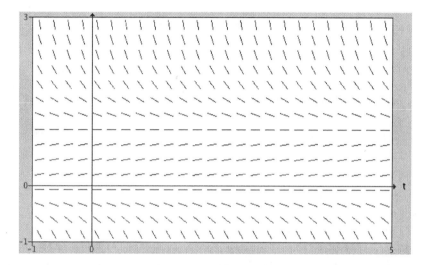

Here are some solution graphs. Note that they do exactly as we expected; they all tend to the equilibrium $y = 1$, the carrying capacity (assuming, of course, that the population is nonzero to start).

Figure 2.4

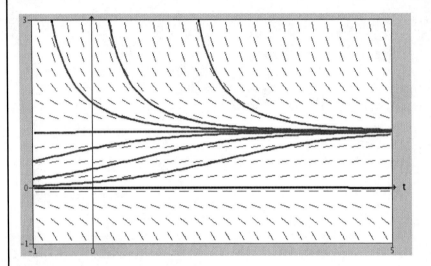

With an eye toward what comes later—when we deal with differential equations whose solutions live in higher dimensional spaces—we will plot the **phase line** for this ODE. This is a picture of the motion of a particle along a straight line where the position of the particle at time t is the value of $y(t)$.

Figure 2.5

This course deals primarily with autonomous differential equations. These are ODEs of the form $y' = F(y)$ (i.e., the right-hand side does not depend on t). We simply find all the equilibrium points by solving $F(y) = 0$. Between 2 such equilibria, the slopes are either always positive or always negative (assuming $F(y)$ is a continuous function), so solutions either always increase or always decrease between the equilibrium solutions. For example, for the ODE $y' = y^3 - y$, we have 3 equilibria at $y = 0$, 1, and -1. Above $y = 1$, we have $y' > 0$. Between $y = 0$ and y = 1, we have $y' < 0$. Between $y = -1$ and $y = 0$, we have again $y' > 0$. And below $y = -1$, we have $y' < 0$. So we know what the solutions will do, at least qualitatively. The phase line and some representative solution graphs are shown below.

Figure 2.6

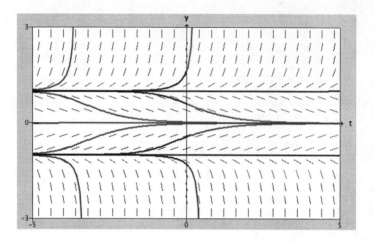

We can use the graph of the right-hand side of the differential equation to read this behavior off. Consider $y' = y^2 - 1$. Here is the graph of $dy/dt = F(y)$.

Figure 2.7

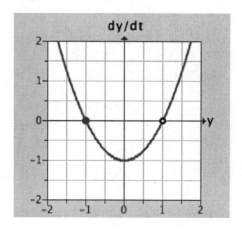

A glance at this graph tells us that if $y > 1$ or $y < -1$, then $y' = y^2 - 1 > 0$, so solutions must increase in this region. If $-1 < y < 1$, then we have $y' < 0$, so solutions decrease. And if $y = 1$ or $y = -1$, we have equilibrium points.

With this information, we know that the slope field looks as follows.

Figure 2.8

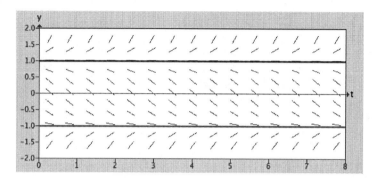

And our solutions behave as follows.

Figure 2.9

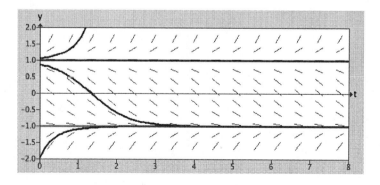

As another example, for the ODE $y' = y^4 - y^2 = y^2(y^2 - 1)$, we see that there are 3 equilibrium points: at 0, 1, and −1. The graph of the expression $F(y) = y^2(y^2 - 1)$ looks as follows.

Figure 2.10

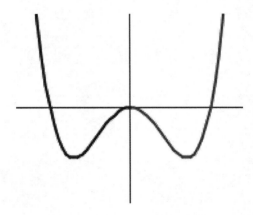

This graph tells us what happens between the equilibria: We have $y' > 0$ if $|y| > 1$, while $y' < 0$ if $0 < |y| < 1$. So the phase line and the solution graphs are as follows.

Figure 2.11

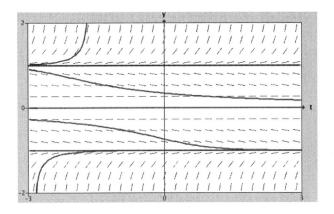

Note that there are 3 different types of equilibrium points for this ODE. The equilibrium point at $y = 1$ is called a **source** since all nearby solutions move away from this equilibrium. The equilibrium point at $y = -1$ is called a **sink** since all nearby solutions move closer to this point. And the equilibrium point at $y = 0$ is called a **node** since it is neither a sink nor a source.

Important Terms

carrying capacity: In the limited population growth population model, this is the population for which any larger population will necessarily decrease, while any smaller population will necessarily increase. It is the ideal population.

node: An equilibrium solution of a first-order differential equation that has the property that it is neither a sink nor a source.

phase line: A pictorial representation of a particle moving along a line that represents the motion of a solution of an autonomous first-order differential equation as it varies in time. Like the slope field, the phase line shows whether equilibrium solutions are sinks or sources, but it does so in a simpler way that lacks information about how quickly solutions are increasing or decreasing. The phase line is primarily a teaching tool to prepare students to make use of phase planes and higher-dimensional phase spaces.

sink: An equilibrium solution of a differential equation that has the property that all nearby solutions tend toward this solution.

source: An equilibrium solution of a differential equation that has the property that all nearby solutions tend away from this solution.

Suggested Reading

Blanchard, Devaney, and Hall, *Differential Equations*, chap 1.3.

Guckenheimer and Holmes, *Nonlinear Oscillations*.

Hirsch, Smale, and Devaney, *Differential Equations*, chap 1.2.

Roberts, *Ordinary Differential Equations*, chaps. 2.1 and 2.3.

Strogatz, *Nonlinear Dynamics and Chaos*, chap 2.1.

Relevant Software

Blanchard, Devaney, and Hall, *DE Tools*, First Order Solutions, HPG Solver, Phase Lines, Target Practice.

Problems

1. Let's begin with some ideas from calculus.

    a. What is the integral (antiderivative) of the function $y(t) = t^2 + t$?

b. Solve the equation $y^4 - 4y^2 = 0$.

c. For which values of t is the function $y(t) = t^3 - t$ positive?

d. Sketch the graph of $y(t) = t^2 - 1$.

e. Sketch the graph of $y(t) = t(2 - t)$.

2. Sketch the phase line for $y' = 1$.

3. Sketch the phase line for $y' = -y$.

4. What are the equilibria for the previous differential equation?

5. What are the equilibrium points for $y' = -1$?

6. What is the phase line for the differential equation $y' = y^n$, where n is some positive integer? Does the answer depend on n?

7. What is the phase line for the differential equation $y' = y^2(1 + y)$?

8. Find all equilibrium points for the differential equation $y' = \sin(y)$ and then sketch the phase line.

9. Find an example of a differential equation that has equilibria at each integer value and each of these equilibria is a node.

10. Find an example of a differential equation that has exactly 2 equilibrium points, each of which is a node.

Exploration

Consider the differential equation $y' = y^2$. We know what the slope field and phase line look like, but can you find actual solutions? How about for $y' = |y|$? Notice that the phase lines are the same for these differential equations, but the solutions are very different. Can you find other differential equations that have the same phase lines but different solutions?

A Limited-Growth Population Model
Lecture 2—Transcript

Hello again. In the last lecture, we introduced probably the simplest of all differential equations, the unlimited population growth model. In this lecture I'm going to introduce a slightly more complicated population growth model that will still look kind of naïve although we'll see a little bit later that in fact it's very interesting.

We'll not solve this differential equation as we did with the unlimited growth model. We'll actually solve it analytically later. That involves a lot more calculus. Today we'll concentrate on some of the stuff that we began with last time, the qualitative behavior, the various pictures that we have of solutions of differential equations. This often but not always gives us a good idea of what happens.

Remember last time when we did the slope fields. Let me go back and do another slope field. Here's a differential equation. This is the differential equation $y' = y^2 - t$. Unlike the differential equation last time, the right hand side here depends both on y and on the time variable t. So over here we see the slope field for this differential equation. Remember the slope field is a bunch of little lines for which are given by the right hand side of the differential equation. The slope of those lines is where all of our solutions are tangent.

Usually the slope field tells us a lot about the solutions to differential equations, but not always. Look at this slope field here. Suppose I put a little target right about there. Can you tell me where I should start down here in the lower part of the y/t plane so that I would get a solution that would hit that target?

It looks like if I would start someplace maybe right around here, looks like I go up and to the right. Let's see. No, not quite. I've missed. Clearly, I have to go a little bit further to the left, click here. No, I missed again. Pretty clearly to hit that target I've got to start out someplace in the middle, click here. I missed again. Let me move a little closer. Start here, I missed again. I'll move up a little closer. Do I hit the target now? No, I missed it again. Over here, I finally got it. That's actually a world record for me. Usually I have to move within a millimeter of that target. The point is that sometimes these slope fields tell you exactly what's going to happen to the solutions of your differential equations, but other times that does not happen.

Now let's move on to the more complicated population model. This is what we call the limited growth population model. Biologists and mathematicians would call it the logistic population model. So again assume we have our population, say some species the population of which is $y(t)$. This species is living alone, no predators. Now let's put in the possibility that the population can get so large we have overcrowding and that causes problems.

Again, as with all of our models, let me begin with an assumption. Here we'll have two assumptions. The first assumption is that the rate of growth of our population is proportional to the current population if the population is small. That was more or less the exact same assumption we made with our unlimited population growth model. Now let's put in a second assumption. Let me assume that there's sort of an ideal population for this species, what we call a carrying capacity. Let me call it N. We assume that carrying capacity is such that if your population is larger than the carrying capacity, you've got a population that's too large. The population must decrease.

On the other hand, if you have a population that's smaller than the carrying capacity, and bigger than 0, then there's room for growth and your population increases. Those are our two assumptions. Let's go ahead and convert them to a differential equation.

Assumption number one says that the rate of growth of population is proportional to the current population if the population is small. The population grows like the unlimited growth model if there's very few of the species around. That should say, in terms of a differential equation, that the derivative of y, the derivative of our population is equal to some constant times y. The rate of growth is proportional to the current population if the population is small.

Let me start out by writing a differential equation as y' is equal to k times y times something. That something will take into account our second assumption. What is the something? The something should be approximately equal to 1, if the population is small. If the population is small, our differential equation, and our rate of growth should be proportional to the current population. So something should be essentially 1 if $y(p)$ is small.

On the other hand, our something should be negative if the population is larger than this carrying capacity. The population is larger than the carrying capacity, there are too many of the population, the population should decrease, y' should be negative. On the other hand, if the population is less than the carrying capacity, then there's room for growth, so the population should increase. We want that something to be positive if y is less than N. Finally, if the population is exactly at that carrying capacity, we want the something to be 0. That would say that if the derivative is 0, that would give us a constant solution.

One possibility for the something is the quantity $1 - y/N$. Notice that if y is larger than N, than $1 - y/N$ is $1 - $ something bigger than 1, that's negative. On the other hand, if y is less than N, $1 - y/N$ is now positive. If $y = N$, $1 - y/N = 0$. So that's one possible model for this differential equation that governs the limited growth or logistic growth model. Our differential equation here is $y' = ky(1 - y/N)$. What you see is we have some parameters in here. We have that proportionality constant k and we have the carrying capacity N.

For simplicity, let me assume that both k and N are 1. Now say in a carrying capacity is 1, I don't really mean that there's 1 person is the ideal population. I mean maybe we're measuring y in terms of the percentage of the maximum or the carrying capacity. So that N would be 1. Of course my students always say oh, that would be a great carrying capacity for math faculty, but we won't go there.

In any event, what I'd like to do is now take that differential equation and go back over the slope field, see some solutions, and then add in some other kinds of geometric pictures, some qualitative methods that will help us understand solutions to differential equations as we go along.

The first observation is, remember an equilibrium solution is a constant solution for this differential equation. In our case, y' is now $y(1 - y)$. What are the constant solutions? Well if $y = 0$, then our derivative y' is equal to 0, we know $y = 0$ is a constant solution. Similarly, the right hand side is $y(1 - y)$, so if the population is equal to 1 carrying capacity, then again $y' = 0$, so we've got a constant solution.

Let me look on the computer at what these and the rest of the solutions look like. Here is a picture of the slope field for this differential equation. I've drawn in two of those constant solutions, the constant solution at 0 and the constant solution at 1. What you notice is the slope field between 0 and 1 is always positive. Yes, that's exactly what we built in to our assumption. That means when the population starts out between 0 and 1, the population increases. It increases directly to the ideal carrying capacity, population $y = 1$. On the other hand, when y is larger than 1, by our assumptions the slope field is always negative. That means solutions decrease, and indeed that's what happened. They decreased down to the equilibrium solution at 1.

Here's another picture that we're going to draw of this behavior. This is what we call the phase line. Remember our differential equation depends on time t. Very often, that's the case. When we go to higher dimensions, again, our differential equations will depend on the time t, but drawing the graphs of those solutions as we're doing here will be much more complicated. So what we'll do is go into the phase plane, or the phase space.

Here's what the phase line is in this case. The phase line is the simplified version of the phase plane or space. What I'll do is I'll think of y as a particle moving along the y-axis, moving with respect to time. Our solution as time goes on moves. That particle moves, and I'll draw that motion. So for example, here is the phase line here, and what I'll do is watch the motion of a particle as it moves along this phase line.

For example if I start at a population that's close to 0, as we know, that population increases, and the particle races up the y-axis, slowly coming to rest at the equilibrium point at 1. If I start out with a population that's larger than 1, as we've seen, the population decreases, along the phase line, you see the particle race down and slowly come to rest at the equilibrium point at 1.

We also have a negative y-axis here. That of course doesn't measure population, but you can also draw that in the phase line. In this case, our population just races off to negative infinity. There's a second way that we're going to display the solutions of differential equations. Again, this becomes much more important when we go into higher dimensional differential equations.

Let me back off a bit and give a couple more examples of differential equations together with some other concepts that will arise. I'm going to for the most part concentrate now on what we call autonomous differential equations. These are differential equations y' equals some function of y. I usually call it $F(y)$. That is to say, there's not t on the right hand side. You think of the target I put up before. That differential equation was $y' = y^2 - t$. That was non-autonomous.

So what we're going to do when we solve differential equations or try to solve a differential equation is first find those most important points, the equilibrium points. How do you do that? Well, to find these equilibrium points, what we've got to do is set the right hand side of our equation equal to 0. If we find those roots, then those roots are places where the derivative equals 0. So those roots will be constant solutions. Then more importantly if we find a pair of constant solutions between those equilibrium solutions, assuming there's no more equilibria there, then the slopes are either positive or negative. So we know what happens between those solutions. Either solutions go from on equilibrium to the other, or reverse. Or if you're coming from infinity, they come down from infinity to one of the equilibrium.

Let's start with an example. Let's take the autonomous differential equation $y' = y^3 - y$. First step, find the equilibrium solutions for that differential equation, $y' = y^3 - y$. Equilibrium solutions are factor the right hand side into y, factor of $y^2 - 1$. Equilibrium solutions are where the right hand side equals 0. The right hand side equals 0 when y is either 0 or ± 1. So we found our equilibrium solutions. Incidentally I should say that this is also a differential equation that we can solve. We can find the other solutions, but boy solving this is a horrendous mess of calculus. That's why these qualitative methods will help us a lot more.

We've got our equilibrium points. It's 0 and ±1. We know between the equilibria or above and below them, the slopes must either be always positive or always negative. So let's find out what the slope field is in these other regions.

First case, suppose that y is larger than 1, say 2. Then $y' = y^3 - y$, $2^3 - 2$ is positive. So everywhere above the equilibrium point at 1, we know that y' is positive. That means slope fields stick straight up. That means solutions must increase, must increase off to infinity and so forth. If y is between the equilibrium point 0 and 1, then you can check pretty easily. Check say $y = \frac{1}{2}$, $\frac{1}{2}^3 - \frac{1}{2}$ is now negative. So between 0 and 1, y' is negative. Solutions must go downhill to the equilibrium point at 0.

Also between -1 and 0, similar computation says that y' is positive. Solutions must go uphill. Finally, when y is below -1, $y = -2$, then the right hand side of the differential equation is -2^3. That's $-8 - (-2)$, that's negative; y' is negative, solutions go downhill. So we know what's going to happen to our solutions at least qualitatively.

Let's turn to the computer and see that actually happen. Here again is the slope field for that differential equation. I've highlighted the three equilibrium points, at ±1 and at 0. Let me put in the phase line. By what we just saw, we know where the solutions tend in each of the regions between the equilibrium. When y was larger than 1, we see that the slope field is always positive, solutions go off to infinity. On the phase line, the particle runs off to infinity. But below $y = 1$, now the slope field is negative. We've calculated that. We know that solutions must go in forward time down to the equilibrium point at 0. In backward time, it must increase to the equilibrium point at 1.

Again, if we go below 0 but above -1, there we calculated that the slope field was positive. We know what happens to solutions. Solutions must continue to increase and eventually level out at the equilibrium point at 0. The final case is when y is less than -1. There the slope field is negative. There our solutions run off to minus infinity. So again, the slope field and whether it's positive or negative and the location of the equilibria tells us everything about the solutions from a qualitative point of view. We don't know the exact position of $y(t)$ at any time, but we know what's going to happen in the future. We know what happens as times goes to infinity.

Now let's move to another picture of what's happening for differential equations. Let's start with again an example. Let's say $y' = y^2 - 1$. $y' = y^2 - 1$. As always, find the equilibria. There are two places where the right hand side vanishes at ± 1. Those are our equilibria. We could also look at the sine of the right hand side to determine where the slope field is positive or where it is negative. For example if y is greater than 1, $y^2 - 1$ is positive. Solutions must increase. If y is less than -1, $y^2 - 1$ is also positive, solutions must increase. But between ± 1, the right hand side of the differential equation is negative, so solutions must decrease.

By the same calculations we did before, we can read that off. But there's another way to do that. Remember as I said, look at the graph of the right hand side of that differential equation. What I'm displaying here on the computer is, first off, the slope field for this differential equation. Secondly I'll incorporate the phase line. Thirdly I'll draw the graph of the right hand side of this differential equation namely $y^2 - 1$. There it is. There are three different pictures that will allow us to understand what is happening to solutions of differential equations.

As we just said, for this differential equation, if y is larger than 1, you can look at the graph of the right hand side of that differential equation, the graph of $y^2 - 1$. Here it is down here. The graph of $y^2 - 1$ when y is larger than 1—incidentally, be careful here—notice that y is my horizontal axis, and $F(y)$ or the right hand side of the differential equation is my vertical axis—in any event, when y is larger than 1, just looking at the graph shows us that slopes must be positive, solutions must increase. Indeed there they do, both in the graph of the solution and the phase line. Between ± 1 the graph is now below the horizontal y-axis. That means y' is negative, solutions must decrease, and there they go.

Again when y is less than -1, now again the graph of $y^2 - 1$ is positive, slopes are positive, solutions go uphill. There they go. So there's another way of understanding what's happening to solutions of differential equations, all together three pictures already and we've got a long way to go.

There's another little thing that's interesting here. Notice that we have two different types of equilibrium points for this differential equation. The equilibrium point at 1 has the property that nearby solutions move away from it. Solutions leave the environment of the constant solution $y = 1$. That's what we call a source.

On the other hand, when $y = -1$ at that equilibrium point, nearby solutions move toward it. That equilibrium point is a sink. Here again are our solutions. Let me start them off again. Near the upper equilibrium point all solutions run away. That's a source. On the other side of that equilibrium point, all solutions move away. Solutions are repelled from that equilibrium point. It's what we call a source or a repelling equilibrium point. Meanwhile down below, the green equilibrium point is a sink. Nearby solutions tend to that equilibrium point. Whether I'm above or below $y = -1$, solutions come in. That's an equilibrium point that we call a sink.

Later on in this course when we move to higher dimensional differential equations, we'll see many, many other types of equilibrium points. Here basically although you only see two. Well that's not quite right. There's one exceptional case. That's what we call a node.

Let me look at the differential equation $y' = y^2$. Here on the computer I've drawn the graph of y^2. The graph is a parabola that just touches the y-axis at 0. So we have a single equilibrium point at 0. Now y^2 everywhere is positive except at 0. That means the slopes are always positive everywhere except at 0. That means when y is larger than 0, solutions run away. Whereas when y is less than 0, solutions come in to 0. Solutions are always increasing. That's an example of a node. From above it looks like a source; from below it looks like a sink. Any equilibrium point that is neither a sink nor a source is what we call a node. As we'll see as we go on, there are many different kinds of nodes.

Let me finish now with one final example. A little more complicated differential equation that would really be horrendous to solve. The differential equation is $y' = y^4 - y^2$. Or factor that into $y' = y^2(y^2 - 1)$. Let's do it all quickly. What are the equilibrium points for this differential equation? Right hand side is $y^2(y^2 - 1) = 0$ at 0 or ± 1. There are our equilibrium points.

What else happens? We could draw the graph of $y^4 - y^2$ or we could just check what happens to various y values between or above and below the equilibrium points.

It's pretty easy to check that y' is positive, $y^4 - y^2$ is positive. If either y is greater than 1, or y is less than -1, there y' is positive, slopes are going up, solutions are increasing. On the other hand, from the graph of this function or just by computing what happens at various values you can see that y' is negative if y is either between -1 and 0, or between 0 and 1.

Let's look at the solutions of this differential equation and read off whether we have sinks, sources, or nodes, so back to the computer. Now here are the graphs of the solution, not the phase lines or anything else. Let me first show you the slope field. There is the slope field for this differential equation. We see pretty easily that you have an equilibrium point at -1. That's where the slope is 0, it's horizontal. Now if I choose a y value below that, solutions come in to that equilibrium point. Or if I choose a y value above, solutions also come into that equilibrium point; $y = 1$ is therefore a sink for this differential equation.

On the other hand, at $y = 1$, that's right about there. We see that above $y = 1$, solutions run away. Below $y = 1$ again solutions run away. So somewhere in between here there's a solution, constant solution, $y = 1$ which is a source. Then finally right in the middle our equilibrium point at 0, from above solutions are converging down to it, whereas from below, solutions are running away. So the equilibrium point at 0 turns out to be a node.

Just to summarize, using all of these qualitative techniques, we know what's happening to solutions. We don't know if solutions race down to -1 very quickly or take a long time to get there. We don't know the exact value of $y(t)$, but we have a good comprehensive idea of what's happening to solutions of differential equations.

There are several other techniques from calculus that are going to come up that help us understand what's happening here. That's what I'll turn to in the next lecture when we talk about the first derivative test for types of equilibrium points. See you then.

Classification of Equilibrium Points
Lecture 3

In this lecture, we see that slope field allows us to completely understand the behavior of solutions of first-order, autonomous differential equations. Recall that there are 3 different types of equilibria: sinks, sources, and nodes.

Figure 3.1

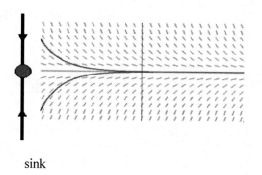

sink

Figure 3.2

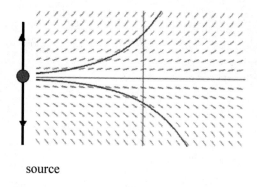

source

Figure 3.3

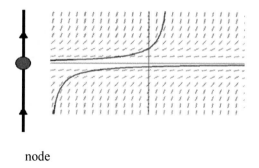

node

There are actually lots of different things that can happen when an equilibrium point is a node. For example, the differential equation $y' = y^2$ has a node at $y = 0$. But for any nonzero value of y, solutions tend away from 0 if the initial value $y_0 > 0$, whereas solutions tend to 0 if $y_0 < 0$. The differential equation $y' = 0$ also has an equilibrium point at 0, and this equilibrium point is also a node, since all solutions are equilibria. And there are examples where an infinite collection of equilibrium points can accumulate on a given equilibrium point, again giving us a node.

We can use calculus to determine whether a given equilibrium point y_0 for the ODE $y' = F(y)$ is a sink, a source, or a node. Since y_0 is an equilibrium point, we know that the graph of F crosses the y-axis at y_0. If the graph of F is increasing as it passes through y_0, then y' is positive when $y > y_0$ and y' is negative when $y < y_0$. This says that solutions move away from y_0 whenever y is close to y_0, so y_0 is a source. On the other hand, if the graph of F is decreasing through y_0, then similar arguments show that y_0 is a sink.

But we know from calculus that if $F'(y_0) > 0$, then $F(y)$ is increasing, and if $F'(y_0) < 0$, then $F(y)$ is decreasing. So this gives us the **first derivative test for equilibrium points**: Suppose we have the differential equation $y' = F(y)$, and y_0 is an equilibrium point.

- If $F'(y_0) > 0$, then y_0 is a source.

- If $F'(y_0) < 0$, then y_0 is a sink.

- If $F'(y_0) = 0$, then we get no information.

Consider $y' = F(y) = y - y^2$. If $y = 1$ or $y = 0$, we have equilibrium points. Moreover, $F'(y) = 1 - 2y$. This means that $F'(1) = -2 < 0$, so 1 is a source. Similarly, $F'(0) = 1$, so 0 is a sink. Therefore we know that the phase line is as follows.

Figure 3.4

And our solutions behave as follows.

Figure 3.5

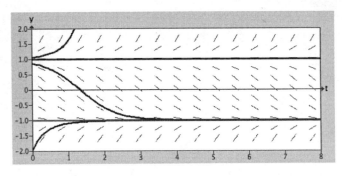

As another example, consider $y' = y^3 - y$. We can factor the right-hand side into $y' = y(y + 1)(y - 1)$, so we have equilibrium points at $y = 0$, 1, and -1. Because $F'(y) = 3y^2 - 1$, we know that

$F'(0) = -1$, so 0 is a sink;

$F'(1) = 2$, so 1 is a source; and

$F'(-1) = 2$, so -1 is also a source.

The phase line is therefore as follows, and we again know the behavior of all solutions.

Figure 3.6

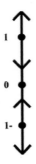

Let's take another example: $y' = y^4 - y^2$. We have equilibrium points at 0 and at ± 2. Since $F'(y) = 4y^3 - 2y$, we have $F'(1) = 2 > 0$; so 1 is a source. Because $F'(-1) = -2$, -1 is a sink. The derivative at the 0 vanishes, so we get no information. But what we have about the 2 equilibria at ± 1 is enough to tell us that 0 is a node.

Figure 3.7

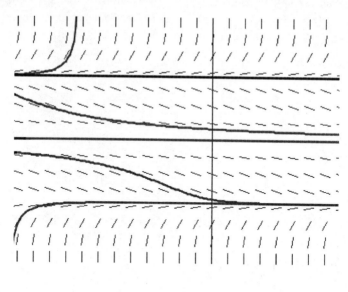

We now turn to perhaps the most important theorem regarding first-order differential equations, the existence and uniqueness theorem. Suppose we have a differential equation $y' = F(y)$, and suppose the function $F(y)$ is nice at the point y_o. (For our purposes, "nice" means that the expression is continuously differentiable in y at the given point y_0.) Then the theorem states that

- there is a solution to the ODE that satisfies y(t0) = y0, and
- that this is the only solution that satisfies y(t0) = y0.

Let's work a couple of examples. First, $y' = \sin^2(y)$. We have equilibrium points at $y = 0, \pm \pi, \pm 2\pi, \ldots$. At all other points, the slope field is positive. So all other solutions are always increasing. However, any solution that starts between $y = 0$ and $y = \pi$ can never cross these 2 lines. The solution is trapped in this region, and since it is always increasing, the solution must tend to $y = 0$ in backward time and to $y = \pi$ in forward time. We have similar behavior between any 2 adjacent equilibrium points.

Figure 3.8

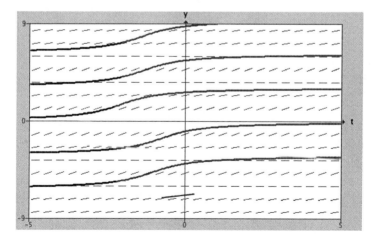

Now let's consider $y' = y/t$. Here we have problems when $t = 0$. The right-hand side of the ODE is not even defined at $t = 0$, never mind continuously differentiable there. It is easy to check that $y(t) = kt$ is a solution of this ODE for any value of k. For each such k, we have $y(0) = 0$, so we now have infinitely many solutions that satisfy $y(0) = 0$. But there are no solutions that satisfy $y(0) = y_0$ for any nonzero value of y_0. Even worse, when the right-hand side of the ODE is not nice, the numerical methods we are using to plot solutions may fail. For example, the differential equation

$$y' = \frac{1+y}{1-y}$$

is not defined at $y = 1$. Look at what happens when we try to use the computer to plot the corresponding solution.

Figure 3.9

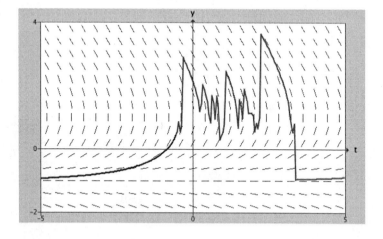

Something is clearly wrong here; the numerical method the computer is using has failed. We will describe why this happens in Lecture 6.

Important Term

first derivative test for equilibrium points: This test uses calculus to determine whether a given equilibrium point is a sink, source, or node. Basically, the sign of the derivative of the right-hand side of the differential equation makes this specification; if it is positive, we have a source; negative, a sink; and zero, we get no information.

Suggested Reading

Blanchard, Devaney, and Hall, *Differential Equations*, chaps. 1.5 and 1.6.

Hirsch, Smale, and Devaney, *Differential Equations*, chap. 7.2.

Roberts, *Ordinary Differential Equations*, chap. 2.2.

Strogatz, *Nonlinear Dynamics and Chaos*, chap. 2.5.

Relevant Software

Blanchard, Devaney, and Hall, *DE Tools*, HPG Solver, Phase Lines.

Problems

1. First let's do some calculus review problems.

    a. For which values of t is the function $y(t) = 1 + t^2$ increasing?

    b. For which values of t is the function $y(t) = |t|$ decreasing?

    c. What are the roots of the function $y(t) = t^2 + 2t + 1$?

    d. Where is the previous function increasing?

    e. Sketch the graph of $y(t) = t^2 + 2t + 1$.

    f. Sketch the graph of $y(t) = \sin(t)$.

    g. Compute the derivative of $\cos(3t + 4)$.

2. Find the equilibrium points for $y' = y + 1$, and determine their type.

3. Find the equilibrium points for $y' = -y^2$, and determine their type.

4. Sketch the phase line for $y' = -y^2$.

5. Find the equilibrium points for $y' = y^2 - 1$ and determine their type.

6. Classify the types of equilibrium points for $y' = y^3 - 1$.

7. Does the existence and uniqueness theorem apply to the differential equation $y' = |y|$?

8. Consider $y' = y^2 - A$, where A is a parameter. Find all equilibrium points and determine their type (your answer will, of course, depend on A).

9. Consider $y' = Ay(1 - y)$, where A is a parameter. Find all equilibrium points and determine their type.

10. What is the actual behavior of solutions of

$$y' = \frac{1+y}{1-y}?$$

Exploration

We saw that the differential equation in problem 10 causes problems when we try to solve it numerically. Can you come up with other examples of first-order ODEs that break the numerical algorithm your computer uses to solve them?

Classification of Equilibrium Points
Lecture 3—Transcript

Hello again. In the last lecture, we saw several different ways to visualize solutions of autonomous first order differential equations. Remember they were differential equations that performed y' is some function of y. No ts on the right hand time. We saw the slope field and solution graphs. We then saw the motion on the phase line. We also used the graph of the right hand side of the differential equation.

In today's lecture, what I do is revert back to some more calculus, use some more calculus for determining the types of equilibrium points. This will be the first derivative test for equilibrium points. Then I'll turn to the main theorem in differential equations. This is the so-called existence and uniqueness theorem, and we'll see how that helps solidify all these qualitative notions.

First remember we saw three types of equilibrium points. There were sinks, sources and nodes. Let's look at the phase lines again to see what they look like. Here's the differential equation we saw last time, $y' = y^2 - 1$. Down here, we see the graph. The graph crosses the y-axis at two points ± 1. They are our equilibrium points. We see from the solutions graphs that $y = +1$ is a source, nearby solution curves move away. Whereas $y = -1$ is a sink, nearby solutions curves come closer to it. There are two other types of equilibrium points.

A third type was the node. Let me turn to the differential equation $y' = y^2$. We see a single equilibrium point at 0. Since the right hand side of the differential equation y^2 is always positive, all other solutions move away. All other solutions move up, so from below solutions converge up to the equilibrium point at 0. But above when y is positive, solutions curves move off to infinity. So some solutions come in, others go away. That's a node.

There could be other types of nodes. For example if you take y' is not y^2 but $-y^2$ then again we have an equilibrium at 0. But now the slopes are always negative, so solutions would always decrease. Another example is we could have the trivial differential equation $y' = 0$. There all solutions are constant.

S all of our solutions are equilibrium points, so arbitrarily close to any equilibrium point you have nothing else but equilibrium points.

Still other things could happen. For example, you could have a differential equation that has an equilibrium point and arbitrarily close to it are infinitely many other equilibrium points that accumulate on it, but not everything is an equilibrium point. In between, they could be doing something.

Let's now turn to the first derivative test for equilibrium. This will be a test from calculus that allows us to determine immediately whether our equilibrium point is a sink or a source. Then with that information we can immediately sketch the phase line. So here's what we'll do. Suppose we have an equilibrium point at y_0. Our differential equation is $y' = F(y)$, so that means $F(y_0) = 0$. So the graph of F crosses the y-axis at your point y_0. Let's assume that the derivative of your function at y_0 is positive. What does that say? That says that our graph increases through y_0. So if y is less than y_0, then $F(y)$ is negative. So for y below our equilibrium point, slopes are negative. That means solutions are moving away.

On the other hand, if y is greater than y_0, then $F(y)$ is greater than 0. That means above y_0 slopes are positives, solutions increase. So that says that if the derivative at y_0 is positive, our equilibrium point must be a source.

Now the exact opposite is true. What happens if the derivative at y_0 is negative? Then that says that $F(y)$ decreases through y_0, so below y_0 of $F(y)$ is positive. That means the slopes are increasing, solutions are coming in to y_0, below y_0. Similarly, above y_0 now $F(y)$ is negative, so solutions are decreasing toward y_0. That says that if the derivative at y_0 is negative, then y_0 is a sink.

The final case, the intermediate case, is where the derivative at y_0 happens to be 0. In that case, we get no information. For example, we could have a node. Remember the equation we just looked at, $y' = y^2$. We have an equilibrium point at 0, but the derivative of y^2 is $2y$, so that's 0 at 0, and we get a node.

On the other hand, look at $y' = y^3$. Again, 0 is an equilibrium point. Again, the derivative at 0, $3y^2$ is 0, so the derivative is 0 at the equilibrium point. But for y^3 if y is positive, then y^3 is positive, so above 0 our slopes are increasing. On the other hand, if y is less than 0, if y is below 0, again y^3 is negative. So again, solutions now are going the other way. That means that for $y' = y^3$ we have a source. Similarly if you look at $y' = -y^3$, again an equilibrium point at the origin where the derivative is 0, but now we compute a sink.

Let's do a couple examples of this. Let's return to our logistic differential equation $y' = y(1 - y)$ or $y' = y - y^2$. We see we have equilibrium points at 0 and 1. Now let's use the first derivative test to determine their type. Take the derivative of the right hand side, the derivative of $y - y^2$ of course is $1 - 2y$, and then plug in our equilibrium points. Take the derivative at the equilibrium point 0. We have $F'(0) = 1 - 2(0)$, that's 1. That's positive, so 0 is a source. The derivative at 1 now is $1 - 2(1)$, that's -1, that's negative. So the equilibrium point at 1 is a sink. So effectively, we know everything about the behavior of solutions just from that information.

Here are our pictures for the differential equation $y' = y(1 - y)$. From the graph of that quadratic function, we know we have roots at 0 and 1 as we just saw. We see the derivative at 0 is positive. The derivative at 1 is negative. By the first derivative test then 1 is a sink. Solutions tend to the equilibrium point at 1. The equilibrium at 0 is a source, solutions tend away. Now because there are no equilibrium points in between, we know the behavior everywhere. Solutions between equilibrium points must always increase. Solutions above $y = 1$ must always decrease, and same thing below $y = 0$. The first derivative test tells us everything.

Here's one more example, how about $y' = y^3 - y$, or factor a y there, $y' = y(y^2 - 1)$. From that equation you see that we have equilibrium points at 0 and ± 1. So let's again use the first derivative test. The derivative of $y^3 - y$ is $3y^2 - 1$. So just plug in all our equilibria and determine their types. For example at 1, we get $3(y^2 - 1)$, that's two. Greater than 0, 1 is a source. At -1 the derivative is again $3(1 - 1)$, 2. That's again a source. But at 0, the derivative is $3(0 - 1)$, that's -1. Less than 0, that's a sink. So with just that little bit of calculus, we now know everything that's happening for solutions to that differential equation.

Let's look at those solutions. Here again is the graph of our differential equation. The right hand side has equilibrium points at −1, 0, and 1. The derivative at 1 as we just saw is positive as it is at −1, whereas at 0 the derivative of the slope is negative. So from the first derivative test we know that 1 is a source, solutions move away; −1 is also a source, the derivative is positive there, solutions move away. And the equilibrium point at 0 derivative less than 0 is a sink. Again, there are no equilibrium points in between these equilibrium points, so we know everything. We know what's happening to all other solutions.

Here's one final example that we visited earlier. Let's take $y' = y^4 - y^2$ or factor out a y^2, that's $y^2(y^2 - 1)$. Again we have equilibrium points at 0, 1, and −1. Now use the first derivative test and we have a little bit of difference here. For example, the derivative here is $4y^3 - 2y$, that's the derivative of $y^4 - y^2$. Again, evaluate that derivative at the equilibria. At 1 we just get the derivative is $4 - 2$, that's 2. So 1 is a source. At −1 the derivative is now $-4 + 2$, that's −2, so −1 is a sink. But at 0, the derivative vanishes, so we get no information.

On the other hand, just having the information about those two flanking equilibria tells us everything again. So here is the slope field for this differential equation. We know we have an equilibrium point at 1, right about there, an equilibrium point at −1, right about here. We know that 1 is a source, so solutions run away. We know that −1 is a sink, so solutions tend in. Just by looking at this situation, we see what the equilibrium point at 0 must be. Above 0 and below 1 solutions are decreasing. Below 0 and above −1 solutions are also decreasing. That says that our equilibrium point at 0 must be a node. Sometimes you get no information from the first derivative test, but sometimes the information at other equilibrium points help you out.

Here's one other example, how about $y' = \sin^2(y)$. Here come some trigonometric functions. Sin is the most basic trigonometric functions. You know that the graph of $\sin(y)$ is periodic with period 2π. And that $\sin(y)$ has zeroes at 0, and $\pm\pi$ and $\pm 2\pi$ and plus or minus any integer times π. So the right hand side of this differential equation has infinitely many equilibrium points at $\pm N\pi$ for each N. Now also from calculus we know that the derivative of $\sin(y)$ is $\cos(y)$.

57

So let's try the first derivative test here for sin², $y' = \sin^2(y)$. The derivative by the chain rule is $2(\sin(y))$ then times the derivative of sin, that's cos. So $F'(y) = 2\sin(y)\cos(y)$. Remember $\sin(y)$ vanishes at all of our equilibrium points. So the derivative is equal to 0 at each of these equilibrium points, so no information at all from the first derivative test. But we're kind of lucky. Look at the right hand side of the differential equation. It's $\sin^2(y)$, so at any nonequilibrium point, $\sin^2(y)$ is positive. So solution curves must all go uphill. We know just by looking at the right hand side that all of those equilibrium points are nodes.

Let's just look at those solutions. So here is the slope field for $y' = \sin^2(y)$. As we just saw, all solutions tend to increase between the equilibrium points. Each of the equilibrium points is a node. So even though the first derivative test tells us nothing, we still know exactly what is happening. Now here's a question. Why can't we have solutions, say here's 0, here's π, why can't we have solutions that increase for a while, climb up the slope field, and then level off at the equilibrium point at 0, run along 0 for a while and then start increasing up to $y = \pi$, cut across that equilibrium solution for y for a while, and then decides to just keep increasing again? Why can't we have solutions that cross those equilibrium points?

That's governed by the existence and uniqueness theorem for differential equations, probably the most important theorem in the field. Here's what it says for first order differential equations. Suppose you have a first order equation $y' = F(t, y)$, may depend on t and y. Suppose that $F(t, y)$ the right hand side is nice at the point (t_0, y_0). I'll explain what I mean by nice in a minute. Then the existence and uniqueness theorem says two things. First, there is a solution that satisfies the initial condition $y(t_0) = y_0$. Given that point (t_0, y_0) where the right hand side is nice, there is a solution that satisfies that initial condition. More importantly, that solution is unique. There's only one solution that passes through the point t_0, y_0.

What does nice mean? Nice means that right hand side F is continuously differentiable at the point (t_0, y_0). Remember what that means. That means that you can differentiate F with respect to t and with respect to y at that point. What you get is a nice continuous function. If that condition holds, then you have first of all a solution through that point, and secondly that solution is unique.

So going back to $y' = \sin^2(y)$ we now see that solutions in fact cannot cross. $\sin^2(y)$ is a perfectly nice right hand side. It's differentiable at every point. In fact, it's continuously differentiable at every point. It's derivative is 2sin cos, so that means solutions can never cross. So we could never have a solution come up and merge with an equilibrium point for a while, and then go on.

There's a slight problem. Sometimes your differential equations on the right are not nice. That causes problems. Let me give a couple examples of that. Let's look at $y' = y/t$. Where is there a problem here? We have a t in the denominator. So when $t = 0$, the right hand side isn't even defined, never mind continuously differentiable. On the other hand, we know some solutions to that differential equation. You look at just the simple function $y(t) = kt$ for any constant k, that is a solution. Take the derivative of kt you just get k. Plug it into the right hand side, you get kt/t, the ts cancel, you get k. So yes, kt is a solution to that differential equation.

Now notice infinitely many of those solutions pass through 0 at time 0, $y = kt$ always satisfies $y(0) = 0$ no matter what k is. In fact, you have infinitely many solutions passing through a point where the existence and uniqueness theorem did not hold. Now there are no solutions whatsoever that satisfy $y(0) = y_0$ when y_0 is a nonzero number. So you have infinitely many solutions through 0 at time 0, but no solutions through any other point y_0 at time 0.

Something else is a little funny here. Let me turn back to the computer. So here is the slope field for $y' = y/t$. I've used the computer to calculate two of those solutions, because we know what they are, they're just kt. But look at what the computer gave me. In one case it gave me a nice straight solution that went right through the origin $y = kt$. There was something funny going on there when we went through 0.

In the other case, the solution just gave me half of that line. What is happening here? Well, near that bad point $t = 0$ the computer's numerical method sometimes gets confused. Let me see if it gets confused again. You go down here, no all the way through, all the way through, all the way through. Oops,

59

that one didn't go all the way through. The numerical methods given by a computer sometimes go bad when you have a bad point on the right hand side of your differential equation.

But it gets even worse. Let's look at another differential equation. Let's say $y' = y/t^2$. Where does that go bad? Again we have a t in the denominator, so when $t = 0$, the right hand side again is not defined. We could go and solve this differential equation, but let's not worry about that. Let's see what happens on the computer here.

Here we see the slope field for $y' = y/t^2$. Think of what solutions should do when t is positive. $T = 0$ is the bad place, but when t is positive and y is positive, then y/t^2 is positive. Our slopes to the right should be going uphill, and yes, they are. But look at what the computer has done to this solution here. Yes, as t goes on, that solution should be increasing. But wow, back here near 0 suddenly it's decreasing.

Again, the numerical algorithm has gone a little bit crazy at this bad point. Other cases, it looks fine there. Oh, look at that. This solution should be going downhill everywhere, slope's a negative to the right of the y-axis, but somehow our numerical method made an error. What we see is numerical methods again can go wrong at bad points on the right hand side of the differential equitation.

Let's actually look at another one. Here's another differential equation. This time it's autonomous. There are no ts. It's $y' = (1 + y)/(1 - y)$. Where's the bad point here? In the denominator we have $1 - y$, so when $y = 1$, we have 0 in the denominator. That means our slope field becomes infinite at $y = 1$. We also have a $1 + y$ in the numerator, so $1 + y$ vanishes at $y = -1$. We have an equilibrium point at -1.

Here's our slope field. Notice that the slopes are pretty nice down here near -1. But as we get closer to $y = +1$, the slopes get more and more vertical. The slopes are infinite at $y = 1$. Watch what the computer does when we try to find solutions now. Whoa, is that the solution to the differential equation? Another one, is that the solution? This numerical method is really going crazy in this case because we have a place again where the denominator is 0.

With an eye toward what's coming later, this is actually chaotic behavior. If I start out very close say to this point, I get one solution. If I move a little bit over and start out, my numerical solution gives me a completely different solution. What's happening is the so-called iterative method is behaving chaotically. This is something that we'll see over and over again in this course when you have first of all a numerical method that goes crazy. Sometimes they go chaotically crazy. Later like we saw with the Lorenz equations, often differential equations themselves behave chaotically. So this is the beginning of that other subtheme in this course.

Here are a couple more examples. Let's look at $y' = 3y^{2/3}$. $y' = 3y^{2/3}$. Is that a nice right hand side? There's nothing in the denominator. It looks pretty good. But remember, what's the derivative of $y^{2/3}$? The derivative of $y^{2/3}$ is $2/3(y^{-1/3})$. So when $y = 0$, you get a 0 in the denominator of the derivative, the derivative doesn't exist there. Remember our right hand side had to be continuously differentiable. So we have a bad point when $y = 0$.

We know some solutions here. Obviously $y = 0$ is a solution. You plug $y = 0$ into the right hand side, you get 0. So $y = 0$ is a constant or equilibrium solution. Do you know any other solutions? Well, we'll see how to do this in a couple of lectures, but another solution is given by, look at the right hand side. It's $3(y^{2/3})$. The function $y(t) = t^3$ solves that equation. Differentiate t^3, you get $3t^2$ and that equals $3(t^3)$ quantity to the 2/3 power. So yes, t^3 is a solution to that differential equation.

But so is y identically equal to 0. So $y(t) = t^3$ crosses $y = 0$ at 0. So we don't have uniqueness for this differential equation. This is very different from the previous ones where all of the sudden we had points where the differential equation wasn't defined. Here the differential equation is everywhere defined, but it's not differentiable.

There's my equilibrium solution, $y = 0$, and now other solutions look like that. It looks just like the graph of tangent. It crosses my equilibrium solution. That's forbidden if the right hand side is nice, but if it's not differentiable, it happens. In fact, infinitely many of these solutions cross our equilibrium solution because $y = 0$ is a bad point.

Here's one more differential equation. Let's take the differential equation $y' = e^t \sin(y)$. That's a nice differential equation. Right, the derivative of e^t and $\sin(y)$ are all nice. So that differential equation satisfies the existence and uniqueness theorem. But now let's go to the computer and see what happens here.

There's the slope field for $e^t \sin(y)$. $\sin(y)$ vanishes at $y = 0$, so we have an equilibrium point at $y = 0$. It's kind of hard to see, but if you look over to the left you see it. $\sin(y)$ also vanishes at $y = \pi$. You can see an equilibrium solution right up about here at $y = \pi$. Now $et(\sin(y))$ for y between 0 and π $\sin(y)$ is always positive. So the slope fields are always increasing. So this is a nice right hand side. We can't have solutions cross. We know we have an equilibrium point at π and at 0. We know that solutions are increasing everywhere in between.

Let's see what happens when we use the computer to find those solutions. Whoa, what happened there? What happened there? And what happened there? Clearly, the computer is going crazy. Make a small change in my initial condition, get a very different change in the behavior of solutions. Notice that all of these solutions are fine for a certain time. The solutions just climb up to the equilibrium point if $y = \pi$. Then eventually they all diverge. Eventually they go crazy. Again, it's chaotic behavior.

One thing to notice is this differential equation has a very nice right hand side. The exponential of $t(\sin(y))$, that's perfectly differentiable, continuously differentiable. So that differential equation satisfies the existence and uniqueness theorem. What's happening here? Why are we seeing this chaotic behavior? Why is the numerical method breaking down? The fact is that there are lots of different numerical methods out there, and given any numerical method you can always find a differential equation, usually very easily, that will break that numerical method.

These numerical approximations usually you find, but often they can break down. We'll return to this idea of numerical methods in Lecture 6, but in the meantime, I'm going to return to our other subtheme, namely bifurcations. That will be the topic of the next lecture.

Bifurcations—Drastic Changes in Solutions
Lecture 4

We now introduce the concept of bifurcation, one of the subthemes in this course that will arise over and over. To bifurcate basically means to split apart or to go in different directions. In the theory of differential equations, to bifurcate means to have solutions suddenly veer off in different directions. In order to have such changes, our differential equation must depend on some sort of parameter. Sometimes, when we change this parameter just a little, we find a major change in the behavior of solutions; this is a bifurcation.

Let's start with a simple example: $y' = y^2 + A$. We know that for $A > 0$ there are no equilibrium points; for $A = 0$ there is 1 equilibrium point, a node; and for $A < 0$ there are 2 equilibrium points, at $\pm(-A)^{1/2}$. Because $F'(y) = 2y$, by first derivative test, the positive equilibrium point is a source and the negative is a sink. Look at the function graphs.

Figure 4.1

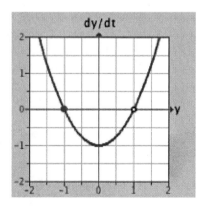

We see a bifurcation at $A = 0$.

In our limited population growth model

$y' = ky(1 - y/N)$

we have 2 parameters: the growth constant k and the carrying capacity N. If we vary these parameters, not much happens to our solutions (as long and both k and N remain positive). Now let's introduce a new parameter into this model to account for harvesting. Suppose our population is a fish population that is subject to a certain amount of fishing governed by the amount of fishing licenses that have been issued. We assume that the rate of growth of this fish population goes down depending on the amount of licenses issued, which is essentially the harvesting rate. Let h be this harvesting rate, so $h \geq 0$. Then one differential equation that illustrates this is the **limited population growth model with harvesting**:

$y' = ky(1 - y/N) - h.$

For simplicity, let's revert to our easier model:

$y' = y(1 - y) - h.$

First let's find the equilibrium solutions of this equation. Look at the graph of the right-hand side of the ODE (i.e., the graph of $y(1 - y) - h$) to see where this graph crosses the y-axis. When $h = 0$, there are only 2 places where $y' = 0$, namely $y = 0$ and $y = 1$. But as h increases, the graphs move downward and the 2 roots move closer together. At some h-value, they merge into a single root. Then, for slightly higher values of h, there are no roots at all. So our equilibrium points have disappeared. This is the drastic change in the behavior of solutions—this is our bifurcation point.

Figure 4.2

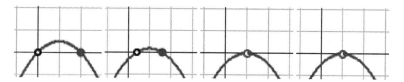

Two questions arise: When does this bifurcation occur? And what are the ramifications in terms of the solutions of our equation (the fate of the fish population)? To answer the first question, we must find the h-value for which our equation

$$y(1-y) - h = 0$$

has a single root. Rewriting this equation as

$$-y^2 + y - h = 0$$

allows us to solve it using the quadratic formula. We find that the roots are given by

$$q_\pm = \frac{1 \pm \sqrt{1-4h}}{2}.$$

So there is a unique root when $h = 1/4$. In this case, we get a single equilibrium point, namely $y = 1/2$. That is, at $h = 1/4$, our population levels off at exactly half the carrying capacity. When $h < 1/4$, we have a pair of equilibria at the 2 points q_+ and q_-. But when $h > 1/4$, these equilibria have disappeared.

By looking at the graphs above, we see that the equilibrium point given by q_+ is a sink when $h < 1/4$, while q_- is a source. Looking at the solution graphs, we see that all solutions that start out above the value q_- tend to the equilibrium solution at q_+. So as long as the population is high enough to begin with, harvesting at this rate is fine—the population survives. Even when $h = 1/4$, all is fine as long as our initial population starts out above the equilibrium point. But as soon as h goes below 1/4, disaster strikes: The population goes extinct.

Figure 4.3

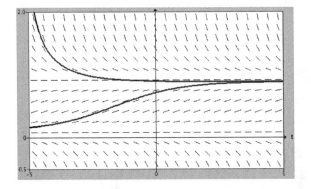

$h < 1/4$; population survives

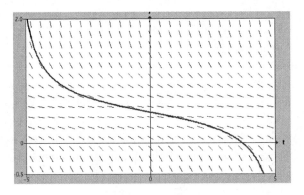

$h > 1/4$; population dies out

To summarize all of this, we can record all of our phase lines in a **bifurcation diagram**. In this plot, we let the harvesting parameter h lie along the horizontal axis. Over each h-value, we superimpose the corresponding phase line as well as all possible equilibrium points. This enables us to see the dramatic bifurcation that occurs when h passes through 1/4.

Figure 4.4

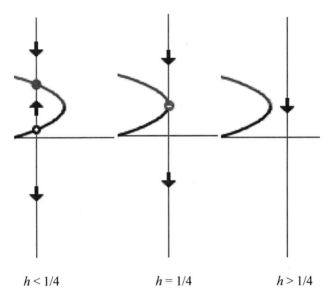

$h < 1/4$ $h = 1/4$ $h > 1/4$

This type of bifurcation is called a **saddle-node bifurcation**.

As another example of a bifurcation, consider the differential equation $y' = ay + y^3$. This equation has equilibrium points at $y = 0$ and $y = \pm\sqrt{-a}$. So there is only one equilibrium point when $a \geq 0$, but there are 3 equilibria when $a < 0$. We see this easily from the graphs of $ay + y^3$ below.

Figure 4.5

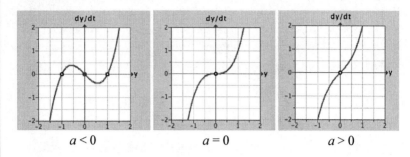

By the first derivative test, we see that 0 is a source when $a < 0$ and a sink when $a < 0$. Meanwhile, when $a < 0$, the other 2 equilibria are sources. We see this in the bifurcation diagram below.

Figure 4.6

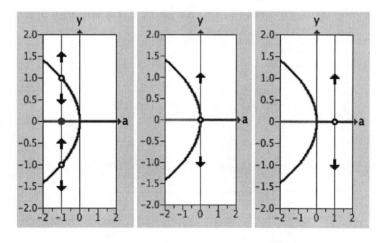

This type of bifurcation is called a **pitchfork bifurcation**.

Important Terms

bifurcation diagram (bifurcation plane): A picture that contains all possible phase lines for a first-order differential equation, one for each possible value of the parameter on which the differential equation depends. The bifurcation diagram, which plots a changing parameter horizontally and the y value vertically, is similar to a parameter plane, except that a bifurcation diagram includes the dynamical behavior (the phase lines), while a parameter plane does not.

limited population growth model with harvesting: This is the same as the limited population growth model except we now assume that a portion of the population is being harvested. This rate of harvesting can be either constant or periodic in time.

saddle-node bifurcation: In an ODE, this is a bifurcation at which a single equilibrium point suddenly appears and then immediately breaks into two separate equilibria. In a difference equation, a fixed or periodic point undergoes the same change. A saddle-node bifurcation is also referred to as a tangent bifurcation.

pitchfork bifurcation: In this bifurcation, varying a parameter causes a single equilibrium to give birth to two additional equilibrium points, while the equilibrium point itself changes from a source to a sink or from a sink to a source.

Suggested Reading

Blanchard, Devaney, and Hall, *Differential Equations*, chap. 1.7.

Guckenheimer and Holmes, *Nonlinear Oscillations*, chap. 3.1.

Hirsch, Smale, and Devaney, *Differential Equations*, chap. 1.3.

Strogatz, *Nonlinear Dynamics and Chaos*, chap. 3.

Relevant Software

Blanchard, Devaney, and Hall, *DE Tools*, Phase Lines.

Problems

1. Find the equilibrium points of the differential equation $y' = y - A$, and determine their type.

2. Are there any bifurcations in the previous differential equation?

3. Find the equilibrium points of the differential equation $y' = A$, and determine their type.

4. At which A-values does a bifurcation occur in $y' = A$?

5. Find the equilibrium points for $y' = y + Ay^2$.

6. Describe the bifurcation that occurs in the family $y' = Ay$.

7. Plot the bifurcation diagram for $y' = Ay - y^2$.

8. At which type of equilibrium points would you expect a bifurcation to possibly occur in the equation in problem 7?

9. Consider more generally $y' = B + Ay - y^2$. Fix a value of B, and plot the bifurcation diagram that depends only on A. For which values of B do you see a different type of picture?

10. Consider $y' = B + Ay - y^3$. Fix a value of B, and plot the bifurcation diagram that depends only on A. For which values of B do you see a different type of picture?

Exploration

Unlike in our example in this lecture, harvesting is not always constant. For example, harvesting of fish in certain locales is seasonal. Hence it makes more sense to allow the harvesting term to be a periodic function. Toward that end, use a computer to investigate the limited population growth model with periodic harvesting given by

$y' = y(1 - y) - h(1 + \sin(2\pi t))$.

Explain the behaviors and bifurcations that you see.

Bifurcations—Drastic Changes in Solutions
Lecture 4—Transcript

Welcome back again. As I've mentioned several times in the past, one of the subthemes in this course will be bifurcation, and that's what I'd like to explain today. To bifurcate means to split apart, to go in different directions. Let me bifurcate right now. But what is a bifurcation in differential equations? A bifurcation is similarly a splitting apart of solutions. Solutions go in different directions. More specifically though, in differential equations, a bifurcation is a sudden dramatic change in the behavior of solutions.

How do they arise? As we've seen, very often differential equations depend on parameters. I think Newton's law, force is mass times acceleration of mass, is a parameter. We talked about the damping constant in the mass-spring system. We saw the carrying capacity in the logistic differential equation. All of those were parameters. Usually when you change parameters, not much happens. If you move your parameters a little bit, the solutions look more or less the same. But sometimes very small changes in those parameter values can create drastic changes in the behavior of solutions. That's a bifurcation.

Let me begin with a simple example. Let's take the differential equation $y' = y^2 + A$. Here A is the parameter. So we have a whole family of differential equations. What are the equilibrium points? Well, when $A = 0$, we just have $y' = y^2$, we know we have a single equilibrium point at 0. In fact, we've seen that equilibrium point is a node. What about when A is greater than 0? Then the right hand side is $y^2 + A$. Both y^2 and A are now positive, so there are no equilibrium points when A is greater than 0.

What about when A is negative? Well, then we have $y^2 + A$. Set that equal to 0, you'll see you find two equilibrium points at $\pm\sqrt{(-A)}$. Remember, that makes sense. A is negative, $\pm\sqrt{(-A)}$. In fact by what we've done in the last lecture, the first derivative test for equilibria. You take the derivative of that differential equation, you get just $2y$, so what you see is the derivative at $+\sqrt{(-A)}$ is a source. And the derivative at $-\sqrt{(-A)}$ is negative, you get a sink. So we have a bifurcation at $A = 0$. So when A is negative, two equilibrium points, a sink and a source. When A is greater than 0, no equilibrium points at all, everything has changed. There's a drastic change in the behavior of solutions.

So let's see that by looking at what happens on the computer. Here we see our usual three pictures, the graph of $y^2 + A$, the phase line and our slope field. Here I'm drawing a case where $A = 0$. We see that we have a node. Nearby solutions tend up from below $y = 0$, and they also tend up above $y = 0$. But now when A goes negative, we suddenly see the birth of two equilibrium points. As we just said, we see a sink below and a source up above.

On the other hand, when A gets larger than 0, all of a sudden there are no equilibrium points. Notice what happens to solutions here. There are no equilibrium points, there's no way for the solution to stop increasing. This solution just goes off to infinity. So when $A = 0$ or below, you see that solutions that start out very low can't go even above the t-axis. They're bounded above by that equilibrium solution, but as A increases through 0, suddenly we get a dramatic change. Now solutions can go all the way from minus infinity to plus infinity. That's a dramatic change, that's a bifurcation.

Let's do another example. Remember we talked a while ago about the logistic population model. That seemed like a very simplistic model for population growth. But just wait until you see the parameters. Right now our logistic model was $y' = ky(1 - y/N)$. Here we had two parameters, one was k, that proportionality constant, the other was N, the carrying capacity. If we change k and N, not much happens when we vary them. On the other hand, let's add a new parameter. Let's add a new parameter to account for harvesting.

Suppose our population is fish, whatever. Let's assume that the rate of growth of the population of fish goes down depending on say the number of fishing licenses that are issued. Let's say the harvesting done is a constant rate. So let me let h be the harvesting rate. So the rate of growth of our population also goes down as h goes up. This is a constant harvesting rate. You don't often have a constant harvesting rate. Where I'm from in New England obviously the fishing is high in the summer and low in the winter. So the harvesting rate is periodic. But if you go to the South, you probably find a more constant harvesting rate.

So what is the differential equation? Let's incorporate that harvesting parameter in there. Our differential equation before was $y' = ky(1 - y/N)$. We want our rate of growth of the population to go down as the harvesting goes up, so let's subtract off h. We have three parameters in this differential equation. To keep it simple, let me just assume as before that k and N were 1. So our differential equation is $y' = y(1 - y) - h$. Let's look at the computer to see if we see some bifurcations.

Here is our previous picture of the slope field for the logistic equation. Let me also add in the phase line and the graph of the right hand side. This is the picture when $h = 0$ for $y' = 1 - y$. We've seen what happens here. We have a sink up above at $y = 1$, and $y = 0$ is a source. Now as I raise the harvesting rate, I raise h, I lower $-h$. The graph of the right hand side is going to change. There is the graph of $y(1 - y) - 0$, but as I lower that parameter, you see that that graph gets smaller and smaller, and at some point those two equilibrium points merge, and then again disappear.

So the first question is what is that merger point when suddenly a single equilibrium point appears and then splits into two. Let's work on that. Our differential equation is $y' = y(1 - y) - h$. We want to find our equilibrium points and specifically we want to find the h value where there's a single equilibrium point.

Let's take the right hand side, that's $y - y^2 - h$, set it equal to 0. You rewrite the right hand side as $y^2 - y + h = 0$, and solve and find the roots. What are the roots? Well that's a quadratic equation. We can find the roots by the quadratic formula. The roots of $y^2 - y + h = 0$ are by the quadratic formula it's $-(-1) \pm \sqrt{1 - 4(1(h - 4h))}$ all divided by 2. So we get two roots, I'll call them q_+ and q_-. So q_+ is 1 + that square root over 2. Q_- is 1 − that square root over 2.

And now notice under the square root sign we have $1 - 4h$. If $h = ¼$ we get 0 under the square root, we get that q_+ and q_- are both equal to ½. We have a single root when $h = ¼$. That's where that bifurcation occurred. We can go a little bit further. We can use the first derivative test and check that q_+ is a sink when h is less than a quarter, and q_- is a source again when h is less than ¼. But now when h is greater than ¼, $1 - 4h$ is negative, we have no roots at all.

So that's what we see. We go from a sink and a source when h is less than ¼, to a single root when $h = ¼$ and then to no equilibrium points when h is greater than ¼.

Let's look at that bifurcation again. Again, when h is just slightly below let's say ¼ we see that if we start with a sufficiently large population, that population is just going to tend to equilibrium. Well yes, if we harvest at that rate, and the initial population is very small, then that population is going to go extinct. But as long as we have a sufficiently large population, that population is fine at that harvesting rate. If we harvest at the rate ¼ where those two equilibrium points merge, still that's fine. As long as we have a sufficiently large population, that population survives. It tends to that limiting value. No matter what initial population we start with, as long as it's sufficiently large, harvesting at that rate is fine.

Now let's give out one more fishing license. Let's harvest at a higher rate. All of the sudden mathematically that equilibrium point disappears. All of a sudden, all of our solutions go off to minus infinity. All of the sudden, our population has gone extinct. Think about it, when we're harvesting at a rate of ¼ everything is fine as long as we have a sufficiently large population. But if we change that harvesting rate just a little bit, equilibrium points disappear and the population goes extinct.

In fact, I have a major story in my life concerning that. Back when I was in high school, in fact all the way through college I used to spend my summers working on a fishing boat in New Hampshire. Back then what we'd fish for mostly was haddock. Back then when you had fish and chips in New England, your fish would be haddock, and it was great. We used to love to fish for haddock. There were plenty of them.

But curiously as time went on, people, ecologists started saying to the fisherman, you'd better stop fishing at this rate because someday the haddock population is going to disappear. Well, all the captains that I was working with would say these guys are crazy. We go out every year and we always catch the same number of fish. We fish a little bit more, same number of fish. This will never happen. Well luckily, I went off to college and a few years later, it happened. They went out fishing and there were essentially no haddock there. The haddock population had just disappeared. The population went through a bifurcation. I'm glad I went on in mathematics.

In any event, what you see is this rather simplistic differential equation, this logistic population limited growth model. When we add in a simple parameter for bifurcations, can predict what's really going on in the real world. You take much more sophisticated differential equations and yes, they're telling you the same things too. There's a real connection between these simple differential equations and what's going on.

This differential equation is an example of a saddle node or tangent bifurcation. What I mean by saddle node is well we'll see this in higher dimensions. That's where saddles merge with nodes. But for some reason mathematicians call this bifurcation a saddle node or a tangent bifurcation.

So let me look again at this bifurcation, but this time displaying it in a different way. So here's our logistic population model again. As we've seen, as we lower A, the graph becomes tangent to the y-axis and then disappears. In each case, we get a different phase line. Before the bifurcation, we see a sink and a source. At the bifurcation, we saw a node. Beyond the bifurcation, we saw everything run off to extinction.

Let me record all of these behaviors in what we call the bifurcation plane. The bifurcation plane will be a record of all of the phase lines in terms of parameter. That's what I'm plotting over here. Over each of those parameters h, what I'll draw is a copy of the phase line. So when A is small, we see a pair of equilibria. I've actually also sketched in the equilibrium points, the green equilibrium, which is a sink, and the red equilibrium point which is a source. As I change my parameter, you see that those two equilibria come closer and closer together, then merge at the bifurcation value and then disappear.

So the bifurcation diagram is a record of all possible phase lines, and in fact, a picture of where your bifurcations occur. This is our first example of a picture of a parameter plane. Actually, it's a parameter line here. We're plotting the parameter horizontally and the y-axis vertically. Now we've seen 4 different pictures, 4 recording the behavior of solutions of first order differential equations. Our solution graphs and slope fields, our phase lines, the graph of the right hand side, and finally the bifurcation diagram. As we go on to more complicated systems, we'll see more and more such pictures of different phenomena occurring with differential equations.

Let's do one more example of a differential equation that depends on a parameter. Let's look at $y' = Ay + y^3$. So A is our parameter here, y' is the parameter $y + y^3$. Where are the equilibrium points? Well, pretty clearly the right hand side always vanishes when $y = 0$, so we always have an equilibrium point at $y = 0$. Let's factor out that y term. We get $y(A + y^2)$, so we see we have additional equilibrium points at $\pm\sqrt{(-A)}$. So we have no equilibrium points of the form $\pm\sqrt{(-A)}$ if A is greater than 0. Then we have the square root of a negative number.

So when A is greater than 0, we have a single equilibrium point at $y = 0$. If A is equal to 0, again a single equilibrium point at $y = 0$. But if A becomes less than 0, suddenly we have three equilibrium points at 0 and at $\pm\sqrt{(-A)}$.

So let me figure out what's happening at those equilibrium points. We can do that using the first derivative test. The right hand side of our differential equation is $Ay + y^3$. The derivative of the right hand side is therefore $A + 3y^2$. So what are the types of our equilibrium points? The equilibrium point at 0 is, the derivative at 0 is $A + 3(0)$, so 0 is a source. If A is greater than 0 the derivative is positive. If the equilibrium point at 0 is a sink if A is negative. If A is negative the derivative is A, that's negative, we have a sink.

Now you can just plug in $\pm\sqrt{(-A)}$ to that equation and to see that when we have those two new equilibrium points, they're both sources. So we have another bifurcation. As A changes, we go from having a single equilibrium point when A is positive, to three equilibrium points when A is negative, one sink and two sources. This is what we call a pitchfork bifurcation.

Here is the picture of the slope field for the differential equation $Ay + y^3$. This is the case where A is pretty large. Let me go to the case where A is 0. That's where we're going to have a bifurcation. When A is 0 our differential equation is just $y' = y^3$. What we see is the equilibrium point at 0 is now a source. If A is larger than 0, same thing happens. The equilibrium point at 0 remains a source. We see in the graph of $Ay + y^2$ there's only one root, one place where that graph crosses the y-axis.

But as I lower A, you see that the graph of that function suddenly pierces the y-axis in two more points. We get two additional equilibrium points. The derivative at each of these equilibrium points is positive as we saw by the first derivative test, those equilibria are sources. On the other hand, the derivative at 0 is negative. The equilibrium point at 0 has become a sink. We see that in the graphs of the solutions. Solutions tend in to 0, but tend away from the upper equilibrium point and tend away from the lower equilibrium point.

So let's record what's happening here again in the bifurcation plane. Here's a picture of what's going on as we vary A. Remember in the bifurcation diagram we're plotting A horizontally, that's our parameter. Over each given A value, we erect the corresponding phase line. So when A is 0 we see a single equilibrium point, which is a source. When A becomes positive, suddenly nothing happens. We still have a single equilibrium point, that's a source. But when A becomes negative, we see the appearance of two equilibrium points, a sink in the middle and a source on either side.

So again, this bifurcation diagram is a nice record, a nice animation of what's happening in terms of the bifurcation. As I said, you see why this is a pitchfork bifurcation. In the bifurcation diagram, I've plotted the locations of all of the equilibrium points and you see that these equilibrium points seem to form a pitchfork.

So this is a second example of a bifurcation. We've seen the saddle node where suddenly we go from having no equilibrium points to having a pair of equilibrium points. Here with a pitchfork bifurcation, we've gone from having a single equilibrium point, which persisted, and two more branched off. But at the same time, that single equilibrium point that persisted changed its type. It went from a source with no equilibria nearby to a sink with a pair of sources nearby, a second type of equilibrium. We'll see as we go into higher dimensional differential equations that many, many more kinds of bifurcations that can occur, and many of them are very interesting.

Here's another family of differential equations. Let's look at the family $y' = r + ay - y^2$. We have two parameters in this differential equation. Here's the picture of what's happening. Let me fix an r-value and vary a. Then what I'll do is plot those same 4 pictures, slope field and solution graphs, phase line, graph of the right hand side and finally the bifurcation diagram depending on a.

So if I fix r equal to 0, and let a vary, we see another bifurcation. Initially when both r and a are equal to 0, we have a single equilibrium point at the origin. It looks like a node. If I let a get larger, then we see that single equilibrium point splits into a pair of equilibria, a sink and a source. As we let a decrease, those two equilibrium points merge, and then suddenly reappear on the other side as two equilibrium points, again a sink and a source. So we go from having two equilibrium points to a single equilibrium point and then back to two equilibrium points, a bifurcation, a kind of a minor bifurcation.

Now let r be something different. Let the parameter say r be negative. What do we see here? Well, now as we vary a, initially when a is large, we see a pair of equilibrium points, one a sink and one a source. As we lower a, suddenly these two equilibrium points emerge and they disappear. We have gone through a saddle node bifurcation. The sink and the source come together and then suddenly they disappear. All solutions now run off to minus infinity. All solutions, the graphs of the solutions, tend off to minus infinity. It went off to minus infinity.

So these families of differential equations that depend on parameters can have multiple bifurcations. We can analyze them both visually, as we're doing now, or we could sit down and compute those equilibrium points, find out whether they're sinks, sources etcetera.

Now if I let r be something positive, we see there are no bifurcations. As I vary a, we see a sink up above, and a source down below and nothing happens. There are no bifurcations in this region here. It's when we change r that we see these bifurcations happen. So it's an interesting way to animate and I think enliven the study of differential equations.

Let me just show you a couple more kinds of bifurcations that occur in differential equations. Here's another family. The family is y', it now depends on two parameters, $r + ay - y^3$. You have two parameters r and a. Let me fix r say at 0, and let a vary. If I let a vary, what you see is when r is 0 we're going through essentially our previous pitchfork bifurcation. On one side, you see a single equilibrium point, a sink. That sink suddenly splits into two equilibria. This time the sink becomes a source and the two new equilibria are both sinks. But if I change that r-value, then something else happens. When I vary a, you see for a while, I have a pair of equilibrium points that suddenly come together and merge in a saddle node bifurcation, and I'm left with just one equilibrium point.

One equilibrium point persists and suddenly we see the appearance of two more, a saddle node bifurcation with an accompanying other equilibrium point. On the other hand, when r is positive, what you see is the exact opposite. If I vary a, now the two lower equilibrium points come together in a saddle node bifurcation and disappear leaving me with just one equilibrium point. That's another way that bifurcations occur in differential equations.

Here's one other example. Let me take the differential equation $y' = r + ay - y^2$. So $r + ay - y^2$ instead of y^3, again, two parameters r and a. Let me fix r and vary a. What you see is initially we have a node at the origin, but when a becomes positive, two equilibrium points emerge, one is a sink and one is a source. As a goes through 0, those two equilibrium points merge, and then reemerge as a pair of new equilibrium points. So we go from having two equilibria when $a = 0$ a single equilibrium, and then when a is larger than 0, again a pair of equilibrium.

Now if I let r be negative, what happens is sometimes we lose those equilibrium points. When a is large, we see a pair of equilibria. They come together in a saddle node bifurcation, and suddenly the solutions all go off to minus infinity. There are no equilibrium points. You have a little avenue where we can go off to minus infinity. It takes a while to get there, but they do go off to minus infinity.

Now let a be more negative, and then more suddenly again you see the emergence of a saddle node bifurcation, a pair of equilibrium points, a sink and a source. Finally, if r is positive, you see there are no bifurcations at all. As I vary a, my sink lives alone and my source lives alone. So what you see is when we vary parameters in differential equations, you can get all sorts of different behaviors, but sometimes you can get dramatic changes in those behaviors. Those are bifurcations that play a real big role in real life.

So that's our first encounter with bifurcations. We'll see many more of them as we go along. Next time though I'm going to back up and go back to not the old method, but the tried and true method of actually solving differential equations. Finding the solutions of differential equations you get by explicit formulas. Not qualitative methods, we'll blend those in. But we'll find explicit formulas for solutions of certain kinds of differential equations.

Methods for Finding Explicit Solutions
Lecture 5

In this lecture we describe several standard methods for solving certain types of first-order differential equations, namely, linear and separable first-order equations. To find these explicit solutions, we need to invoke another tool from calculus, integration (or antidifferentiation). Finding an (indefinite) integral or antiderivative is basically the opposite of finding the derivative; the integral of a given function is just a function whose derivative is equal to the given function.

For example, the integral of the function t^n for $n > 0$ is the function $t^{n+1}/(n + 1) + C$, where C is some constant, since the derivative of $t^{n+1}/(n + 1) + C$ is $(n + 1)t^n/(n + 1) = t^n$. We denote the integral of the function $y(t)$ by

$$\int y(t)\,dt.$$

The dt indicates that the independent variable is t.

Other examples from calculus that we will use are the following, in which $\ln(t)$ is the natural logarithm function.

$$\int e^{kt}\,dt = e^{kt}/k + C$$

$$\int \frac{dt}{t} = \ln|t| + C$$

$$\int \frac{dt}{1-t} = -\ln|1-t| + C$$

A first-order linear differential equation (with constant coefficients) is a differential equation of the form

$$y' + ky = G(t).$$

The function $G(t)$ may be nonlinear, however, and this expression is sometimes called the forcing term. If $G(t) = 0$ (i.e., we have no forcing term), then our equation is simply $y' = ky$. This is called a first-order linear homogeneous equation, and we know its general solution, $y(t) = Ce^{kt}$. To solve these equations, we use the "guess and check" method. We know how to solve the homogeneous part of this equation; the solution is $y(t) = Ce^{kt}$. We then make an appropriate guess to find the general solution of the nonhomogeneous equation.

As an example of this method, consider $y' + y = e^{3t}$. The solution of the homogeneous equation $y' + y = 0$ is $y(t) = Ce^{-t}$. So we would make a guess of the form $y(t) = Ce^{-t} + Ae^{3t}$. The question is, what is A? Plugging this into the ODE yields $A = 1/4$, so our specific solution is $y(t) = Ce^{-t} + (1/4)e^{3t}$. The solution resembles e^{-t} as time goes backward and e^{3t} as time goes forward.

Figure 5.1

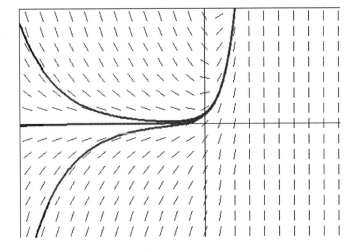

Another example of a first-order linear ODE is **Newton's law of cooling**. Suppose you have a cup of coffee whose temperature at the moment is 170°. Suppose also that the ambient temperature is 70°. Can you find a function $y(t)$ that gives the temperature of your coffee as a function of t? Newton's law of cooling says that the rate of cooling is directly proportional to the difference between the current temperature and the ambient temperature. As an ODE, this says that $y' = k(y - 70)$. If we assume that the coffee is initially cooling at a rate of 20° per minute, our equation at time $t = 0$ reads $-20 = y'(0) = k(y(0) - 70)$. So $k = -0.2$, and the differential equation is $y' = -0.2(y - 70)$ or $y' + 0.2y = 14$, which is first-order, linear, and nonhomogeneous.

We know how to solve the homogeneous equation $y' = -0.2y$; the general solution of this equation is $y(t) = Ce^{-0.2t}$. So plugging this into the left-hand side of the ODE yields 0. We want to find a function that we can plug into the left-hand side to get out the constant 14. So let's guess a solution of the form $y(t) = Ce^{-0.2t} + A$, where A is some constant. Plugging this expression into the differential equation yields $-0.2Ce^{-0.2t} + 0.2Ce^{-0.2t} + 0.2A = 14$, so $A = 70$. Therefore our explicit solution is $Ce^{-0.2t} + 70$. We want the solution that satisfies $y(0) = 170$, so we put this initial value into our solution, which yields $170 = Ce^0 + 70$, or $C = 100$. Thus our specific solution is $y(t) = 100e^{-0.2t} + 70$.

Another method for solving this equation is **separation of variables**. This equation is separable because we can get all the y's on the left and all the t's on the right. That is, we can rewrite the differential equation given by

$$\frac{dy}{dt} = -0.2(y - 70)$$

in this manner:

$$\frac{dy}{y - 70} = -0.2 \, dt \, .$$

Now we integrate both sides of this equation, the left side with respect to y and the right with respect to t:

$$\int \frac{dy}{y-70} = \int -0.2 \, dt.$$

The integral of the left is just $\ln(y - 70)$ + constant while on the right we find $-0.2t$ + constant. Lumping both of the constants together and calling them D, we find that

$$\ln(y - 70) = -0.2t + D.$$

Exponentiating both sides, we find

$$y - 70 = e^{-0.2t+D} = Ce^{-0.2t},$$

where we have written the term e^D as the constant C. Thus, exactly as above, we find the solution $y(t) = Ce^{-0.2t} + 70$, and this is easily seen to be the general solution.

Now let's return to the limited population growth model given by $y' = y(1 - y)$. This equation is separable, so we are left with doing the 2 integrals:

$$\int \frac{dy}{y(1-y)} = \int dt.$$

The right-hand integral yields t + constant, whereas the left-hand integral is more complicated. We can break up this complicated fraction into 2 "partial fractions" this way:

$$\frac{1}{y(1-y)} = \frac{1}{y} + \frac{1}{1-y}.$$

So our integrals on the left reduce to

$$\int \frac{dy}{y} + \int \frac{dy}{1-y} = \ln|y| - \ln|1-y| + const.$$

Let's assume that $0 < y < 1$, so both of the terms inside the absolute values are positive. Therefore we are left with

$$\ln(y) - \ln(1-y) = t + C$$

for some constant C. Exponentiation plus some algebra yields

$$y(t) = \frac{De^t}{1 + De^t}.$$

These functions have graphs, below, that we have seen before.

Figure 5.2

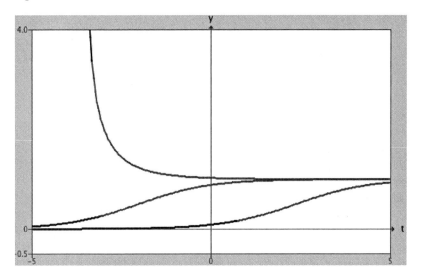

If we were to assume $y > 1$, we would find the same solution. So is this the general solution? Not quite. If we try to find a value of D that solves the initial value problem $y(0) = y_0$, we must solve the equation

$$y_0 = y(0) = \frac{De^0}{1+De^0} = \frac{D}{1+D}.$$

Doing the algebra yields

$$D = \frac{1}{1-y_0},$$

so we have found such a D-value as long as y_0 is not equal to 1. But we know the solution to the initial value problem $y(0) = 1$; this solution is just the equilibrium solution $y(t) = 1$. So we do in fact have the general solution once we tack on this solution.

Taking the limit as t goes to infinity of

$$y(t) = \frac{De^t}{1+De^t} = \frac{D}{e^{-t}+D}$$

yields 1, so all solutions tend to the equilibrium solution $y(t) = 1$, just as we saw earlier.

Important Terms

Newton's law of cooling: This is a first-order ODE that specifies how a heated object cools down over time in an environment where the ambient temperature is constant.

separation of variables: This is a method for finding explicit solutions of certain first-order differential equations, namely those for which the dependent variables (y) and the independent variables (t) may be separated from each other on different sides of the equation.

Suggested Reading

Blanchard, Devaney, and Hall, *Differential Equations*, chap. 1.2.

Hirsch, Smale, and Devaney, *Differential Equations*, chap. 1.2.

Roberts, *Ordinary Differential Equations*, chap. 1.3.

Strogatz, *Nonlinear Dynamics and Chaos*, chap. 2.3.

Problems

1. First let's review some integral calculus. Find the integrals of the following functions.

   a. $y(t) = t + 5$

   b. $y(t) = t^2 + 2t + 1$

   c. $y(t) = 1/t^2$

   d. $y(t) = e^{-t}$

   e. $y(t) = 0$

2. Solve the equation $\ln(t) - 4 = 0$.

3. Solve the equation $e^t - 4 = 0$.

4. Find the general solution of $y' - y = 0$, and sketch the graphs of some solutions.

5. Find the general solution of $y' - y = 1$, and sketch the graphs of some solutions.

6. For the unlimited population model with $y > 1$, check that solutions assume the form discussed in the lecture.

7. Find the general solution of the ODE $y' = y^2$.

8. Suppose you have a cup of coffee whose temperature at the moment is $200°$, and the ambient temperature is $80°$. Suppose that 1 minute later the temperature of the coffee is $180°$. Find the function $y(t)$ that gives the temperature of this cup of coffee as a function of t.

9. Find explicit solutions of the differential equation $y' = 1 - y^2$ satisfying the initial conditions $y(0) = 0$, $y(0) = 1$, and $y(0) = 2$.

10. What is the general solution of the equation $y' = 1 - y^2$?

Exploration

If you are familiar with other techniques of integration, you can find solutions of some more complicated differential equations (though most such equations do not have solutions that can be found explicitly). Refresh your knowledge of integration techniques to find the general solutions of the following differential equations.

1. $y' = 1 + y^2$
2. $y' = e^t y/(1 + y^2)$
3. $y' = y + y^3$

Methods for Finding Explicit Solutions
Lecture 5—Transcript

Here we go again. In this lecture, we're going to take a slightly different tact, actually a major different tact. We're going to describe several methods for finding explicit solutions of differential equations. As I mentioned earlier in the old days back when I was a fisherman in New Hampshire, that was the only way we could solve differential equations. Now of course we have so much more. But that's not to diminish the importance of being able to solve these differential equations. Yes, if you have a differential equation and you can find an explicit formula, you can predict exactly what's going to happen for all future times.

So to understand how to solve these differential equations, we need several more tools from calculus. First we need integration. Remember what the integral is. It's the anti-derivative. I'll denote the integral of a function $y(t)$ by $\int y(t)\, dt$. Then the derivative of the integral of $y(t)$ is just $y(t)$. The derivative undoes what the integral does. That's why the integral is the anti-derivative.

So let's start with a couple of examples. First, what's the $\int t^2\, dt$? Well the $\int t^2\, dt$ is $t^3/3$ plus some constant. Because if you differentiate $t^3 + 3$, you get $3t^2$ divided by 3, that's just t^2, and the derivative of the constant is just 0. So the derivative of $t^3/(3 + C) = t^2$. The $\int t^2\, dt$ is $t3/(3 + C)$.

Here are a couple more. What's $\int t^n\, dt$? Well that's just $t^{n+1}/(n + 1) + C$. Since the derivative of $t^{n+1}/(n + 1)$ is just tn, and again the derivative of the constant is 0. Whenever you do the integral of some function, you always get a new function plus a constant. Another example $\int e^t$, the exponential function. The $\int e^t\, dt$ is just $e^t + C$. Of course, the derivative of e^t is itself.

Back to trig, the integral of sin function dt. The derivative of sin is cosign, but what's the integral? It's not cosign because the derivative of cosign would be −sin, the interval of sindt would just be −cos + C. And similarly the $\int$ cos with respect to t is the sin function plus a constant.

Here's one more topic that'll come up. As we've seen in differential equations, the exponential comes up all the time. Going backwards the inverse of the exponential will come up all the time. That's the natural logarithm function. I'll denote it by $ln(t)$. Now the natural logarithm is only defined for t greater than 0. So the $ln(t)$ is only defined for t greater than 0.

The natural logarithm is the inverse of the exponential function. What that means is if you take the exponential applied to the natural logarithm function of t, you get back just t. That is the exponential undoes what the natural logarithm does to t, $\exp(ln\ t) = t$. So what's the derivative of the natural logarithm function? Let's just differentiate that formula we just wrote down. The derivative of $\exp(ln\ t) = t$ is the derivative of the left hand side is just $e^{ln(t)}$, but now by the chain rule times the derivative of $ln(t)$. All of that is equal to the derivative of t on the right.

So put that together we get $e^{ln(t)}$, that's just t, times the derivative of the $ln(t) = 1$. Or, solving that equation you get the derivative $ln(t) = 1/t$. So the derivative of a natural logarithm function is this kind of interesting rational function $1/t$.

Here are a couple more examples now of integrals. What's $\int dt/t$? Well we just saw that the derivative of the $ln(t) = 1/t$, so the $\int dt/t$ is the $ln(t) + C$. Here again we're assuming that t is greater than 0. If t happened to be less than 0, we'd get that it would be equal to $ln(-t)$. You're taking the integral of $1/t$ where t is negative you get $ln(-t)$. You'll make that t positive.

Here's another example, how about the $\int dt/(1-t)$. It's the integral of the function $1/(1-t)$. You might think that might be just $ln(1-t) + C$, but be careful. The derivative of $ln(1-t) = 1/(1-t)$ times a derivative of $(1-t)$, that's -1. We want to just $1/(1-t)$, not $-1/(1-t)$. So the integral of $1/(1-t)$ is actually $-ln(1-t) + C$. Here too we're assuming that $1-t$ is positive.

What I'd like to do now is solve two kinds of differential equations. The first is one that's going to come up a lot in the future. These are linear differential equations. We'll solve first order linear differential equations. These are differential equations of the form $y' + ky = G(t)$. This could be hopelessly nonlinear. In any event, this is a linear differential equation because the y terms stand alone, $y' + ky = G(t)$.

We're going to solve this in two steps. The first thing we're going to do is solve the so-called homogenous linear equation. That's when we get rid of $G(t)$, where $G(t)$ is 0. We have just $y' + ky = 0$. We know how to solve this. We've done this before. We have $y' = -ky$, so we know one solution is e^{-kt}. In fact we know the general solution of this homogenous is some constant, call it Ce^{-kt}.

Now we have to move to the solution of the nonhomogeneous equation where we put that $G(t)$ term back in. This is how we're going to do it. The easiest way is the guess and test method. We're just going to guess one solution of the nonhomogeneous equation. Then if we guess correctly, if we have one solution of the nonhomogeneous equation, then our guess plus that previous solution a Ce^{-kt} is also a solution for the nonhomogeneous equation. In fact, it's the general solution.

Why is that? Think about it. We plug our guess into the differential equation, $y' + ky$ applied to our guess we get out on the right the right hand term. We get out $G(t)$. On the other hand, if we plug that second term in Ce^{-kt} we get out 0. So when we plug in both of these, we get out $G(t) + 0$, we get a solution of the differential equation. Since Ce^{-kt} is the general solution of the homogeneous, this guess plus Ce^{-kt} is the general solution of the entire equation, the nonhomogeneous equation.

So how are we going to guess? Let me do an example. Let's take $y' + y = e^{-3t}$. Our first step would be to solve the homogeneous equation. No problem, that's $y' + y = 0$ where $y' = -y$ we know that the general solution is Ce^{-t}. So we have to guess some solution of $y' + y = e^{3t}$.

What are you going to guess? Would you guess a sin function? No way to plug in sin you're going to get y' sin, derivative sins is cosign + y. You're going to get sin and cosigns. Would you guess t^5? No, you'd get back polynomials. Would you guess tangent? No, you wouldn't. What you'd clearly guess here is some constant times e^{3t}. That would be the only way to get an e^{-3t} out when you plug it in to the differential equation.

So let's do that. Let's guess a constant, say Ae^{-3t}. Throw that into the differential equation. What do you get? On the left you have to take $y' + y$, so you get the derivative of Ae^{3t} is $3Ae^{3t} + y$, that's $+ Ae^{3t}$. All of that has to be equal to on the right e^{3t}. But on the left we really have $3A + A(4Ae^{3t}) = e^{3t}$ or $A = ¼$. So we guessed correctly. We have one solution of our nonhomogeneous equation. That's $¼e^{3t}$. So by what we saw earlier the general solution is $Ce^{-t} + (¼)e^{3t}$. That's the general solution of the nonhomogeneous equation.

Here's the typical graph of a solution of that differential equation. As time goes backwards this solution looks like Ce^{-t}, e^{-t} increases in backward time. But in forward time this solution looks like $¼e^{3t}$. E^{3t} increases in forward time whereas Ce^{-t} goes to 0. Similarly, all these solutions look like e^{-t} in backwards time and e^{3t} in forward time. So there's the picture of the graphs of these solutions of the differential equation. Luckily, we've been able to find them explicitly.

Let me give another example of that. Here's Isaac Newton again. Here's Newton's Law of Cooling. Suppose you have a cup of coffee whose temperature at the moment is 170°, and suppose you're sitting out on a deck someplace and the ambient temperature is 70°. What you'd like to know is what is the temperature of the coffee as time goes on. How does the coffee cool down?

We're looking for the temperature of the coffee. Let me again call that $y(t)$. So we know that $y(0) = 170$. So what is Newton's Law of Cooling? As usual with a model, we have to make an assumption. Here's the assumption. Newton's Law of Cooling says that the rate of cooling, that's the derivative of y, is directly proportional to—read is equal to a constant to—the difference of the current temperature and the ambient temperature. The difference between $y(t)$ and our ambient temperature namely 70°. So the differential equation here is $y' = k(y - 70)$. The rate of cooling, the rate of change of temperature of the coffee is proportional to the difference of the temperature and the ambient temperature.

Now let me assume that the coffee is initially cooling at a rate of say 20° per minute. I could also assume that the temperature was 70° at the starting time and 65° two minutes later and get the same result, but it's easier to do it this way. So I assume that the coffee is initially cooling at a rate of 20° per minute.

Now remember $y(0)$, the initial temperature is 170°. So what we have is the initial cooling rate is -20, that's $y'(0)$. That must be equal to k, that constant, times $y(0) - 70$. But we know that our initial temperature is 170, so that must be equal to $k(170 - 70)$, that's 100. So $-20 = k(100)$, so $k = -0.2$. So we have the differential equation for this specific cup of coffee $y' = -0.2(y - 70)$ or $y' + 0.2y = 14$. That's a first order, linear, nonhomogeneous term. That function $G(t)$ is now 14. So let's go ahead and solve that.

Remember what to do. First thing we use is to find the solution of the homogeneous equation. That's $y' + 0.2y = 0$, or $y' = -0.2y$. y' again is equal to a constant times y. The general solution is some constant $Ce^{-0.2t}$. There's the general solution of the homogeneous. The next thing is to guess the solution of the nonhomogeneous equation.

What will you guess here? How would you get out that right hand side term 14? Again, would you guess sin or cosign? There's no way. The only way you'd get that out is by guessing a constant. So let me make the guess $y(t) = A$, and determine what A is. So I plug that into the equation. The derivative of $A = 0 + 0.2y$, that's $+ 0.2A$ must be equal to 14, $0.2A = 14$, so $A = 70$. So we have our general solution. The general solution is $70 + Ce^{-0.2t}$.

Since $y(0) = 170$, we know that 170 equal to $y(0)$ must be equal to $70 + Ce^0$. So we find that $C = 100$. $y(t) = 100$, that's C times $e^{-0.2t} + 70$ is the solution of our initial value problem.

What does this mean for your cup of coffee? Well the solution is $100e^{-0.2t} + 70(0)$. We're starting out at temperature 170, and what happens to that first term? It's a decaying exponential. $E^{-0.2t}$ goes to 0 as time goes on. As that term goes to 0, you're left with just 70. So guess what, your coffee will cool off and end up at exactly the ambient temperature. You won't get a cup of ice coffee unfortunately.

Let me now turn to the same differential equation but solve it a different way. I'm going to solve this by the method of separation of variables. So here was our differential equation again. Let me write it as dy/dt or y prime, $= -0.2(y - 70)$. The method of separation variable says if you can get all the

ys on the left and all the ts on the right, then provided you can do an integral, you're done. You've got the solution.

So look at our differential equation. We've got dy/dt. I'm going to write it as dy/dt now, $= -0.2(y - 70)$. I can divide both sides by $(y - 70)$, get all the ys on the left. I can multiply both sides by dt, you get $-0.2dt$ on the right. Oh boy, yes, I have all the ys on the left, but how can you multiply dy/dt by dt? Well, this is a trick. You'll see that it works. It gives us the same solution. So I'm done. I have all of the ys on the left, and all the ts on the right. That was a separable differential equation.

Now what I'm going to do is integrate both sides of that equation. On the left I have $\int dt/(y - 70)$, and on the right I have $-0.2dt$. Integrating them is easy. Right, the integral of the left hand side $1/(y - 70)$ is just the $ln(y - 70)$. Remember y is our temperature. It's bigger than 70, so $(y - 70)$ is positive. So on the left we get $ln(y - 70)$. On the right, we had to integrate $-0.2dt$. But the integral of $-0.2dt$ is just $-0.2t$ plus some constant, and let me now call it D.

So we are almost there. We have $ln(y - 70)$ on the left. We now want to solve for y. Let me exponentiate both sides, $e^{ln(y-70)}$ is just $(y - 70)$. The exponential of $-2t + D$ is $e^{-0.2t} + D$. I can combine, I can use the properties of exponential to say $e^{-0.2t + D}$ is $e^D e^{-0.2t}$. But if D is an arbitrary constant, then e^D is an arbitrary constant. So let me just call e^D C. What I get on the right is $Ce^{-0.2t}$. Then now bring the 70 to the right, we get the exact same solution as before, $y(t) = 70 + Ce^{-0.2t}$. Then go solve the initial value problem to determine that $C = 100$. So again, your coffee cools off to the ambient temperature.

Let's finally go back to that differential equation that has come up several times in the previous lectures, the logistic or limited population growth model, $y' = ky(1 - y)$. I'm not going to include any harvesting here and I'm setting k to be 1 and the carrying capacity to be 1 as usual. So how do we solve this differential equation? How do we find a specific solution?

Notice that this differential equation is also separable. We can get all of the ys on the left and all the ts on the right. Dy/dt on the left divide by $y(1 - y)$

and multiply both sides by t/dt on the right, we get $dy/y(1-y) = dt$. So we've separated this equation so now let's integrate.

So we have two things to integrate. We have to integrate dt on the right hand side, but that integrates to just t plus a constant. Then we also have to integrate $1/(y(1-y))$. But $1/(y(1-y))$ can be broken up into two simpler fractions. The integral of $1/(y(1-y))$, that looks pretty hard to do. But I can break that up into $1/y + 1/(1-y)$. So all I have to do on the right is integrate $1/y$ and $1/(1-y)$.

Let me assume that the population is between 0 and 1. So the integral of $1/y$, y is positive. The integral of $1/y$ is just $ln(y)$. As we saw earlier the integral of $1/(1-y)$ is $-ln(1-y) + C$. So when we integrate, $1/y + 1/(1-y)$ we get $ln(y) - ln(1-y) + C$.

So all together our integrals yield $ln(y) - ln(1-y)$ on the left, and $t + C$ on the right. Now we have to solve for y. How are we going to do that? Let's exponentiate both sides of that equation. When we exponentiate the right we get e^{t+C}. That again is $e^t(e^C)$. But if C is an arbitrary constant, e^C is an arbitrary constant, call it D. When we exponentiate the right we get De^t.

Now what happens when we exponentiate the left? We have $e^{ln(y) - ln(1-y)}$. That breaks up into, by properties of the exponential, $e^{ln(y)}(1/e^{ln(1-y)})$. Remember the minus in front of $ln(1-y)$. But $e^{ln(y)}$ is just y, and $e^{ln(1-y)}$ is just $(1-y)$. So we end up on the left with $y/(1-y) = De^t$. So now we can solve that equation.

Let me multiply both sides by $1 - y$. We get $y = De^t - yDe^t$. Or bringing the y back to the left we get $y + yDe^t = De^t$. Or dividing both sides by $(1 + De^t)$ we find our solution $y(t)$. $y(t)$ is just De^t divided by $(1 + De^t)$.

This was what happens when y was between 0 and 1. If y happened to be bigger than 1, then that $(1-y)$ term would be negative. When it involves logarithms, you have to involve a negative there. What you'll end up with is a different constant when you've solved that differential equation. Call that constant C. It doesn't matter what it is, you'll get the same analytic expression. So we found the solution to the logistic population model. It's $De^t/(1 + De^t)$. Let's go back and see what this actually looks like graphically.

Here again are our solutions for the logistic population model. Just as we saw using graphical techniques, solutions increase from the equilibrium point at 0 up to the equilibrium point at 1. Each of these that I'm drawing is the graph of some function $De^t/(1 + De^t)$ for different values of D.

Now we have a very different situation. We have graphs of actual functions. If we wanted to predict the behavior of the solutions, we now have a formula for it. We can tell exactly what's going to happen for future times. You want to know what the population is at time 50, you just plug it into the $De^t/(1 + De^t)$.

Of course the question is what is D? To find out what D is you just plug in your initial value. If you knew your initial population that would determine what the value of D is. Instead of doing that, let me answer the question is this the general solution of the differential equation, of the limited growth population model. Remember what the general solution is. The general solution is a family of solutions, that's what we have $De^t/(1 + De^t)$. It's a family of solutions. But that family of solutions must enable us to find a solution that satisfies any initial condition.

So can we find a solution of the form $De^t/(1 + De^t)$ that satisfies $y(0)$ is some given number, say $y(0)$ whatever? What do we have to do? We have to solve the equation $y(0)$ = the given number $y(0) = De^t/(1 + De^t)$ with 0 plugged in. But when $t = 0$ that right hand side is just $D/(1 + D)$. We have on the left y_0 our given number. On the right, we have $D/(1 + D)$. Multiply through by $(1 + D)$. Solve that equation, you'll find that $D = 1/(1 - y_0)$. So D, that parameter in our family of solutions, is equal to $1/(1 - y_0)$.

Can we then determine all solutions of this differential equation? Can we find the solution to any initial value problem? We can, sort of, as long as y_0 is not equal to 1, then we have a D value that gives us the solution to the initial condition $y_0 = y(0)$. If our initial population is 1, then $D = 1/(1 \quad 1)$ or 0, not good.

We have a solution to the initial value problem as long as $y(0)$ is not equal to 1. So what do we do? We know a solution that satisfies the differential equation when $y(0) = 1$, when the initial condition is 1. That's just our friend, the equilibrium solution y identically equal to 1. So we're done. We have the general solution to this differential equation. Either it's of the form $De^t/(1 + De^t)$ or it's the constant function 1.

Just notice if we take this expression $De^t/(1 + De^t)$, what happens to this expression as time goes to infinity? We have De^t up top. As time goes to infinity, that goes to infinity. Then on the bottom, we have $(1 + De^t)$. That also goes to infinity. You get infinity over infinity, not good. But if I just divide top and bottom by e^t I get D divided by $1/e^t(D)$. I get $D/(e^{-t} + D)$. Now what happens as t goes to infinity here? As t goes to infinity here, our e^{-t} goes to 0, and we get just D/D or 1.

So as time goes on, these solutions always tend to the equilibrium solution at 1 just as we saw before. On the other hand, now we have explicit solutions to this differential equation. We can tell exactly what the population is at any time. So as I said earlier, knowing an explicit solution is great. On the other hand, we can't always do that. That's why we use these qualitative techniques.

Next time we're going to veer off, take another little bifurcation. Next time I'm going to talk about the numerical methods that we're using to solve these differential equations. We'll investigate one specific numerical method, and you'll see why some of those earlier differential equations that behave chaotically that broke the numerical method do exactly that. So stay tuned.

How Computers Solve Differential Equations
Lecture 6

We have used the computer several times to plot the solutions of differential equations. But how does a computer produce these solutions? Actually, there are many different numerical algorithms for generating these solutions. Perhaps the most widely used method is called Runge-Kutta 4. For simplicity, we'll describe a much easier to understand though less accurate numerical method, **Euler's method**. This algorithm is constructed in much the same way as the more accurate methods: It uses a recursive procedure to approximate the solutions.

Here is the idea behind Euler's method. We wish to approximate the solution to the differential equation $y' = F(y, t)$ starting at the initial value (y_0, t_0). We first choose a step size Δt. This step size will usually be very small. Then we "step" along the slope field, moving by Δt units in the t-direction at each stage. The resulting conglomeration of little straight lines will be the approximation to our solution.

More precisely, starting at the given initial point (y_0, t_0), we will recursively construct a sequence of points (y_n, t_n) for $n = 1, 2, 3, \ldots$ and join each pair of points (y_n, t_n) and (y_{n+1}, t_{n+1}) by a straight line. This straight line is constructed as follows. We start at (y_0, t_0) and draw the slope line out to the point whose t-coordinate is $t_0 + \Delta t$. This is the value of t_1. The corresponding y-value over t_1 is y_1. Then we do the same at our point (y_1, t_1): Draw the slope line and move along it to (y_2, t_2), where $t_2 = t_1 + \Delta t$.

Figure 6.1

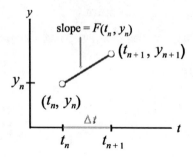

We therefore have recursively $t_{n+1} = t_n + \Delta t$. So the only question is how do we generate the value of y_{n+1} knowing both y_n and t_n? The answer comes from algebra. We have the straight line segment passing through the known point (y_n, t_n) and having slope given by the right-hand side of the ODE—that is to say, the slope is $F(y_n, t_n)$, whose value we know. Therefore our little slope field line has equation

$$y = Mt + B,$$

where $M = F(y_n, t_n)$ is the slope and B is the y-intercept. To determine the value of B, we know that the point (y_n, t_n) lies on our line. So we have

$$y_n = F(y_n, t_n)t_n + B.$$

Therefore

$$B = y_n - F(y_n, t_n)t_n.$$

So the equation for y_{n+1} reads

$$\begin{aligned} y_{n+1} &= F(y_n, t_n)(t_n + \Delta t) + y_n - F(y_n, t_n)t_n \\ &= y_n + F(y_n, t_n)\Delta t. \end{aligned}$$

Thus the recursive formula to generate the values y_{n+1} and t_{n+1} is given by

$$t_{n+1} = t_n + \Delta t$$

$$y_{n+1} = y_n + F(y_n, t_n)\Delta t .$$

Below is an image of Euler's method applied to the differential equation $y' = y^2 - t$ with inordinately large step sizes of 0.25 and 0.125. Also displayed is the actual solution starting at (y_0, t_0). The arrows are little segments of the slope field. Note how when the step size is 0.25, our approximate solution is way off, but when we lower the step size to 0.125, we get a better approximation. Lowering the step size further and further usually gives better and better approximations.

Figure 6.2

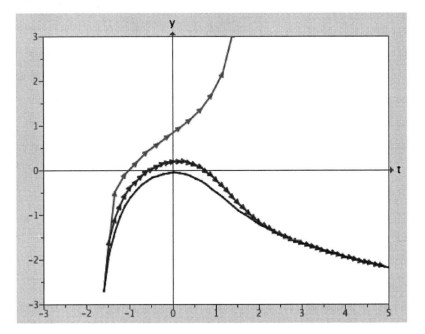

Perhaps the easiest way nowadays to invoke Euler's method is to use a spreadsheet. Here, for example, is a spreadsheet calculation of the solution to the differential equation $y' = y - t$ starting from the initial position $y(0) = 0.3$.

Figure 6.3

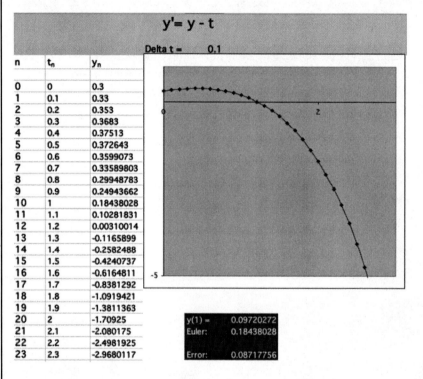

Above we have used the step size $\Delta t = 0.1$ and displayed the corresponding sequence of straight line segments.

Unfortunately, no numerical algorithm is flawless; we can always find a differential equation that breaks a given numerical method. For example, consider the differential equation $y' = e^t \sin(y)$. Clearly there are equilibrium points at $y = 0, \pm \pi, \pm 2\pi, \ldots$. So a solution that starts at $y(0) = 0.3$ should simply increase up to $y = \pi$ as time moves on. But you can see below a time series for what Euler's method yields with a step size of .08. Clearly, the results are bad. What you are actually seeing is chaotic behavior, another theme we will return to later.

Figure 6.4

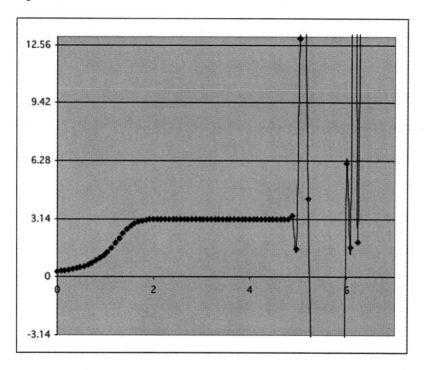

If we change the step size just a little bit to .079, we get very different results for Euler's method. This is sensitive dependence on initial conditions, the hallmark of the phenomenon known as chaos.

Figure 6.5

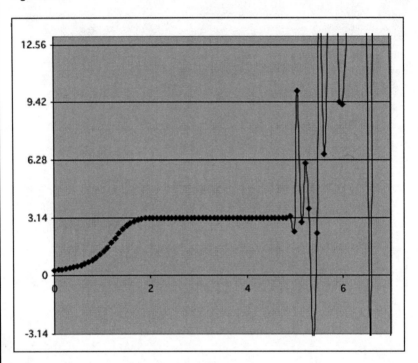

Important Term

Euler's method: This is a recursive procedure to generate an approximation of a solution of a differential equation. In the first-order case, basically this method involves stepping along small pieces of the slope field to generate a "piecewise linear" approximation to the actual solution.

Suggested Reading

Blanchard, Devaney, and Hall, *Differential Equations*, chaps. 1.4 and 7.1–7.2.

Hirsch, Smale, and Devaney, *Differential Equations*, chap. 7.5.

Roberts, *Ordinary Differential Equations*, chap 2.4.

Strogatz, *Nonlinear Dynamics and Chaos*, chap. 2.8.

Relevant Software

Blanchard, Devaney, and Hall, *DE Tools*, Euler's method.

Problems

1. First let's try some refreshers from algebra in the t-y plane.

   a. What is the slope of the straight line in the plane connecting $(1, 1)$ to $(3, 4)$?

   b. What is the slope of the straight line connecting (t_0, y_0) to (t_1, y_1)?

   c. What is the equation of the straight line connecting $(1, 0)$ to $(0, 1)$?

   d. What is the equation of the straight line connecting $(1, 0)$ to $(3, 0)$?

   e. What is the equation of the straight line connecting $(1, 0)$ to $(1, 3)$?

2. What is the equation of the tangent line to the graph of $y = t^2$ at the point $t = 1$?

3. Consider the differential equation $y' = y$. Given the initial condition $t_0 = 0$, $y_0 = 1$ and step size 0.1, use the formula for Euler's method to compute by hand t_1 and y_1.

4. Repeat the above to compute t_2 and y_2 as well as t_3 and y_3.

5. Sketch what you have done via Euler's method for the approximate solution so far.

6. **a.** For the differential equation $y' = y$, find the solution satisfying $y(0) = 1$.

   b. Then use Euler's method with a step size of 0.1 to approximate the value of $y(1)$.

   c. Then recompute using step sizes of 0.05 and 0.01. How do these compare to the approximation above with step size 0.1?

   d. Using the value of e as approximately 2.718281, how does the error change as we vary the step size?

7. Use Euler's method to approximate the solution of the differential equation $y' = (1 + y)/(1 - y)$ with any initial condition $y(0) \neq 1$. What happens here?

Exploration

A slightly better numerical approximation is given by the so-called improved Euler's method. This method also uses a step size Δt. However, instead of using the slope field at the point (y_n, t_n) to approximate the solution, this method uses a line with a slightly different slope. To get this slope, first use the Euler's method approximation at the point (y_n, t_n) to find the endpoint of the slope line at this point whose t-coordinate is $t_n + \Delta t$. Then take the average of the slopes at these 2 points.

This is the slope of the straight line for the improved Euler's method. Find the formula for this slope line. Now use the improved Euler's method to solve the above differential equation with step sizes 0.1, 0.05, and 0.01. How does the improved Euler's method compare to the ordinary Euler's method?

How Computers Solve Differential Equations
Lecture 6—Transcript

Here we go again. Over the past few lectures, you've seen that I've used the computer numerous times to display the solutions of differential equations. The question is how does the computer do that? Just like humans, computers cannot solve most differential equations. What they do is use what we call a numerical method to approximate the solutions of these differential equations.

In this lecture what I'm going to introduce is the simplest possible such numerical method, so called Euler's method. To be honest, in practice you would never use this simple method to solve a differential equations. There are many others out there. Most commonly used I think is Runge-Kutta 4. I'm going to stress Euler's method first of all because it's simple. It's easy to describe. Secondly, all other numerical methods like Euler's method use an iterative algorithm to produce the approximation. This iterative algorithm for Euler's method is easiest to define.

Here's what we're going to do. We want to approximate the solution of the differential equation say $y' = F(t,y)$. We can have both t and y on the right hand side. We're given some initial condition. Say y at time t_0 is equal to some number y_0. What we'll do is construct a series of straight lines, very small straight lines that approximate the solution. How do we do that?

We fist pick what we call a step size, I'll call it Δt that's very small. Then what we'll do is starting at our point t_0, y_0 we'll draw the slope field but we'll step along that slope field exactly Δt units in the t direction. That's the beginning of our numerical algorithm. That takes us from t_0 to a point Δt units away along the slope field. That's our new point. I'll call it (t_1, y_1). Then what I'll do is the exact same thing. I'll start at (t_1, y_1), draw the slope field and step along at exactly Δt units. I'll get to (t_2, y_2). Then we'll continue all the way up producing this sequence of numbers (t_n, y_n). Those little straight lines will be our approximation of the differential equation.

Let me just show you what this looks like. Here is a differential equation. It's $y' = y^2 - t$. Let me start out by using a step size of 0.5. Let me start at this initial point and I'll draw the slope line right there, and I'll step along it in an inordinately large step size of 0.5. Usually you'd use something like 0.001. Anyway, that's my first slope line. I get to the point (t_1, y_1) and I step again. Pretty clearly, I'm going very far away.

Let me change the step size to be 0.25. Again start at the same place, draw the slope line, but only step half as far, go to 0.25. Then at that point, draw the next slope line and step 0.25. Draw the next slope line and step 0.25. The next slope line and so forth. We continually draw these little slope lines and that gets us an approximation to the solution to this differential equation.

Again, 0.25 is pretty large. Let me take a step size of 0.125. Again, start at the same point. Now we get smaller pieces of lines that are presumably a better approximation to our solution of the differential equation. As you can see these different little slope lines that I'm drawing are actually different. They look like they're moving in different directions. As Δt gets smaller, presumably we're getting a better and better approximation.

Let me compare all of these little slope lines. There's what you see. The final curve I drew, the purple curve, is the actual solution. What you see is, as Δt get smaller, we get closer and closer to our actual solution. That's the approximation to the real solution of our differential equation.

The question is given a point (t_n, y_n) in that sequence, how do we produce the next two points? How do we produce (t_{n+1}, y_{n+1})? That's the recursive procedure. Given (t_n, y_n) producing t_{n+1} is easy. T_{n+1} is just take our number t_n and add to it the step size, Δt. Then what do we do? Then we draw the slope line through the point (t_n, y_n) and precede out to the point Δt units away, that's t_{n+1}. The point on that slope line directly above t_{n+1} is our point y_{n+1}. That's now what we have to find. How do we find the point y_{n+1}?

We've got at the point (t_n, y_n) our slope line. We know the slope there. The slope is the right hand side of the differential equation $F(t_n, y_n)$, that's a number. We have that. We know the equation for that slope line is y is slope times $t + B$. In our case we have the slope is $F(t_n, y_n)$, so our line is $y = F(t_n, y_n)t + B$ where B is the y intercept.

How do we find B? We know one point on this line. We know that the point (t_n, y_n) lies on that straight line. We can throw that point into the equation for our line. When we do, we get that $y_n = F(t_n, y_n) t_n + B$. That gives us B. B is just $y_n - F(t_n, y_n) t_n$. Now all of those letters are actually numbers. We know t_n, we know y_n and we know the slope at that point.

We have the equation for our straight line. Our line is $y = F(t_n, y_n) t + y_n - F(t_n, y_n) t_n$. There is the equation for our straight line. We can use that to generate y_{n+1}. y_{n+1} is gotten by putting our point t_{n+1} that is $t_n + \Delta t$ into that equation. So $y_{n+1} = F(t_n, y_n) t_n + \Delta t$, that's our t coordinate plus the y intercept, that's $y_n - F(t_n, y_n) t_n$.

All together, some of those terms cancel. We see that y_{n+1}, our new y-value is our old y-value plus the slope $F(t_n, y_n) \Delta t$. That gives us our formula for Euler's method. The recursive formula for Euler's method is the new t-value t_{n+1} is given by the old t-value $t_n + \Delta t$. The new y-value y_{n+1} is given by the old y-value $y_n + F(t_n, y_n)\Delta t$.

Let me give a couple of examples of how to use Euler's method to approximate the solutions of differential equations. My first example will be pretty simple. It'll be $y' = 2t$ with the initial condition $y_0 = 1$. Of course we know the solution to $y' = 2t$ starting at 1. That solution is just $t^2 + 1$. Let me use Euler's method to see how well I can approximate the solution $t^2 + 1$.

How am I going to do this? It turns out that in my classes in differential equations, I always assigned a number of different labs, several of which involve numerical methods, and almost all of my students used spreadsheets to solve these differential equations. Spreadsheets I think are one of the most amazing mathematical tools out there. Most people don't realize how many different things in mathematics you can do with spreadsheets.

In any event, let me use a spreadsheet to approximate the solution of that differential equation. Here I've put in the data for the differential equation $y' = 2t$. I've started at t-value 0 and y-value 1. I used again a very large step size of 0.2. Below you see all the data that's generated. Here I've just plotted the straight lines generated by Euler's method going from 0 out to 1. You see very few straight lines.

On the other hand, let's lower this step size from 0.2 to 0.1. What you see is the data changes and so does my plot. Right up here is the y-value 2. Remember our solution is $t^2 + 1$ at time 1; $y_1 = 2$. Notice that my approximation is getting kind of close to 2, but it's still kind of far away. Let me lower the step size again. Let me lower the step size from 0.1 to 0.05, cut it in half. Now when I do that, again the data changes and again you see my approximation comes closer to the actual value.

If I go down to step size 0.01 and evaluate the data, now you hardly see those straight lines. They're packed so closely together, the step size is 0.01, but you see we're almost there. We have a very good approximation to our solution of the differential equation.

Let's do another one. This is the differential equation $y' = y - t$. Again, this is a first order differential equation, now linear and nonhomogeneous. We've seen how to actually solve this kind of differential equations. Let's just do it with a spreadsheet. Again I'm using a step size of 0.2, very large. Let me change that to 0.1. When I do, you see the straight lines get smaller and my approximation changes. If I go to step size 0.05 and change the step size, again the straight lines get smaller still. We're presumably getting closer and closer to our approximation. If I go to 0.01, again the step sizes get very small and presumably this is even closer to our actual solution.

Let me show you this in a slightly different way. Let me vary these step sizes continuously. Okay, so I'm starting here with a pretty large step size of 0.2, and now as I change that step size continuously, notice that my lines are getting shorter, they're indeed moving to the left, and they're getting closer and closer to that magenta line. That's our actual solution. As Δt gets down to close to 0.01 you see, we really hone in on our actual solution.

This is one of the wonderful tools I think about spreadsheets in the sense that you can a) display the data of course, b) you can see the graphical representation of the data, but c) you can actually animate this data. You can move continuously through your step sizes and see the actual convergence of your approximation to the actual solution of your differential equation.

So that's Euler's method. Let me just maybe take a brief detour and show you how easy it is to in fact do this spreadsheet analysis. Let me turn again to the differential equation $y' = y - t$ and show you how to enter all of this data. It just takes a couple of seconds.

We're going to start with the differential equation $y' = y - t$. I'll start at the t-value 0, and the y-value 0.3. I'll start with a Δt of 0.05. Here's the formula for Euler's method, the recursive formula that I have to enter. How do I do that in a spreadsheet? I need my t_1 value, but remember my t_1 value is just t_0 plus my step size. I have to enter an equation for t_0 plus my step size. In a spreadsheet, you do that by first inserting equals and then my t_0 value resided in the cell directly above t_1. That's cell B6, so I enter B6 and then $+ \Delta t$; Δt we're taking to be 0.05.

There's the formula for my new p-value. How do I get my new y-value? Basically it's the same thing. Enter a formula by starting with equals. The formula for the new y-value is the old y-value, which resides directly above in cell C6. Then I have to add to that $+\Delta t$. Remember Δt is 0.05 times my old value. My old value, again sits directly above in cell C6 minus my old t-value, which sits over here in cell B6. There is the formula for t_1 and y_1.

There we have our new t and new y-values. We have to now go to t_2 and y_2. How are we going to do that? Am I going to type in these formulas again? No way, it's much easier to do in a spreadsheet. What I'll do is just highlight both of these formulas and actually fill them down. I just fill them down and there is my new data. Notice when I put this formula in, I had cell C6 + 0.05 times C6 − B6. When I filled this down, what went into this cell was now C7 + 0.05 times C7 − B7, etcetera, etcetera. When I fill these formulas down, they just go down recursively.

There's the data for Euler's method. How do I see this data? How do I visualize it? Let me just highlight all this data and insert a chart. In spreadsheets, there are all sorts of different charts you can put in. For this method, I'd prefer a scatter plot where the dots are connected by straight lines. There it is. There's my plot of the numbers that come out via Euler's method applied to this differential equation.

There's one more thing; let me change Δt. How am I going to change Δt here? Let me change Δt to say be 0.01. I have to change Δt in each of these formulas. What am I going to put in here to the formula for t_1? I had 0.05 before; let me get rid of that and well could I enter what's here in the cell for Δt in cell F3? No, if I entered that, when I filled the spreadsheet down F3 would become F4 would become F5. What I have to do is put in a constant reference to cell F3. The way you do that is with \$F\$3. That gives me a constant reference to cell F3. You'll notice that my Δt has changed to 0.01.

I have to do the same thing for my y equation. Instead of 0.05 I have to put in a constant reference to cell F3, \$F\$3, and there it is. Again, let's fill this down. If I want to go out to say $t = 1$ step size of 0.01 I need 100 entries here. Let me fill this down 100 places. There I have it. There's the data for Euler's approximation when step size is 0.01.

Let's highlight all this data, 100 bits of data. Now I'll insert a chart. Again, a scatter plot, and again connect the dots with straight lines, and there it is. Here's our new approximation via Euler's method to that differential equation. You see the dots are much closer together. Our solution is presumably a better approximation. As I said, this is a wonderful way for students at least to do some mathematics, see the data and see it displayed visually.

One thing that we saw last time was that certain numerical methods would fail when you applied them to certain differential equations. Let's look at what happens when we apply Euler's method to one of those differential equations. Let me take the differential equation for $y' = e^t \sin(y)$. What do we know about that differential equation? Well when $y = 0$ or $\pm\pi$ or $\pm 2\pi$ we have equilibrium solutions. There are equilibrium solutions at 0, π, 2π etcetera. Then the solutions must go up and down in between.

What happens between 0 and π? Between 0 and π our differential equation has right hand side $e^t \sin(y)$. E^t for t greater than 0 will be positive. $\sin(y)$ between 0 and π is also positive. So our slopes should be positive. They should be increasing. They should be moving away from our equilibrium point at 0, and up to our equilibrium point at π, and then they should level off there.

Go back to the spreadsheet. Here's the spreadsheet for the differential equation $y' = e^t \sin(y)$. Here I've started with the initial condition $y(0) = 0$. We know that's an equilibrium solution. The spreadsheet gives us that this is an equilibrium solution. Let's now start out someplace between 0 and π, let's say at 0.3. Our initial condition is $y(0) = 0.3$ and here's what the spreadsheet gives us.

We started out okay. We started out at 0, increased gradually up to π, and then all of the sudden things went crazy just as we saw before. Why is that? Well maybe we're using too large a step size. Let me change that step size. We started out with a step size around 0.1. Let me lower that step size. Whoa, a very different behavior. Lower it again, and again.

Every time we change this step size we start out okay, but then our numerical method fails. Our iterative process goes crazy. What you're actually seeing is chaotic behavior, one of the subthemes of this course. Change Δt by just a little bit, change it in the third decimal place, you see dramatically different behavior for the results of this iterative process.

Why is that happening? Why are these solutions going crazy? Think about it, as we get out near the lines $y = \pi$, yes, $\sin(y)$ is pretty small in this $\sin\pi$. A little bit below π. Think about the term e_t. E_t, that's going to be part of our formula for the slope, e_t is getting very large. Out around 4 or 5, e_t is about 500. That's getting pretty big. Even though $\sin y$ is small at my t_n and y_n value, that e^{tn} value is getting very large. Eventually that pokes me up quite a bit. Pokes me up above $y = \pi$, well into the regime between π and 2π, then I have another large e^t term. Boom, that's driving me crazy. That's what we're seeing over here in this approximation.

As I change Δt you see that eventually I go a little bit above the line $y = \pi$, then my slope is huge because it's being multiplied by an e^t term and I'm just sent into various regions between $n\pi$ and $n + 1\pi$.

Here's one last thing. We saw last time that when I took the differential equation $y' = (1 + y)/(1 - y)$, that had the same kind of behavior. Let's go back to that differential equation and see why that happened.

Here is the differential equation $y' = (1 + y)/(1 - y)$. Remember when $y = -1$ we have an equilibrium solution. Below $y = -1$ things are fine, solutions just tend off to minus infinity. Above $y = -1$, our slope fields are increasing, and increasing, and increasing, and then when we get to $y = 1$ the denominator is 0. Then our slope field is vertical, and after that, it's pointing in the opposite direction. What happens to solutions here? Our numerical method has failed.

What has happened is we climb above the line $y = 1$, then our slope field is pointing down, it sends us this way. Then we climb back up until we get above $y = 1$. Our slope field now points us down, up, down, up, down, up, down. That's what happens over and over again. We keep crossing that line, our slope fields are very big pointing in opposite directions. Again, our numerical method fails. Euler's method or whatever, Runge-Kutta 4 method fails dramatically. In fact, no matter which numerical method you use, you can always find a fairly simple differential equation that will behave just like that. That will destroy that numerical method.

This completes the first part of the course dealing with first order differential equations. What we're going to do now is move to systems of differential equations. That's the topic for the next series of lectures.

Systems of Equations—A Predator-Prey System
Lecture 7

We now begin the main part of this course, which deals with systems of differential equations. Our first example is the predator-prey system. Here we assume we have 2 species living in a certain environment. One species is the prey population (we'll use rabbits), and the other species is the predator population (we'll use foxes). Denote the prey population by $R(t)$ and the predator population by $F(t)$.

We assume first that if there are no foxes around, the rabbit population obeys the unlimited population growth model. If, however, foxes are present, then we assume that the rate of growth of the rabbit population decreases at a rate proportional to the number of rabbit and fox encounters, which we can measure by the quantity RF. So the differential equation for the rabbit population can be written

$$\frac{dR}{dt} = aR - bRF,$$

where a and b are parameters.

For the fox population, our assumptions are essentially the opposite; so the differential equation for $F(t)$ is

$$\frac{dF}{dt} = -cF + dRF.$$

Again, c and d are parameters.

Note that both of these equations depend on $R(t)$ and $F(t)$. This is typical of systems of differential equations: We have a collection of differential equations involving a number of dependent variables, and each equation depends on the other dependent variables. So the **predator-prey system** is given by the following equations.

$$\frac{dR}{dt} = aR - bRF$$

$$\frac{dF}{dt} = -cF + dRF$$

This is a 2-dimensional system of ODEs, since we have just 2 dependent variables.

More generally, a system of ODEs is a collection of n first-order differential equations for the missing functions $y_1, \ldots, y_n$. Each equation depends on all of the dependent variables $y_1, \ldots, y_n$ as well as (possibly) t. So we get the following.

$$\frac{dy_1}{dt} = F_1(y_1, \ldots, y_n, t)$$
$$\vdots$$
$$\frac{dy_n}{dt} = F_n(y_1, \ldots, y_n, t)$$

This is an n-dimensional system of differential equations. Later we will spend a lot of time looking at linear systems of equations with constant coefficients. A 2-dimensional version of such a system is

$x' = ax + by$

$y' = cx + dy$

where a, b, c, and d are constants.

A solution of this system of ODEs is then a pair of functions, $(R(t), F(t))$, each of which depends on the independent variable. We will think of this pair of functions as a curve in the plane parameterized by t. We call this curve a **solution curve**. Thus the plane becomes our **phase plane**, in analogy with the phase line viewed earlier.

Note that we know the tangent vector to any given solution curve: This is the vector given by the right-hand side of our pair of differential equations (R', F'). So the right-hand side of our equation determines a **vector field** in the phase plane. Solution curves in the R-F plane are everywhere tangent to the vector field. Since these vectors can often be very large (and therefore cross each other in ways that make viewing the vector field difficult), we usually scale the vector field so that all vectors have the same length; this scaled field is called the **direction field**.

Figure 7.1

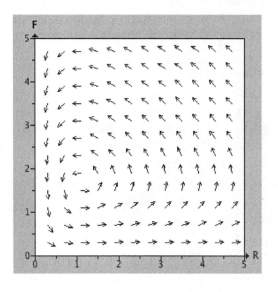

The point (R_0, F_0) is an equilibrium point if the right-hand sides of both differential equations vanish at this point. This means that we have a constant solution for the system given by $R(t) = R_0$ and $F(t) = F_0$. We indicate this by a dot in the phase plane. So the equilibrium points for the predator-prey system are given by solving the pair of algebraic equations simultaneously:

$$R' = aR - bRF = 0, F' = -cF + dRF = 0.$$

Solving this shows that there are equilibria at $(0, 0)$ and $(c/d, a/b)$.

Note that when $R = 0$, we have $R' = 0$, so the rabbit population stays fixed at 0. Also, in this case, the equation for foxes becomes $F' = -cF$. This is just our old population model, this time with population decline rather than growth. So, along the F-axis, we can see the phase line corresponding to this first-order equation. Similarly, when there are no foxes, the differential equation for rabbits becomes $R' = aR$, the unlimited population growth model for rabbits. Thus we see the phase line for this equation along the R-axis in the phase plane.

Figure 7.2

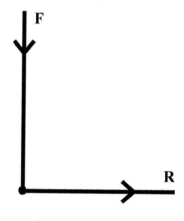

We may then sketch the vector field in the special case where $a = b = c = d = 1$ and use numerical methods to superimpose the plots of various solution curves.

Figure 7.3

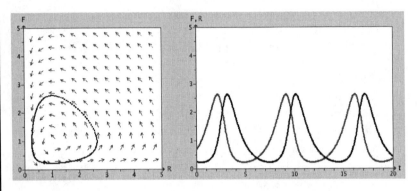

On the left, we see a solution in the phase plane. Here the R-axis is the horizontal axis and the F-axis is the vertical axis. This solution winds repeatedly around the nonzero equilibrium point. On the right, we have superimposed the corresponding graphs of both $R(t)$ and $F(t)$. Now the t-axis is the horizontal axis, while the F- and R-axes are vertical.

We may modify the predator-prey system by assuming that the rabbit population obeys the limited growth model with carrying capacity $N = 1$. For simplicity we'll choose all the other parameters to be equal to 1, except d. So our new system is

$$\frac{dR}{dt} = R(1-R) - RF$$

$$\frac{dF}{dt} = -F + d\,RF.$$

On the R-axis, our phase line is now the limited growth phase line with equilibrium points at $R = 0$ (a source) and $R = 1$ (a sink). On the F-axis, we have (as before) the fox population tending to the equilibrium at 0. Setting the 2 differential equations equal to zero yields 2 equilibrium points, at the origin and (1, 0). But if $d > 1$, we get a new equilibrium point at $(R = 1/d, F = (d - 1)/d)$. If $0 < d < 1$, there is an equilibrium point at $(R = 1/d, F = (d - 1)/d)$, but the F-coordinate here is negative, so we will not consider this in our population model. Thus we have a bifurcation when $d = 1$.

But there is more to this story. When $d \leq 1$, the phase plane appears to show that any solution that begins with a positive fox population now tends to the equilibrium point at (1, 0). That is, the fox population goes extinct. But when $d > 1$, the solutions seem to tend to the equilibrium point $(1/d, (d - 1)/d)$, so both populations seem to stabilize as time goes on. This is another example of a bifurcation.

Figure 7.4

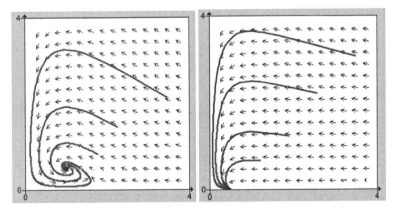

Phase plane for $d = 2$ Phase plane for $d = 1/2$

Many new types of solutions appear in systems of ODEs. For example, there may be periodic solutions as in the original predator-prey system.

Figure 7.5

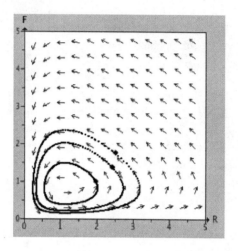

And solutions may spiral toward or away from such a periodic solution. For example, consider the following system of equations.

$$x' = (1-\sqrt{x^2+y^2})x - y$$
$$y' = x + (1-\sqrt{x^2+y^2})y$$

It is very rare to be able to find periodic solutions explicitly for systems of ODEs, but here we can. Note that when $x^2 + y^2 = 1$, the system reduces to $x' = -y$ and $y' = x$. Do you know a pair of functions that satisfies this relation? Sure—from trigonometry we know that $\cos^2(t) + \sin^2(t) = 1$, and from calculus we know that the derivative of $\cos(t)$ is $-\sin(t)$, while the derivative of $\sin(t)$ is $\cos(t)$. So $x(t) = \cos(t)$, $y(t) = \sin(t)$ is a solution of this system, which is the periodic solution we see above.

Solutions spiral in toward a periodic solution (below).

Figure 7.6

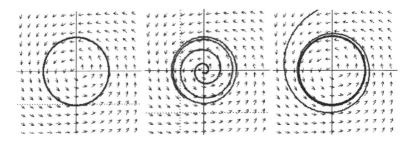

Another type of differential equation that comes up often is a second-order equation. This is an equation that involves both y' and y''. We will deal for the most part with second-order equations of the form

$$Y'' + ay' + by = F(t),$$

where $F(t)$ is considered a forcing term. For example, in the next lecture we will consider the mass-spring system given by $y'' + by' + ky = 0$. We can write this equation as a system by introducing the new variable $v = y'$ (the velocity). Then our system becomes

$$y' = v$$
$$v' = -ky - bv.$$

Note that this is a linear system of ODEs. This will enable us to see solutions both as graphs of $y(t)$ and also as curves $(y(t), v(t))$ in the phase plane.

Important Terms

direction field: This is the vector field each of whose vectors is scaled down to be a given (small) length. We use the direction field instead of the vector field because the vectors in the vector field are often large and overlap each other, making the corresponding solutions difficult to visualize. Those solutions and the direction field appear within the phase plane.

phase plane: A picture in the plane of a collection of solutions of a system of two first-order differential equations: $x' = F(x, y)$ and $y' = G(x, y)$. Here each solution is a parametrized curve, $(x(t), y(t))$ or $(y(t), v(t))$. The term "phase plane" is a holdover from earlier times when the state variables were referred to as phase variables.

predator-prey system: This is a pair of differential equations that models the population growth and decline of a pair of species, one of whom is the predator, whose population only survives if the population of the other species, the prey, is sufficiently large.

solution curve (or graph): A graphical representation of a solution to the differential equation. This could be a graph of a function $y(t)$ or a parametrized curve in the plane of the form $(x(t), y(t))$.

vector field: A collection of vectors in the plane (or higher dimensions) given by the right-hand side of the system of differential equations. Any solution curve for the system has tangent vectors that are given by the vector field. These tangent vectors (and, even more so, the scaled-down vectors of a corresponding direction field) are the higher-dimensional analogue of slope lines in a slope field.

Suggested Reading

Blanchard, Devaney, and Hall, *Differential Equations*, chap. 2.1.

Edelstein-Keshet, *Mathematical Models in Biology*, chap. 6.2.

Hirsch, Smale, and Devaney, *Differential Equations*, chap. 11.2.

Roberts, *Ordinary Differential Equations*, chap. 10.5.

Relevant Software

Blanchard, Devaney, and Hall, *DE Tools*, Predator Prey.

Problems

1. Find all of the equilibrium points for the system below.

   $x' = x + y$

   $y' = y$

2. Find the equilibrium points for the system below.

   $x' = x + y$

   $y' = x + y$

3. **a.** Sketch the direction field for the system below.

   $x' = x$
   $y' = y$

   b. What happens to solutions of this system?

   c. Find a formula for the solutions of this system.

4. Write the second-order differential equation $y'' = y$ as a system of differential equations.

5. **a.** Sketch the direction field for the following system of differential equations.

   $x' = y$
   $y' = -x$

   b. Can you find some pairs of solutions $(x(t), y(t))$ of this system of differential equations?

6. **a.** Sketch the direction field for the following system of differential equations.

   $x' = x(1 - x)$
   $y' = -y$

   b. What are the equilibrium points for this system?

   c. This system decouples in the sense that the equation for x' does not depend upon y, and the equation for x' does not depend on x. Use this fact to describe the behavior of all solutions in the phase plane.

Exploration

It is impossible to find explicit solutions of the predator-prey equations described in this lecture. However, there is a technique from multivariable calculus that can give explicit formulas for where solutions lie. That is, we can find a real valued function $H(x, y)$ that is constant along each solution. So plotting the level curves of this function shows us the layout of the solutions. Your job is to use calculus to find such a function. Hint 1: Find a function H such that dH/dt is identically equal to zero. This will involve the chain rule. Hint 2: A function of the form $H(x, y) = F(x) + G(y)$ works.

Systems of Equations—A Predator-Prey System
Lecture 7—Transcript

We're back. Now we move on to the main part of the course where we'll deal with systems of differential equations. I call this the main part of the course because most of the differential equations you encounter in science and engineering are actually systems of differential equations.

As with all other parts of the course, let me begin again with a model. This time I'll take a model from biology. This will be the predator-prey system. Let me assume we have two species out there, one predator (think foxes) and the other prey (think rabbits). I'll let $F(t)$ be the fox population at time t, and $R(t)$ be the population of rabbits at time t, the population of the prey.

What are the assumptions? Let's assume that if there are no foxes, then the rabbit population obeys our old unlimited population growth model. The rabbit population just explodes; but if there are foxes around, then let's assume that the rate of growth of rabbits goes down at a rate proportional to the number of rabbit and fox encounters.

How are we going to write down these differential equations? If there're no foxes, then the unlimited growth law says that R' is just aR for some constant, a. Now, if there are foxes around, the growth goes down, dR/dt will be negative, proportional to the number of rabbit-fox encounters.

There're a number of ways to measure the number of rabbit-fox encounters. Probably the easiest is to measure it as some constant times R times F. If the fox population is large, F is large, and you are going to have lots of rabbit-fox encounters. If both R and F are large, you are going to have even more rabbit-fox encounters, but if the rabbit population is very small, there won't be as many rabbit-fox encounters; RF will be smaller. There's the differential equation for rabbits, for prey, R', the derivative of R with respect to t, is $aR - bRF$.

What about for foxes? The fox population will be essentially the opposite. Let me assume if there are no rabbits, then the fox population exponentially declines. $F' = -cF$, but if there are rabbits around, there's food, there's prey, so the fox population will go up at a rate proportional to the number of rabbit-fox encounters. I can write the predator differential equation for the population of foxes as $F' = -cF$, c is a constant, plus dRF. Here, d is positive because the number of rabbit-fox encounters only helps the fox population to survive.

Note that both of these differential equations depend on each of R and F. This is what we call a two-dimensional system of differential equations. R' is $aR - bRF$, and F' is $-cF + dRF$. More generally, a system of differential equations is a collection of n differential equations for the missing functions, y_1, y_2, all the way up to y_n. All of these differential equations will involve y_1', y_2', etcetera. We'll have n different differential equations, and each of them will depend, more or less, on all of the y values and possibly time. So y_1' will be a function of $y_1, y_2 \ldots$ up to y_n, then time, and similarly for the rest of the differential equations.

Later we'll spend a lot of time on a particular kind of system of differential equations, namely linear systems, in fact, linear systems with constant coefficients. These are systems of the differential equations of the form $x' = ax + by$, and $y' = cx + dy$. Note that both of these differential equations depend on x and y, what we call our dependent variables, and the independent variable again is t. Here, all of the letters, a, b, c, and d, are constants. We have a constant coefficient differential equation.

Now let's go back to the predator-prey equation. What is the solution going to be? A solution is going to be functions that measure the rabbit population and the fox population, so it'll be a pair of functions, $R(t)$ and $F(t)$, and we're going to think of this as a parametrized curve in the plane. We're going to draw the R-axis horizontally, the F-axis vertically, and as time goes on, our rabbit and fox population will change. This will trace out some sort of curve called the parametrized curve that depends on time.

That plane has a name. You can probably guess what it is; it is the phase plane, again, the R-axis horizontal, the F-axis vertical. First step, what do we do? We know what happens if there are no rabbits or no foxes, if the rabbit or fox population is 0.

For example, if the rabbit population is 0, then looking at the rabbit differential equation we see R' is 0, so there are no rabbits at all and R stays identically 0. Of course if you have no rabbits to begin with, you are not going to have rabbits appearing out of nowhere, but then our fox population differential equation reads just $F' = -cF$, so we get a phase line for F. We know the solutions to F' as $-cF$ just exponentially decay down to 0. We can draw that in the phase line.

Similarly, when $F = 0$, we have no foxes to start, there will be no foxes later, so we only have the rabbit differential equation which reads $R' = aR$. The rabbit population is exponentially increasing. In the phase line for rabbits, we see solutions shooting off to infinity.

Second step, as with first-order equations, let's find the equilibrium solutions. What are the equilibrium solutions? Those are the points where both of the right hand sides equal 0. Both of our right hand sides of the differential equation equal 0 at some R and F population, then that function, R identically equal to that number and F identically equal to that number, is a constant solution of the differential equation. To find the equilibrium solutions, what we have got to do is solve the right hand side of the differential equations equal to 0 simultaneously, so we have the equations $aR - bRF = 0$, and $-cF + dRF = 0$. Pretty clearly, if R and F are 0 (we have no rabbit and no fox population), that will stay that way. That is an equilibrium solution.

What about other solutions? Let's remove the R from the first equation. We're left with a $-BF = 0$, and if I remove the F from the second equation, I get $-c + dR = 0$. That says that we have another equilibrium solution when $R = c/d$. When $F = a/b$, there's another equilibrium solution for this differential equation.

Third step, what is going on in the phase plane? Our differential equation is R' is something and F' is something. What is that something? If we take a point in the RF plane, whatever point, then at that point, the right hand side of the differential equation is really giving us a vector. We think of the right hand side of this differential equation as giving us a vector field in the RF plane. At each point (R, F), the right hand side is a vector that I'll then translate to start at the point (R, F), so this is the analog of the slope field that we saw for first-order differential equations.

What do we know about our solutions? Our solutions must have derivative, given by the right hand of this differential equation, so our solution curves, these curves in the phase plane, must be everywhere tangent to the vector field. This is again, exactly what went on with first-order differential equations, the right hand side of our differential equation tells us not the slopes, but the tangent vectors to all of our solutions.

Here's an example of the predator-prey equations. I have assumed that all of my constants, for simplicity, are equal to 1. Over here, I am drawing the phase plane. If you look at these vectors, you see they are all the same size. This is what we call a direction field. Often these vectors that we plot in the vector field will be huge and they will be crossing each other, so we just simplify them so that they are basically the same size.

What is the solution to this differential equation? Let me start a solution right here, and what I am plotting is that parameterized curve in the RF plane, and over here the corresponding graphs of $R(t)$ and $F(t)$. Let's think what is going on here. We started here at a particular point in the RF plane. Over here, I am plotting the graphs of rabbit population in green, and fox population in red. What happens to this solution? As time went on, we saw that this solution moved in this direction. What is happening? Remember, rabbits are plotted horizontally, so the rabbit population starts going down, and we see this over here in the green graph, the graph of rabbits with respect to time.

Meanwhile, as we run along this parametrized curve, our fox population, measured here on the vertical axis, is increasing. Our rabbit population is going down, but our fox population is thriving. It is thriving until we get right about here; that is where the fox population reaches its maximum. Meanwhile, the rabbit population has gone way down; that is where we are right when the red graph reaches its maximum. You see the green graph, the rabbit population, is pretty small. Now there are so few rabbits, the fox population starts dying out. You see that F starts to decrease in time. Similarly, rabbits are decreasing in time, but eventually, right about here, there are so few foxes that the rabbits start to increase in population.

That is what is happening over here. We have reached, right about here, the place where the green graph has its minimum. The rabbit population suddenly rekindles and starts growing, and thereafter, we keep going along the parametrized curve; right around here, finally the fox population reaches its minimum. That is right here on the red graph. You see that the rabbit population has started to increase, and what our solution to the differential equation says is we come back to, in fact, the original starting place. Then our rabbit and fox populations behave periodically.

Let's modify this a little bit to have a slightly different system of differential equations. We saw that the unlimited population growth was not so natural when we did the one-dimensional case. We saw a more natural one was the limited population growth, a logistic population growth model. Let's modify this system so that we assume that the rabbit population obeys the logistic population growth model if there are no foxes. Let's assume again, for simplicity, that the constants in the logistics models, k is 1 and carrying capacity is 1.

Our new system of differential equations says R' is logistic $R(1 - R)$ minus some constant times R and F. Let's just let that constant be 1, and then the fox population, let's assume that does not obey a logistic population model. Let's assume that the fox population dies out if there are no rabbits, so dF/dt, fox population F', is $-F$ plus—now let me put in one constant—let's say dRF. Now what we get is the logistic population phase line on the R-axis. Before we had a constant times R; the solutions just went off to infinity. Now we have, when $F = 0$, $R' = R(1 - R)$, so we have two equilibria at the origin, and at the point $R = 1$ and $F = 0$.

What about other equilibrium points? We have one at the origin, and 1,0; to get other equilibrium points, let's set the equation $-F + dRF$ to be 0. Of course $F = 0$. We already know what is happening there, so get rid of the F. We get, then $-1 + dR = 0$, so in order for that equation to vanish, we must have $R = 1/d$. Then go back to the first equation, plug that in, $R = 1/d$, set the first equation equal to 0, you get R, $(1/d)(1 - 1/d) - F/d$ to be 0. If you solve that, multiply through by d, you see that F is $1 - 1/d$ or taking a common denominator, $(d - 1)/d$.

What we see is if d is less than 1, that equilibrium point, $F = (d - 1)/d$ is negative, so that is not part of our population model. In that case, we have only the equilibrium point at the origin and 1,0. On the other hand, when d is greater than 1, we have a new equilibrium point at $1/d$ and $(d - 1)/d$. Notice that when $d = 1$, this new equilibrium point is just 1,0. It is merged with one of the previous equilibrium points that we found on the rabbit axis, so we have a bifurcation when $d = 1$. Again, that subtheme of bifurcations arises.

So here is the direction field in the case where $d = 0.5$. I have set $d = 0.5$, so we only have two equilibrium points, at the origin and 1,0. What we see is all of our solutions just tend to that one equilibrium point. That is, the fox population dies out, F becomes 0, and the rabbit population stabilizes at its carrying capacity.

Now let me change that constant d; it was 0.5. Let me change that to 1.3, and this is what happens. We now know we have another equilibrium point, and all solutions seem to be leveling off at that other equilibrium point. All these rabbit and fox populations seem to be limiting on a constant value as time goes on. That is a bifurcation. We have moved away from the equilibrium point at 1,0, off into another equilibrium point in the phase plane.

Let me take another value of d, let d be 4, and what we see is something a little different. Notice solutions still seem to be coming in to that equilibrium point. Let me let t run a little bit further out. Solutions still seem to be coming in to that equilibrium point, but now they seem to be spiraling in. Solutions look like they are spiraling in to that equilibrium point, whereas before, solutions just tended directly toward it. It looks like something else is happening here; it looks like we have another kind of bifurcation.

There are many kinds of new solutions that show up in systems of differential equations. We have already seen for the predator-prey equation that there can be what we call periodic solutions. Here's the predator-prey equation. We saw that a solution there behaved periodically. A solution here behaved periodically. Come closer to the equilibrium point that we're running around here, we get another periodic solution, so often we see these periodic solutions in systems of differential equations.

Let me just modify this a little bit. Remember we have parameters a, b, c, and d. Let me lower the parameter b, and let me raise the parameter d. Now what happens? Whoa, look at that. The fox population goes way up, the rabbit population goes way down, and then it repeats, another periodic solution. If I go the opposite way, let me raise the parameter b and lower the parameter d, then exactly the opposite happens. Now the fox population goes way down, but the rabbit population goes way up. Again, these solutions behave periodically. We have a periodic solution to this differential equation and lots of other such periodic solutions.

That was kind of a very special case because all of our solutions behave periodically. All of our solutions repeated themselves. Most often, that does not happen. Most often, if you have a periodic solution, it is isolated. It is something that we will later call a limit cycle. What often happens is, yes, you have one solution that is periodic, but all other solutions spiral into or spiral away from that solution. For example, let's look at the system of differential equations,

$$x' = 0.1(1-\sqrt{x^2+y^2})x + y$$
$$y' = -x + 0.1(1-\sqrt{x^2+y^2})y$$

That looks like a strange differential equation, but I do that because for this differential equation, you can actually find a periodic solution explicitly.

Before I do that, let's look and see what the behavior of this system is in the phase plane, so here's the phase plane for that system of differential equation. If I start at $x = 0$ and $y = 1$, there's our solution. It just runs around periodically. Over here I am drawing the graph of x and y with respect to t; you see a periodic solution.

Now let me choose another solution, say right there. Did you see what happened? That solution started very close to the origin, kept oscillating around, and eventually went out and limited on that equilibrium solution, on that limit cycle. Similarly, if we start far away, that solution will spiral into the limit cycle. Initially, the values are very large, but they very quickly come in to that limit cycle.

How do we know that there's a limit cycle? How do we know there's a periodic solution for that differential equation? Remember we had inside some parentheses, $(1-\sqrt{x^2+y^2})$. That says that if $x^2 + y^2 = 1$, that term is 0. That is, for our differential equation, if we're on the unit circle, our equation is just $x' = y$ and $y' = -x$.

Do you know one solution of that system of differential equations, $x' = y$, and $y' = -x$? Do you know a function whose derivative is another function whose derivative is then the negative of the first function? Of course, if $x(t) = \sin(t)$ and $y(t) = \cos(t)$, then x' by some calculus and trigonometry, the derivative of x' is cos and the derivative of cos is $-\sin$. That is, $x' = y$, cosine, and $y' = -x - \sin$; and of course, from trigonometry, you know that $\cos^2 + \sin^2 = 1$. The parametrized curve, $x = \sin(t)$, $y = \cos(t)$ is a solution to this differential equation, so we actually can find the limit cycle in this case.

While I am here, I should mention a couple of things that harkens back to what we have done before. The first is the existence and uniqueness theorem that we had for first-order differential equations holds for systems. For systems of differential equations, as long as the right hand side of your differential equation is nice, then solutions cannot cross, and we have seen that with all of the systems we have looked at already. Solutions never cross; as long as the right hand side is continuously differentiable in all your variables, you are fine. You have a nice differential equation.

Second thing, last time we saw that we could use Euler's method to solve, or rather to approximate, first-order differential equations. You could do exactly the same thing with systems. Now instead of stepping along the slope field, now we'll step along the vector field using a certain step size.

Here's a differential equation, $x' = y$ and $y' = -x$. Remember that? That is the differential equation we just ended up with when we looked at the limit cycle, when we were on the circle of radius 1. What happens to this differential equation? We know the solution, $x' = y$, $y' = -x$, again, we get $\sin(t)$, $\cos(t)$ as our solution, but let's use the numerical method, Euler's method, to approximate this solution.

Again, I'll use pretty large step sizes. Let me start at $x = 0$, $y = 1$, with a step size of 0.5, so we have put the vector field there, and run out 0.5 units. Then we staple in another vector field. We run out a step size of 0.5, we keep doing this, and we're doing exactly what we did for Euler's method for first-order differential equations except now we're stepping along the vector field.

Were we to use a smaller step size, then we go not as far along the vector field, we keep stepping by 0.25, and you get a different approximation to the solution. We know, of course, that the solution is the circle of radius 1. Since our step size is so large, we're veering away from that pretty quickly. The point is, the same stuff we did with Euler's method for first-order differential equations goes over exactly for systems. Let me just see a whole bunch of these things. Here's what we do for Euler's method for various step sizes, and then finally for our actual solution. There it is, so as I said, Euler's method works for these systems of differential equations just as well.

One other thing that is going to come up often in this part of the course is what we call second-order differential equations. A second-order differential equation will be an equation of the form $y' + ay' + by$, where a and b are constants, equal to something on the right hand side, some sort of forcing term. Actually for the most part, we'll be looking at $y'' + ay' + by = 0$.

These second-order equations can also be written as systems. We have y'' is equal to something. Let me introduce a new variable, $y' = v$. So $y' = v$ is our first differential equation, and then what is v'? $V' = y''$, but bring all the other stuff over to the right; we get $y'' = -bv - ky$. We have a system of differential equations, $y' = v$, and $v' = -bv - ky$. That is a linear system of differential equations.

Next time we'll investigate a specific example of such a linear system. We'll introduce the mass-spring system. That is given by the second-order differential equation $y'' + by' + ky = 0$, no forcing, where b is what we call the damping constant, the damping parameter, and k is the spring constant, so two parameters arise here. All of the stuff that we have seen before, like bifurcations, will also arise. In any event, that is the next example that we'll go to in the next lecture.

Second-Order Equations—The Mass-Spring System
Lecture 8

In this lecture, we introduce a slightly different kind of differential equation, second-order (autonomous) linear differential equations. These are differential equations of the form

$y'' + ay' + by = G(t).$

As usual, we begin with a model, this time the **mass-spring system**, also known as the harmonic oscillator. Imagine holding a spring hanging with a weight (the mass) attached. If you pull the mass downward (or push it straight upward) and let it go, the mass will then move along a vertical line. Let the position of the mass be $y = y(t)$. The place where the mass is at rest would be called $y = 0$, whereas $y < 0$ if the spring is stretched and $y > 0$ if the spring is compressed.

To write the differential equation for the motion of the mass, we invoke Newton's law that says that the force acting on the mass is equal to the mass times its acceleration. There are 2 types of forces acting on the mass: One is the force exerted by the spring itself, and the second is the force arising from friction (like air resistance). For the force exerted by the spring, we invoke Hooke's law that says that the force exerted by the spring is proportional to the spring's displacement from its rest position and is exerted toward the rest position. So this force is $-ky$. Here the constant $k > 0$ is the **spring constant**.

For the force arising from friction, we make the simple assumption that the force is proportional to the velocity. So this damping force is given by $-by'$, where b is the **damping constant**. Here we either have $b = 0$ (the undamped case) or $b > 0$ (the damped case). The minus sign here indicates that the damping pushes against the direction of the motion, thereby reducing the speed. This gives us the equation for the damped harmonic oscillator:

$my'' = -by' - ky.$

To simplify matters, we usually assume that the mass equals 1. So the equation for the damped harmonic oscillator is

$y'' + by' + ky = 0.$

Usually such a second-order differential equation comes with a pair of initial conditions: the initial position of the spring $y(0)$ and its initial velocity $y'(0)$.

We can write the equation for the harmonic oscillator as a system of differential equations by introducing a new variable, the velocity $v = y'$. So the system of equations is

$y' = v$

$v' = -ky - bv,$

where $k > 0$ and $b \geq 0$. Since $k > 0$, we see that the only equilibrium point for this system is at the origin (i.e., where the mass attached to the spring is at rest). We now have 3 different views of the mass-spring system: the actual motion of the mass on the left, the solution in the phase plane in the center, and the graphs of both position $y(t)$ and velocity $v(t)$ on the right.

We also get 3 different types of solutions: one where the mass oscillates back and forth as it goes to rest, one where it oscillates forever, and one where it proceeds directly to the rest position.

Figure 8.1

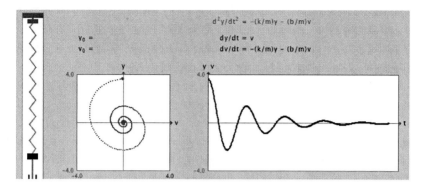

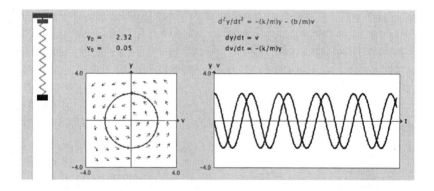

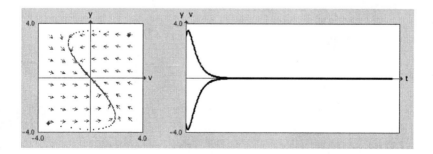

How do we solve such a second-order linear equation? Recall that, for the analogous first-order equation $y' = ky$, we always had the solution $y(t) = Ce^{kt}$. So why not guess a solution of the form e^{st} and try to determine what value of s works? Plug e^{st} into the equation $y'' + by' + ky = 0$. We find

$$e^{st}(s^2 + bs + k) = 0.$$

Therefore any root of the equation $s^2 + bs + k = 0$ yields an exponential function that solves the equation. This quadratic equation is called the **characteristic equation**; we will see many other cases where this equation arises. By the quadratic formula, we find that the roots are

$$\frac{-b \pm \sqrt{b^2 - 4k}}{2}.$$

But there is a problem here. What happens if the quantity $b^2 - 4k$ is negative? Then the roots of our characteristic equation are complex numbers. What is the exponential of a complex number? We'll tackle that problem a little later. Notice also that, if $b^2 - 4k = 0$, then we get only 1 root of the characteristic equation, namely $-b/2$. So we get 1 exponential solution $e^{-bt/2}$. This will not give enough solutions to solve every initial value problem, so more must be done here as well. We'll come back to this situation again later.

Let's suppose that $b^2 - 4k > 0$. Then we have 2 distinct and real roots of the characteristic equation. Let's call them α and β, where the following are true.

$$\alpha = (-b - \sqrt{b^2 - 4k})/2$$
$$\beta = (-b + \sqrt{b^2 - 4k})/2$$

Clearly, $\alpha < 0$. We also have $\beta < 0$ since $b^2 - 4k < b^2$ (remember that k, our spring constant, is always positive). So we have 2 decreasing exponential solutions, $\exp(\alpha t)$ and $\exp(\beta t)$. And you can check that the general solution is

$$k_1 e^{\alpha t} + k_2 e^{\beta t}.$$

Anytime we find 2 such real, distinct, and negative roots of the characteristic equation, our harmonic oscillator is overdamped. This means that the mass slides down to rest without oscillating back and forth.

As an example, suppose the spring constant is 2 and the damping constant is 3. So the equation for this mass-spring system is

$$y'' + 3y' + 2y = 0.$$

The characteristic equation is

$$s^2 + 3s + 2 = 0,$$

whose roots are −1 and −2. So the general solution here is

$$y(t) = k_1 e^{-t} + k_2 e^{-2t}.$$

Here are the phase plane and the $y(t)$ plots for several of these solutions.

Figure 8.2

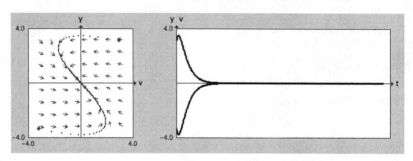

Note that both of these solutions tend directly to the equilibrium point at the origin. The corresponding motion of the mass-spring system is not the usual oscillation back and forth, eventually coming to rest. It's as if we were playing with this mass-spring system in an environment with a lot of resistance, like under water. This is the behavior of the overdamped harmonic oscillator.

So let's now turn to the undamped harmonic oscillator. So $b = 0$, and the equation is $y'' + ky = 0$ or $y'' = -ky$. Do you know a function for which the second derivative is $-k$ times the function? What if $k = 1$? Do you know a function for which $y'' = -y$? Of course, both $\sin(t)$ and $\cos(t)$ have this property. For other k-values, $\sin(k^{1/2}t)$ and $\cos(k^{1/2}t)$ have this property. And we can easily check that the general solution is of the form

$$c_1 \sin(k^{1/2}t) + c_2 \cos(k^{1/2}t).$$

That's why we see the periodic behavior of solutions; any such function is periodic with period $2\pi/k^{1/2}$.

Figure 8.3

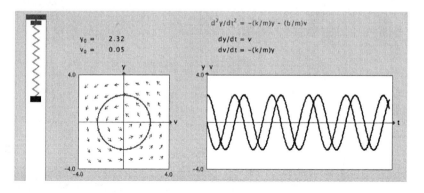

Important Terms

characteristic equation: A polynomial equation (linear, quadratic, cubic, etc.) whose roots specify the kinds of exponential solutions (all of the form $e^{\lambda t}$) that arise for a given linear differential equation. For a second-order linear differential equation (which can be written as a two-dimensional linear system of differential equations), the characteristic equation is a quadratic equation of the form $\lambda^2 - T\lambda + D$, where T is the trace of the matrix and D is the determinant of that matrix.

damping constant: A parameter that measures the air resistance (or fluid resistance) that affects the behavior of the mass-spring system. Contrasts with the spring constant.

mass-spring system: Hang a spring on a nail and attach a mass to it. Push or pull the spring and let it go. The mass-spring system is a differential equation whose solution specifies the motion of the mass as time goes on. The differential equation depends on two parameters, the spring constant and the damping constant. A mass-spring system is also called a harmonic oscillator.

spring constant: A parameter in the mass-spring system that measures how strongly the spring pulls the mass. Contrasts with the damping constant.

Suggested Reading

Blanchard, Devaney, and Hall, *Differential Equations*, chap. 2.1, 3.6.

Hirsch, Smale, and Devaney, *Differential Equations*, chap. 6.2.

Roberts, *Ordinary Differential Equations*, chap. 6.1.

Strogatz, *Nonlinear Dynamics and Chaos*, chap. 5.1.

Relevant Software

Blanchard, Devaney, and Hall, *DE Tools*, Mass Spring.

Problems

1. Write the second-order differential equation for the mass-spring system with mass = 1, spring constant = 2, and damping constant = 3.

2. Convert this second-order equation to a system.

3. Find all equilibrium points for this system.

4. Compute the second derivative of the functions $\sin(2t)$ and $\cos(2t)$.

5. Sketch the curve (e^{-t}, e^{-t}) in the plane.

6. For the differential equation $y'' + 3y' + 2y = 0$, show that $y(t) = k_1 e^{-t} + k_2 e^{-2t}$ is the general solution. Do this by showing that you can generate a solution of this form that solves any initial value problem $y(0) = A$, $y'(0) = B$.

7. a. Find the general solution of the mass-spring system with spring constant 6 and damping constant 5.

   b. Find the solution of the above mass-spring system that starts at $y(0) = 0$ with $y'(0) = 1$.

   c. Sketch both the graph of $y(t)$ and the corresponding curve $(y(t), v(t))$ in the phase plane.

   d. Explain in words what actually happens to this mass as time unfolds.

Exploration

Later in the course, we will investigate what happens to the mass-spring system when we apply an external periodic forcing. As a prelude to this, investigate what happens when we apply other, easier forcing terms. Consider first the mass-spring system in problem 2 above, with spring constant 6 and damping constant 5. What happens if we apply a constant force, say 1, to this system? That is, what happens to solutions of the ODE $y'' + 5y' + 6y = 1$? How about the systems where the forcing is 1 from time 0 to time 5, and then we turn the forcing off? Find the solution to this equation that satisfies $y(0) = y'(0) = 0$.

147

Second-Order Equations—The Mass-Spring System
Lecture 8—Transcript

Welcome back. In the last lecture, we introduced a new type of differential equation, namely, systems of differential equations, and right at the end, I showed a very special case of a system, namely, a second-order, linear differential equation of the form $y'' + ay' + by = 0$. In the next three lectures, we are going to spend our time talking about this special case.

As we showed last time, we can write this second-order differential equation as a system of differential equations and display the corresponding solutions in the phase plane. What we do is introduce the new variable, b, so $y' = v$, and then $v' = y''$ which is equal to $-av - by$. We will think of this second-order differential equation as actually a system of differential equations. Now what is the general solution of such a family or collection of differential equations?

The general solution will be a family of solutions from which we could solve any initial value problem. In the first-order case, the initial value problems were pretty simple. They were just y_0 is some constant, say y_0. In this case, in the case of second-order equations, we'll have another condition. Our initial conditions will be y_0 is some number, but also y'_0 will be some other number or v_0 will be something.

As always, let us begin with a model. This time, not biological, this time we will talk about the mass-spring system, also known as the harmonic oscillator. Here is the mass-spring system. Suppose you take a spring, nail it to the wall, to the ceiling, or whatever; it is hanging downward, and you attach some mass to it. Now we are going to move that spring up and down. The position of the mass will be noted by $y(t)$, the height of the mass along that vertical line. I'll always say that $y = 0$ is the rest position of that particular mass, and then $y > 0$ will mean that the spring is compressed, or above the rest position, and $-y$ will mean that the spring is stretched.

As always, we need to make some assumptions before we write down the differential equation. Our assumptions here will be twofold. We will assume that the mass obeys Newton's law, mass times acceleration is equal to the force on the mass, so $m(y'')$ is equal to whatever the forces are, and there will be two distinct forces.

The first force is the force exerted by the spring, and by Hooke's law, this force is proportional to the spring's displacement from rest. That is to say, Hooke's law says that mass time acceleration is equal to minus a constant times y, $-ky$. Here, k will be called the spring constant, and k will always be positive, so Hooke's law says the force is minus that spring constant times y.

Why do we choose $-ky$? There is a physical reason and a mathematical reason. The physical reason is the force is pushing the mass in the opposite direction of the position. If your mass is raised above the equilibrium position, the spring is forcing it down in the opposite direction. If your mass is way below the rest position, your spring is pulling it back again in the opposite direction.

But there's also a mathematical reason. Suppose at some instant of time, your position is positive, say $y(t)$ is positive, and suppose the derivative of $y(t)$, the velocity, is also positive. That means that your mass is above the rest position and it's going up, so what would the graph look like of this solution? We know $y(t)$ is positive and y' is positive. The mass is instantaneously moving up, but the force of the spring should be slowing it down, so that says that, yes, we're at a positive position, $y(t)$ is positive, slope is positive, but the graph should be concave down because we're slowing down as this mass moves up. That says that then y' should be decreasing, or the concavity of your graph, looking like that, should be downward, so the second derivative should be negative. That's why we choose y'' being $-ky$.

There is another force that's going to enter the play that's due to friction, so what is that force. We'll assume here that the force due to friction, whether it's air resistance, water resistance, a heavier spring, or whatever, the force is proportional to velocity. That is to say that this force is again given by minus a constant times y'. That constant I'll call b. It's what we call the damping constant. Unlike the spring constant, this damping constant can be 0; that means there's no damping. We're doing this in an environment where there's no friction whatsoever, no air resistance or anything. It's the so-called undamped case, but b can be positive when there is friction; that's the damped case.

Altogether, what is our second order differential equation for the mass-spring system? Mass times acceleration, mass times y'', is equal to these two forces. The first is $-by'$, and the second, by Hooke's law, is $-ky$. I'm a mathematician, so I always let the mass be 1. I can do that.

Let's assume that m, the mass, is 1, so our differential equation reads, bringing everything over to the left, $y'' + by' + ky = 0$. That's a second-order linear differential equation with two parameters, b the damping constant greater than or equal to 0, and k the spring constant, always positive.

We're always going to have initial conditions here. Our initial conditions will be first $y(0)$ is something, our initial position, and $y'(0)$ is something, our initial velocity. As we saw before, we can also write this differential equation as a system of differential equations. Again, we always do that by introducing the new variable, v, $y' = v$. Then v', the second differential equation, is y''. But that's $-ky - by'$, but $y' = v$. So the equation for v is $v' = -ky - bv$, and the first equation was $y' = v$. That's a linear system of differential equations.

Let's look at the different possible behaviors we could get for this mass-spring system as we change, say, the damping constant. Here, again, we see several different views. Over on the left, I actually have my mass spring. It's hanging from the ceiling, and there's the mass. Next to it, I'm plotting the phase plane. I'll plot y vertically and v horizontally. Remember y measures the position of the mass. When it's at rest, y is 0. When y is positive, the spring is compressed, our mass moves up, and when y is negative the spring is stretched. Our mass moves down, and then finally, over here in the right, I'll plot the graphs of $y(t)$. We could also plot the graph of v with respect to t, velocity with respect to t, but we're really interested only in the position of the mass.

Here's a particular example. Let me start with spring constant 1, and let me let the damping constant be relatively small, 0.17. Here you see the direction field in the phase plane, and what we see is, as you would expect, solutions just oscillate down to rest. Our solutions go up and down; the mass goes up and down, but with decreasing amplitude, decreasing height above and below, and it slowly goes down to rest. You see solutions spiraling into the point $y = 0$, $y' = 0$. That's an equilibrium point where the position is right at rest, and there's no velocity.

So there's one type of behavior for this mass-spring system, but now let me let b be 0. Let me assume there's no damping. So there is no air resistance, no resistance whatsoever. The only force we have is the force exerted by the spring, and now something different happens. If we start our mass off with the spring compressed, what we see is the solution just behaves periodically, up and down, up and down; there's no friction around to push it back to equilibrium. This is the undamped case.

One more example, now let me let the damping constant be relatively large. This is the case where we're watching our spring. I always say we live in a house filled with oatmeal. You attach your spring to the ceiling, and it's got a lot of oatmeal to go through. What happens here? What happens here is our mass just glides very slowly down to its rest position without ever oscillating back and forth. There's a lot of resistance to this mass-spring system. Start from below, same thing, the spring just glides down to equilibrium without oscillating back and forth, so this is what is called the overdamped case.

Now the question is how do we solve this system of differential equations, or this second-order differential equation? Recall that for first-order linear equations, $y' = ky$ where we didn't have any forcing term, then we just had the solution, $y' = ky$. A solution was e^{kt}. The more general solution was some constant times e^{kt}. So we got exponential solutions for linear first-order differential equations; probably, we'll get exponential solutions here.

First off, let's guess a solution of the form e to the some constant times t, let's say e^{st}, and see if we can determine what value of s gives us a solution. So let's take e^{st}, plug it into the harmonic oscillator equation, and you get $y'' + by' + ky = 0$. If you plug e^{st} in there, differentiate it twice, and each time you differentiate e^{st} and s comes down. So y'' is $s^2(e^{st})$. We've got to add to that b times the derivative, that's $bs(e^{st})$, and then $k(e^{st})$. All of that is equal to 0, so there is our equation with e^{st} plugged in. We can factor out the e^{st}, and we're left with $e^{st}(s^2 + bs + k) = 0$.

Now, e^{st} is never 0, so the only way we can solve that equation is if $s^2 + bs + k = 0$, and that, of course, is a quadratic equation. That's an equation that's going to come up over and over again with systems. It's what we call the characteristic equation. What are the roots of this characteristic equation? The roots will give us the s values for which e^{st} is a solution to this differential equation.

By the quadratic formula, the roots of this equation are $-b \pm \sqrt{b^2 - (4 \times 1 \times k)}$, that's the numerator, , all divided by 2. Our roots are given by $-b \pm$ that square root divided by 2. Two immediate problems jump out at you. We have that square root of $b^2 - 4k$. That can be negative. Uh-oh. If $b^2 - 4k$ is negative, we've got the square root of a negative number. We've got complex roots. What's going to happen here? That's one thing we've got to worry about.

Another thing is $b^2 - 4k$ could be 0. Then we only get one solution of the form e^{st}, in fact, it's $e^{(-b/2)t}$, and we get one solution, whereas if the root, $b^2 - 4k$ is positive, we get two roots. So those two cases, the complex case and what we call the repeated root case, only one solution, are very special. We'll talk about those special situations, in fact, more difficult situations, in the next lecture.

For now on, let's assume that $b^2 - 4k$ is positive. What are our roots? We have two.

$$\alpha = (-b - \sqrt{b^2 - 4k})/2$$
$$\beta = (-b + \sqrt{b^2 - 4k})/2$$

Now, clearly that root, α, is negative. We've got $-b$ and minus a square root over 2; that's a negative number, but β where we have $-b$ plus the square root is also negative because remember, $b^2 - 4k$, k is always positive. So $b^2 - 4k$, we're subtracting something off b^2, so that's smaller, and the square root there is smaller than b. We've got $-b$ plus something smaller than b; that's also negative.

We have two solutions to this differential equation, the exponential of αt, and the exponential of βt where α and β are those roots of the characteristic equation. In fact, as we'll see, this is the general solution of this second-order differential equation because we can find values of constants k_1 and k_2 such that $k_1(e^{\alpha t}) + k_2 e^{\beta t}$ times our second solution, $e^{\beta t}$, solve any initial value problem, which remember is of the form $y(0) = A$, and $y'(0) = B$.

What does that mean? Remember, our initial conditions involved not only our solution, $e^{\alpha t} + k_2 e^{\beta t}$, but also its derivative. Remember the derivative of our solution is k_1. We differentiate $e^{\alpha t}$ you get $\alpha e^{\alpha t}$ and the derivative of the second term is $k_2 \beta e^{\beta t}$. So we've got to find solutions of the equation, take our solution, plug in 0, set it equal to some given number, a, take the derivative of our solution, plug in 0, and set it equal to b. We have to find values of the constants k_1 and k_2 that work. That is, if you plug 0 into the first equation, you get $k_1 e^0 + k_2 e^0 = a$, some given number, a, and to the second equation, you get $k_1 \alpha e^0$ that's $1 + k_2 \beta = B$. So you've got a family of two linear equations, $k_1 + k_2 = A$, and $\alpha k1 + \beta k2 = B$. You have to show that you can solve that system of equations given any values of A and B.

Rather than do that in general, let me do a specific example. Let me take the spring constant to be 2 and the damping constant to be 3, so $y'' + 3y' + 2y = 0$, there is our mass spring linear differential equation. What's the characteristic equation? Exactly as we saw before, if you plug in e^{st}, you're going to get an $s^2 + 3s + 2 = 0$. That's the characteristic equation. Notice that comes right out of the differential equation; we have y'' gives you $s^2 + 3y'$ gives you $3s + 2y$ gives you just $2 = 0$. This characteristic equation factors into $(s + 2)(s + 1)$, so we have the roots. The roots are -1 and -2, so I claim here that the general solution is $k_1 e^{-1t}$, one of our roots of the characteristic equation, $+ k_2 e^{-2t}$.

Let's check this. So the question is, for any given initial conditions, A and B, can you find values of k_1 and k_2 so that $A = y(0)$, that's just $k_1 \exp(0)$; that's $0 + k_2$ times the exponential. That is 0, so $k_1 + k_2 = A$, and similarly, as we saw before, $B = y'(0)$ and that's $\alpha k_1 - k_1 \beta k_2 - 2k_2$.

Can you solve the equations $A = k_1 + k_2$ and $B = -k_1 - 2k_2$ where A and B are given numbers, they are constants; the variables are k_1 and k_2. First, if you add these two equations, notice the k_1 terms disappear, and what you get on the left is $A + B$ and on the right you get $k_2 - 2k_2$. That's just $-k_2$, so we have

$k_2 = -A - B$. We can at least find k_2. Now, go back to the first equation. The first equation says that $A = k_1 + k_2$, but we know that $k_2 = -A - B$. So that first equation says that $A = k_1 - A - B$, or we have k_1; k_1 is $2A + B$, so yes, we can find k_1 and k_2 that solve any initial value problem.

What are these solutions going to look like? Both of these are decaying exponentials, either the $-t$ or the $-2t$, so we know that solutions are going to go directly to the rest position. These things are not going to oscillate back and forth as in some of the cases we saw earlier. I assumed here that the spring constant was 2, the damping constant was 3, and there's the picture in the phase plane and the corresponding graph of $y(t)$. If I start the mass off with the spring compressed, decaying exponentials say I go right to 0. No matter where I take this initial condition, right to the rest position, no oscillations back and forth. All solutions $k_1 e^{-t} + k_2 e^{-2t}$ just go to equilibrium. This is what we call the overdamped harmonic oscillator. This is what always happens when we have a pair of real roots to the characteristic equation.

Let's turn to the undamped case, and that's the case where $b = 0$. So our equation now is $y'' + 0(y') + ky$, or as a system, $y' = v$, and $v' = -ky$. But let's not think of it as a system; let's think of this as just the second order differential equation, $y'' - ky$. Remember the roots of the characteristic equation here where the root α was $(-b - \sqrt{b^2 - 4k})/2$.

In this case, what do we get? The b terms disappear, so α is just $-\sqrt{-4k}/2$. If k is positive $-4k$ is negative, we're taking the square root of a negative number, and we're getting imaginary roots. Oh, boy, looks bad. The same thing with β, the root β is given by, this time $+\sqrt{-4k}/2$. So we've got complex roots. That's something we're going to have to deal with a little bit later.

Before I do that, let me take a very special case. Let's just look back at $y'' = -ky$. Do you know a function whose second derivative is equal to $-k$ times itself? I think you do. Let's ask a simpler question, what if $k = 1$? Do you know a function whose second derivative is equal to minus itself? We've actually seen this a couple of times already as sines and cosines entered the fray. We know that the first derivative of sine is cos, second derivative is minus sine. The second derivative of sine is minus itself, and similarly, the second derivative of cosine is minus itself. So we have some solutions; they involve sines and cosines.

What about other values of k? What happens if k is positive but not equal to 1? Now $y'' = -ky$ has solutions that are the $\sin(k^{1/2}t)$ and the $\cos(k^{1/2}t)$. If I differentiate $\sin(k^{1/2}t)$, I get $\cos(k^{1/2}t)$, but a $k^{1/2}$ comes out. I get $k^{1/2}\cos(k^{1/2}t)$. If I differentiate that again, I have $1k^{1/2}$, I differentiate the cosine, I get $-\sin(k^{1/2}t)$, but a second $k^{1/2}$ comes out. I get $-k\sin(k^{1/2}t)$. So I've found a solution. In fact, that is the general solution. The general solution is $k_1\cos(k^{1/2}t) + k_2\sin(k^{1/2}t)$, and this is a periodic solution. That's why we see those oscillations.

Let me go back to the computer and see a couple of these pictures. Let me first choose my k to be 1 and my damping to be 0. Then we saw sines and cosines were solutions, and sure enough, that's what we have. Solutions run around circles in the phase plane; you see the graphs of a nice periodic function elsewhere. If I choose a larger initial value say $y(0)$ is 0, but $y'(0)$ is a large negative, we get a larger circle. If I go closer to the origin, still, we get circular motions. So these sines and cosines behave periodically. This is the undamped harmonic oscillator.

Let me now change k to be 2, so $k = 2$. We know we have solutions of the form $\sin(2^{1/2}t)$ and $\cos(2^{1/2}t)$. What do these solutions look like? Now these solutions look like they are running around the origin again, but now they are doing so on ellipses. These solutions lie on ellipses. They're still periodic, but they run around on ellipses. Go much further away, and you see the ellipse leaves the phase line, phase plane, and comes back in.

There is the behavior of the undamped harmonic oscillator in this very special case, but now we also can have roots of the characteristic equation that involve not just imaginary numbers, but in fact, complex numbers. The question is what happens there? What happens if our root is a complex number of the form a + ib. That's when things get kind of complicated; that's what we'll come back to next time.

Next time we'll see the case where we've got complex roots of the characteristic equation, and also that very special and kind of difficult case where we have repeated roots, a single root, to the characteristic equation so more to come on the harmonic oscillators. See you then.

Damped and Undamped Harmonic Oscillators
Lecture 9

Now we are ready to complete our investigations of the mass-spring differential equation $y'' + by' + ky = 0$. In the last lecture, we saw that there is an exponential solution of the form e^{st} when s is a root of the characteristic equation $s^2 + bs + k = 0$. But this quadratic equation may have complex roots, so the question is what happens to solutions when this occurs? Keep in mind that we also saw that the general solutions of the undamped equation $y'' + y = 0$ involved sines and cosines.

So how did we end up with trigonometric solutions to this differential equation instead of exponentials? Our differential equation is $y'' + y = 0$, and a guess of e^{st} yields the characteristic equation $s^2 + 1 = 0$. The roots of this equation are the imaginary number $i = \sqrt{-1}$ and its negative, so we have a solution of the form e^{it}. What is this complex expression? Recall from calculus that we can express certain functions using their power series expansions. That is, assuming our function $f(t)$ is infinitely differentiable at $t = 0$, we can write $f(t)$ as below.

$$f(t) = f(0) + f'(0)t + f''(0)\frac{t^2}{2!} + f^{(3)}(0)\frac{t^3}{3!} + f^{(4)}(0)\frac{t^4}{4!} + \cdots$$

In the case of e^{it}, we can write this expression in a power series as below.

$$\begin{aligned} e^{it} &= 1 + it + \frac{(it)^2}{2!} + \frac{(it)^3}{3!} + \frac{(it)^4}{4!} + \frac{(it)^5}{5!} + \cdots \\ &= 1 + it - \frac{t^2}{2!} - i\frac{t^3}{3!} + \frac{t^4}{4!} + i\frac{t^5}{5!} + \cdots \end{aligned}$$

Collecting the terms that involve the constant i and those that do not yields

$$e^{it} = 1 - \frac{t^2}{2!} + \frac{t^4}{4!} - \frac{t^6}{6!} + \ldots + i\left(t - \frac{t^3}{3!} + \frac{t^5}{5!} - \frac{t^7}{7!} + \ldots\right).$$

But look at the right-hand side here: The first power series is just that of the cosine function, while the second series is the sine function. That is, we have one of the most famous and beautiful of all formulas in mathematics, **Euler's formula**.

$$e^{it} = \cos(t) + i\sin(t)$$

Letting $t = \pi$ in this formula, we get every mathematician's favorite formula: $e^{i\pi} + 1 = 0$.

Now back to the differential equation $y'' + y = 0$. We have the complex solution

$$e^{it} = \cos(t) + i\sin(t).$$

But our differential equation was a real differential equation; there were no complex numbers in sight. This means that both the real part of our complex solution, $\cos(t)$, and the imaginary part of the solution, $\sin(t)$, are also solutions. This is what we saw before: Our general solution was the combination of the real and the imaginary parts of our complex solution:

$$y(t) = k_1\cos(t) + k_2\sin(t).$$

More generally, for the differential equation $y' + ky = 0$, we get the general solution

$$y(t) = k_1 \cos(\sqrt{k}t) + k_2 \sin(\sqrt{k}t),$$

whose behavior is also periodic.

Figure 9.1

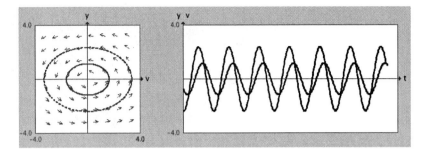

Now consider the mass-spring system with spring constant $k = 1.01$ and small damping constant $b = 0.2$. The characteristic equation is $s^2 + 0.2s + 1.01 = 0$, and the roots are the complex numbers $-0.1 + i$ and $-0.1 - i$. So we have a complex solution of the form $y(t) = e^{(-0.1+i)t}$. We can use properties of the exponential function and write it in the form $y(t) = e^{-0.1t}e^{it}$. So by Euler's formula, we now have complex solutions given by $e^{-0.1t}(\cos(t) + i\sin(t))$. Taking the real and imaginary parts of this solution gives the general solution

$$k_1 e^{-0.1t}\cos(t) + k_2 e^{-0.1t}\sin(t).$$

Now we have a decaying exponential as part of our solution. So the solution does oscillate around its rest position, but now the amplitude of the oscillation (the up and down height of the $y(t)$ graph) tends to zero as time goes on. We see that the mass does tend to its rest position, but it oscillates back and forth as it does so. This is the underdamped harmonic oscillator.

Figure 9.2

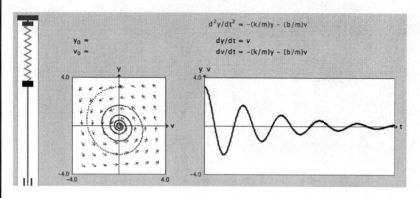

More generally, when we have complex roots of the characteristic equation, say $a + ib$ and $a - ib$ (here a is necessarily negative), we get the general solution

$$k_1 e^{at} \cos(bt) + k_2 e^{at} \sin(bt),$$

which similarly oscillates down to its rest position.

Recall that there was one other case of the harmonic oscillator that we have not yet solved, the case for which we found only one root for the characteristic equation. This is the **critically damped mass-spring** case. For example, consider the mass-spring system

$$y'' + 2y' + y = 0.$$

The characteristic equation is now $s^2 + 2s + 1$, which factors into $(s + 1)^2 = 0$. So we have a pair of real roots given by $s = -1$. We know one solution, namely, e^{-t}. As before, we can multiply this expression by any constant to find other solutions of the form $k_1 e^{-t}$. What about other solutions? We can easily check that $y(t) = te^{-t}$ also solves this differential equation and that, moreover,

$$k_1 e^{-t} + k_2 t e^{-t}$$

is the general solution. What happens to the solution given by te^{-t} as time goes on? Using l'Hopital's rule, we find that the limit of the expression te^{-t} also tends to zero. Thus we see that both terms in the general solution go to zero—and in fact, in the phase plane, they do so just as in the overdamped case.

Figure 9.3

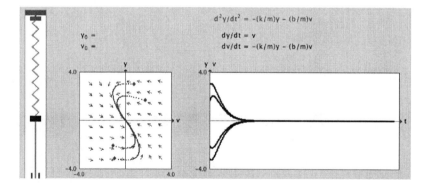

Important Terms

Euler's formula: This incredible formula provides an interesting connection between exponential and trigonometric functions, namely: The exponential of the imaginary number ($i \cdot t$) is just the sum of the trigonometric functions $\cos(t)$ and $i \sin(t)$. So $e^{(it)} = \cos(t) + i \sin(t)$.

critically damped mass-spring: A mass-spring system for which the damping force is just at the point where the mass returns to its rest position without oscillation. However, any less of a damping force will allow the mass to oscillate.

Suggested Reading

Blanchard, Devaney, and Hall, *Differential Equations*, chaps. 2.1 and 3.6.

Hirsch, Smale, and Devaney, *Differential Equations*, chap. 3.4.

Roberts, *Ordinary Differential Equations*, chap. 6.1.

Strogatz, *Nonlinear Dynamics and Chaos*, chaps. 5.1 and 7.6.

Relevant Software

Blanchard, Devaney, and Hall, *DE Tools*, Mass Spring, RLC Circuits.

Problems

1. Use the product rule to compute the derivative of $e^{2t}\cos(3t)$.

2. Use Euler's formula to expand $e^{(2t + 3it)}$.

3. Solve the quadratic equation $s^2 + bs + k = 0$, where b and k are constants.

4. The point $(\sin(t), \cos(t))$ lies on the unit circle in the plane. In which direction does it move as t increases?

5. Is the mass-spring system $y'' + 5y' + 6y$ overdamped or underdamped?

6. Use l'Hopital's rule from calculus to verify that the limit as t approaches infinity of te^{-t} is 0.

7. Find the general solution of the mass-spring system with spring constant 2 and damping constant 2.

8. If we start the mass-spring system in problem 7 off from its rest position but with velocity equal to 1, what is the maximum distance the center of mass will move?

9. Consider the mass-spring systems with spring constant 1 and damping constant b. For which values of b is this system underdamped, overdamped, undamped, or critically damped?

10. We know the general solution of the undamped harmonic oscillator given by $y'' + y = 0$. What happens if we apply a constant force equal to 1 to this system? That is, what is the general solution of the equation $y'' + y = 1$? And what is the motion of the corresponding mass-spring?

Exploration

One of the simplest examples for electric circuit theory is the RLC circuit,

Figure 9.4

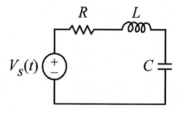

where R is the resistance, L is the inductance, C is the capacitance, and $V_S(t)$ denotes the voltage at the source. The differential equation for the voltage across the capacitor $v(t)$ is similar to our mass-spring example and is given by

$$LCv'' + RCv' + v = V_S(t).$$

Find the solution to this equation when $V_S(t)$ is zero. Compare the behavior of these solutions to those of the mass-spring system.

Damped and Undamped Harmonic Oscillators
Lecture 9—Transcript

So here we go again, time for some more mass-spring systems. Remember, in the last lecture, we introduced the mass-spring system second-order differential equation, $y''+$ damping constant, $b \times y' +$ spring constant, $k \times y = 0$. Here the damping constant, b, could either be 0 or greater than 0 whereas the spring constant, k, was always positive.

Last time we found exponential solutions by guessing a solution of the form e^{st}. When we did that, out came the characteristic equation, $s^2 + bs + k = 0$ that had roots that were real and distinct. As we mentioned, those roots could also be complex or repeated. We also saw, right at the end of the last lecture, that when we took the damping constant, b, to be 0 and k to be 1, we saw solutions that were sines and cosines. Where do they come from? We were always getting exponential solutions before. How did we get trigonometric solutions? Well, stay tuned. Today we're going to see, if you haven't seen it before, the most amazing formula in all of mathematics, at least in my opinion, and I'm sure it will be in yours.

For the differential equation $y'' + y = 0$, the characteristic equation was plug in e^{st}, differentiate it twice, get $s^2(e^{st})$, add it to e^{st}, and you get the equation $s^2 + 1 = 0$, or $s^2 = -1$. The roots here are the imaginary numbers, $\pm i$. We have a solution of the form e^{it} where i is the imaginary number. I know some of you are engineers out there, and you call the imaginary number ij. I am a mathematician. It's imaginary to me; it's i.

So the question is what does e^{it} mean? Here is what e^{it} is. If you remember from calculus that you take any infinitely differentiable function, say $F(t)$, then it has what we call a power series. You can write $F(t)$ as the following infinite sum of functions, $F(t) = F(0) + F'(0)t + F''(0)(t^2/2!) + F^{(3)}(0)(t^3/3!) + F^{(4)}(0)(t^4/4!)$ and on and on and on. The nth term is the $F^{(n)}(0)(t^n/n!)$.

The exponential is infinitely differentiable, so how can we write e^{it}? Using that power series, $e^{it} = F(0)$, that's $e^{01} + F'(0)t$, the derivative of e^{it} is $i(e^{it})$ at 0. That is $i(t)$. The second derivative at 0 is, each time we differentiate e^{it}, another i comes down, so we're going to have i^2, that's -1, remember i is the square root of -1; $i^2 t^2/2$, that's $-(t^2/2!)$ and so on. Differentiate again, we get $it^3/3!$, but i^3 is i^2 is -1 times i is $-i(t^3/3!)$ and on and on. The fourth, the next, derivative is $(it)^4$, but i^4 is 1 because i^2 is -1, divide it by 4!, then $(it)^5$, now we're back to $it^5/5!$, the power series for e^{it} is a conglomeration of terms that involve i and no i. It is $1 + it - t^2/2! - i(t^3/3!) + (t^4/4!) + i(t^5/5!)$. Each term alternates between an i and a no $-i$ term, a real term and an imaginary term.

Let's collect all of the terms that don't involve the imaginary number, i, and all of them that do involve the imaginary term, i. What do we get? We get that e^{it} has real terms, 1, no t because that has an i in front of it, $-t^2/2! + t^4/4! -t^6/6!$ plus all the $t^{2n}/2n!$, sometimes plus, sometimes minus. Now put in the i terms, so plus i times all the terms involving i were $t - t^3/3! + t^5/5! -t^7/7!$.

What we have is the power series for the complex exponential, for e^{it}. It involves a term involving no imaginary numbers, i, and another term involving the imaginary numbers, i, two separate power series.

What is that power series on the left? Let's take the derivative of the left, the derivative of the right, and see what we get. First, the derivative of the right, the terms involving i, you get $t - t^3/3! + t^5/5!$, etcetera. What's the derivative there? The derivative there is derivative of t is 1, the derivative of $-t^3/3!$ is $-3t^2/3!$. The 3s go away, and we get $-t^2/2!$. Hmm, look over to the left and you see that. What is the derivative of $t^5/5!$? That's $5t^4/5!$, again the 5s go away, and we get $+ t^4/4!$ and so forth. The derivative of the right hand power series is the left hand power series.

What is the derivative of the left? The left is $1 - t^2/2! + t^4/4!$. The derivative of 1 is 0; the derivative of $-t^2/2!$, the 2s go away, leaving us with t. The derivative of $t^4/4!$, the 4s go away, we get $t^3/3!$, and then $- t^5/5! + t^7/7!$. The derivative of the left is minus the right. That is, the first derivative of the right hand series is the left, the second derivative of the right is the first derivative of the left, and that's minus the right. The second derivative of each of these power series is exactly minus itself.

You've seen functions that have that property; those are the sines and the cosines. In fact, the left hand power series, if you plug in 0, you get 1. So that left hand power series is the cosine function, $\cos(t) = 1 - t^2/2! + t^4/4! - t^6/6!$, and if you plug in 0 to the right, you get 0; that's the sine function. Think of what that says. This is Euler's formula. It says that $e^{it} = \cos(t) + i\sin(t)$, the exponential function of a complex number somehow involves cosines and sines; e^{it} is cos + $i$sin, an absolutely amazing formula. Like I said, this is the greatest formula in all of mathematics.

Well, there is one better. The one better is this, since e^{it} is $\cos(t) + i\sin(t)$, let me plug in $t = \pi$. What do we get? We get $e^{i\pi}$ on the left is equal to $\cos(\pi) + i\sin(\pi)$. The $\cos(\pi) = -1$ and $\sin(\pi)$ is 0, so we get $e^{i\pi}$ is -1, or bring the -1 to the left, we get $e^{i\pi} + 1 = 0$. There it is. That's the greatest formula in all of mathematics, $e^{i\pi} + 1 = 0$.

Look what you're doing in that equation. You have a sum; you're adding something. You're multiplying i and π, and you're exponentiating $i\pi$, the three basic operations, addition, multiplication, and exponentiation. Look at the numbers that are there. The numbers that are there are 0 and 1, the basis for all integers, all rational numbers. You also see your two favorite irrational numbers, e and π, and then what's the only other great number? The imaginary number i. You put them all together as $e^{i\pi} + 1$, you get 0. They all combine to be in that amazing form. That's just incredible. That's what Euler discovered.

Now a little story, while I'm here, since I'm crazy about this equation. I was on sabbatical a number of years ago, spent some time in Bonn in Germany, had to rent a house for my family, and we found a house on, you guessed it, Eulerstrasse. We lived on Eulerstrasse, loved it. Better than that, when I would walk in to the institute, I'd come down Eulerstrasse, and I'd get on Hausdorffstrasse. You may not know Felix Hausdorff, but my area of research involves a lot of fractals. A lot of fractals involve Hausdorff dimension, so here's my research right there, Eulerstrasse and the corner of Hausdorffstrasse. I'd sit there for days just thinking about mathematics.

Now, we have our solution of $y'' + y = 0$. We have the solution in terms of exponentials; it's e^{it}, but that's $\cos(t) + i\sin(t)$. Wait a minute. We took a real differential equation, $y'' + y = 0$, and got a complex solution out? No, in fact, when you've got a complex solution to a real differential equation, both the real parts, in this case cos(*t*), and the imaginary parts, this time sin(*t*), are also solutions, and in fact, that's what we saw in the last lecture.

The general solution of this system is k_1 times the real part of our complex solution, cos, plus k_2 times the imaginary part sine. As a system of differential equations our solution will be that, $y(t) = k_1\cos(t) + k_2\sin(t)$. And $v(t)$ is the derivative of y, and that's $-k_1\sin(t) + k_2\cos(t)$. Of course, if you let k_1 equal 0 and k_2 be 1, you get the solution that we saw last time, namely sin(*t*) and cos(*t*), which lies on a circle.

More generally, $y'' + ky = 0$, we saw that last time, but now we'll find a solution $e^{i\sqrt{k}t}$. Our solutions will be of the form $y(t) = k_1\cos(\sqrt{k}t) + k_2\sin(\sqrt{k}t)$, and the corresponding derivatives. That's how the sines and cosines come up in the second-order differential equations where we guess an exponential and get trigonometric functions.

So, again, I've shown you these pictures before, but let's go back to the computer and see what the solutions look like when *k* is equal to 1, and *b*, our damping constant, is equal to 0. There we saw solutions, sines and cosines, that run around in circles, periodic solutions of the differential equation.

On the other hand, if we let *k* be something else, let's say *k* is smaller, say *k* is ½, now our solutions run around on ellipses. We still have periodic solutions, but $\cos(\sqrt{k}t)$ and $\sin(\sqrt{k}t)$ give elliptic solutions to this differential equation. No matter where we start again, you see an harmonic oscillator oscillating back and forth without ever turning down to rest. As well called this the last time, this is the undamped harmonic oscillator.

We can also have roots of this characteristic equation that are real complex numbers of the form a + ib. Let me do an example there. Let me, for reasons you'll see in a minute, let *k* not be 1, but 1.01, and let me take a small damping, say *b* = 0.2. Our differential equation is $y'' + 0.2y' + 1.01y = 0$, characteristic equation. As always, $s^2 + 0.2s + 1.01 = 0$. Yikes, it looks kind of complicated. But plug this into the quadratic formula, we get the roots -0.2 ± 0.2^2, that's $\pm$ the square root of 0.2^2, that's 0.04, minus 4 times 1.01, all divided by 2. That's $(-0.2 \pm \sqrt{-4})/2$.

The square root of −4 is $2i$, the real part is −0.2/2, and we get roots $-0.1 + i$ and $-0.1 - i$. Those are the roots of our characteristic equation, so we get exponential solutions of the form, the exponential of that complex number $((-0.1 + i)t)$. What are those solutions? We have $e^{(-0.1 + i)t}$. We can factor that into $e^{-0.1t}e^{it}$, and here comes Euler again. You get $e^{-0.1t}$ times, by Euler, $\cos(t)$ plus $i\sin(t)$, so we've got, again, a complex solution to this differential equation.

Just as with the previous case, we can take the real and imaginary parts of this differential equation, of this solution. The real part is $e^{-0.1t}\cos(t)$ and the other one is $e^{-0.1t}\sin(t)$. Our general solution is the combination of those two, $k_1 e^{-0.1t}\cos(t) + k_2 e^{-0.1t}\sin(t)$. What's going to happen to these solutions?

If you forget about the exponential solutions, there we're going to have cosines and sines, there we're going to get solutions that are periodic in the phase plane. We're multiplying these solutions by $e^{-0.1t}$. We're multiplying it by e to a negative constant times t. When we multiply something by the negative constant times t, that's an exponential that's decaying, so instead of our solutions running around periodically, as t gets bigger, we're multiplying those numbers by a smaller and smaller number, $e^{-0.1t}$. This is our underdamped harmonic oscillator.

More generally, if we have roots of the form $a \pm ib$, we're going to get a general solution of the form $k_1 e^a$ that's the real part of that complex number, t, times $\cos(bt) + k_2 e^{at}\sin(bt)$. It turns out that because our spring constant is always positive, that a value will always be negative unless it is 0 as in the undamped case. We'll always have a negative exponential, a decaying exponential, multiplying our cosines and sines. This is the underdamped harmonic oscillator, so let me let k be something, say 1, and let me let b be 0.2. What happens to this harmonic oscillator? Let me stretch it out, or push the mass way up, let it go, and as we've seen before, it does exactly what you usually see with mass-spring systems. This harmonic oscillator just oscillates down to rest. After a very short time, these oscillations are invisible.

What's going on here is, yes, we have a cosine and sine solution which, as we know, runs around the origin in circles. But as time gets higher, we're multiplying that cosine and sine by exponentials that are slowly forcing this solution into the origin, that's why we see these oscillations, and no matter where we start this mass-spring system, the system always oscillates down to rest, as I said, the underdamped harmonic oscillator.

So that's the next case, now let me turn to the one remaining case. The one case left is where when we take the characteristic equation, we get only one root of that equation. Let me start with an example. This we call, by the way, repeated roots. This is going to come up several times in the future as we go along. It'll be very important, in fact, in terms of bifurcations.

Here's the example. Let's take y'' plus damping constant, 2, times y' plus spring constant $1y = 0$. Our characteristic equation is now $s^2 + 2s + 1 = 0$, and there, we see it. We have a repeated root. That factors into $(s + 1)^2 = 0$, so the only root is -1. We have one solution, e^{-t}. In all of the other cases, we saw that we had two solutions. If we had repeated roots, we had $e^{\alpha t}$ and $e^{\beta t}$. If we had the undamped case, we had a cosine and a sine. Underdamped, we had $e^{\alpha t}\cos$, $e^{\alpha t}\sin$.

So where are we going to get this second solution from? It turns out that this problem is going to come up over and over again, and we're always going to do the same thing. We're going to guess a second solution of the equation $y'' + 2y' + y = 0$. What are you going to guess? This is where things get a little complicated. What's going to give you 0 when you come out there? It is not at all apparent what it should be, but here's the situation. Whenever we encounter a situation where we have to find a second solution, knowing one already, in our case we know e^{-t}, what usually works is to guess t times your one solution, or $t(e^{-t})$.

So let me make that guess, $t(e^{-t})$. Let's check that that's the solution to the differential equation. We have to plug that into the differential equation. That means you have to differentiate $t(e^{-t})$. Here comes some more calculus. How do you differentiate the product of two functions? Remember, you have the product of $u(t)$ and $v(t)$; the derivative of that product is given by, guess what, the product rule. It's given by the derivative of the first, $u'v$, plus the derivative of the second, $v'u$. That is the old product rule from calculus.

So we have $t(e^{-t})$. We have to differentiate that twice. Let's differentiate it the first time. By the product rule, when we differentiate $t(e^{-t})$, differentiate the t term, you get 1 times the other term, e^{-t}, and then we have to add to that the derivative of the second term, that's e^{-t}. That derivative is $-e^{-t}$, but times the first, so $-t(e^{-t})$. So $y' = e^{-t} - t(e^{-t})$.

Now we have to differentiate again. Second derivative is the derivative of $e^{-t} - t(e^{-t})$. Derivative of e^{-t} we know is $-e^{-t}$, and now the derivative of $-t(e^{-t})$, again by a product rule, differentiate the t term. It goes away, you have $-e^{-t}$, and then you have to add to that $-t$ times the derivative of e^{-t}. That gives you $a + t(e^{-t})$. Altogether, what we get is if you take y'' plus $2y'$ plus y, $y'' = -2e^{-t} + t(e^{-t})$, then $2y' = 2e^{-t} - t(e^{-t})$, and then plus, finally, $t(e^{-t})$, and that should be equal to 0 if, in fact, it's a solution. If you add up all of those terms, you see you do, indeed, get 0. So $t(e^{-t})$ is a solution of this differential equation.

Then as we've seen over and over again, so is any constant times $t(e^{-t})$. If you plug in a constant times something that you know is a solution and you differentiate it twice, the constant comes out. Add two times it; the constant comes out. y times the constant comes out. You get a constant times what you put in before. Any constant times $t(e^{-t})$ is also a solution. We've got it; there's the second solution, and then you can easily check, as we've done before, that any constant times that old solution, (e^{-t}) plus any other constant times this new solution $t(e^{-t})$ is the general solution to this differential equation.

This is what we call the critically damped harmonic oscillator. What's happened is, in the case that we did before, the overdamped case, we had two real roots, and we sort of vary things continuously. What's happening is those two real roots are coming together and merging at a place where we have a single root. Then if we keep going, they go off into the complex plane, and we've got complex roots. We're in the underdamped case.

This is kind of a special case. It's where we see bifurcations again, but before getting into that, let's ask what's happening to this solution, say $t(e^{-t})$. What's going on with that solution? Well, t, as time goes on, is going to infinity. Oh, my God. Does that mean that this spring is going off to infinity? The (e^{-t}) is going to 0, but what happens to $t(e^{-t})$? T is going to infinity; (e^{-t}) is going to 0.

Here comes some more calculus. Let's write $t(e^{-t})$ is t/e^t, and now the numerator goes to infinity, and so does the denominator. What's happening here? There's l'Hopital's rule; l'Hopital's rule says that if you have two things going off to infinity, $u(t)/v(t)$, then what's happening to these things in the limit is the same as what's happening to the derivatives. The limit of u'(t)/v'(t) is the same. We have t/e^t; the derivative of t is 1. The derivative of e^t is e^t, so the limit of t/e^t is the same by l'Hopital's rule as a limit of $1/e^t$. As time goes to infinity, that goes to 0, so we know that both of our solutions go to 0 as before. This is the critically damped harmonic oscillator. Let me let b be 2 and k be 1, and we see that, just like in the overdamped case, solutions tend right into equilibrium without oscillating.

Finally, let me summarize everything we've done with the harmonic oscillator system. The harmonic oscillator system, we've seen we have two roots to the characteristic equation. The roots are real and distinct. They'll both be negative; we're in the overdamped case. If those two roots merge together, we've got only one root, we're in the critically damped case. In both cases, everything glides down to 0 without oscillation. If we have complex roots of the form a ± ib, we're in the underdamped case. Assuming that a < 0, now as we've seen, solutions oscillate down to rest. Then the final case is when our roots have no a term, our complex roots have no real part, and our roots are ± ib. That's the undamped case; that's where things oscillate around forever.

With an eye toward what we're going to do a little later, here's another summary of what's happening. Let's look at $y'' + by' + y$. I'm going to set $k = 0$, and let b vary. What we saw was that as long as $b > 2$, then $\sqrt{b^2 - 4}$ was real.

So we were in the overdamped case, and solutions glide down to rest. If, on the other hand, $b = 2$, then in the characteristic equation, $b^2 - 4$ is now 0, so we've got repeated roots. We're in the critically damped case; again, solutions glide right down to 0.

Then, as soon as b becomes less than 2 but bigger than 0, we get oscillations. If b is kind of large, you don't really see those oscillations too well. Maybe you see them on the y' graph a little bit, but they're there. We know we've got solutions oscillating down to rest. If b is very small, like way over here, then you do see the oscillations. That's the visible underdamped case.

Then, the final case is, of course, when $b = 0$, that's our friend the undamped case, and that's where we get perpetual oscillations. So we can draw, we can summarize, everything by recording everything along the b-axis: $b > 2$, overdamped; $b = 2$, critically damped; b between 0 and 2, underdamped; and $b = 0$, undamped. That's a picture of the parameter line; parameter planes will play a role later.

That finishes off the harmonic oscillator equations. Well, not quite, because next time, what we're going to do is look at the harmonic oscillator equations where we add some forcing terms. That will be very different and very interesting. See you then.

Beating Modes and Resonance of Oscillators
Lecture 10

In this lecture, we consider the periodically forced harmonic oscillator. That is, instead of hanging our spring to the ceiling, we now move the spring up and down periodically. So our differential equation becomes

$$y'' + by' + ky = G(t),$$

where $G(t)$ is some periodic function. For simplicity, let us take $G(t) = \cos(wt)$. So the period of the forcing term is $2\pi/w$. There are 2 very different cases to consider. The first is when the damping constant is nonzero. In this case, if there were no periodic forcing, our mass-spring system would settle down to rest. But with the periodic forcing present, the mass cannot end up in its rest position; rather, the mass eventually begins to oscillate with period $2\pi/w$.

As an example, consider the differential equation

$$y'' + 3y' + 2y = \cos(t).$$

Solving the homogeneous equation (the equation with 0 replacing $\cos(t)$) yields the solution

$$y(t) = k_1 e^{-2t} + k_2 e^{-t}$$

since the characteristic equation is $s^2 + 3s + 2 = (s + 2)(s + 1)$. So the unforced oscillator tends to rest just as we expected.

To find one solution of the forced mass-spring system, we need to make a guess. Clearly, we cannot guess just $A\sin(t)$, since the y' term will give us a cosine term. Similarly, we cannot guess $B\cos(t)$. So the appropriate guess is $A\sin(t) + B\cos(t)$. Plugging this guess into the equation and doing a little algebra yields the particular solution with $A = 3/10$ and $B = 1/10$. So any solution of this equation is of the form

$$y(t) = k_1 e^{-2t} + k_2 e^{-t} + \tfrac{3}{10}\sin(t) + \tfrac{1}{10}\cos(t).$$

No matter which values of k_1 and k_2 we take, the corresponding solution ends up looking like

$$y(t) = \tfrac{3}{10}\sin(t) + \tfrac{1}{10}\cos(t).$$

This equation has period 2π, which is exactly the period of our forcing term. This is called the **steady-state solution**.

Here are the graphs of 3 different solutions, together with the solution curves in the phase plane. Note how all solutions quickly approach the steady-state solution.

Figure 10.1

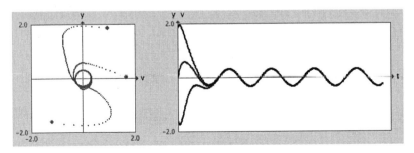

Now let's consider the case where there is no damping term and our forcing term is cos(wt), so we have a parameter present. For simplicity, consider the equation

$$y'' + y = \cos(wt).$$

Since we have no damping, the homogeneous equation is $y'' + y = 0$. So the general solution is $k_1\cos(t) + k_2\sin(t)$. This solution does not tend to rest. To get the full solution, we make the guess $A\sin(wt) + B\cos(wt)$. Plugging this into the differential equation, we find

$$B(1 - w^2)\cos(wt) + A(1 - w^2)\sin(wt) = y'' + y = \cos(wt).$$

So we must have $B = 1/(1 - w^2)$ and $A = 0$. That is, our particular solution is

$$\frac{1}{1-w^2}\cos(wt) + k_1\cos(t) + k_2\sin(t).$$

But wait a moment—this is fine as long as w is not equal to 1; if $w = 1$, this is no longer the solution. That is, if the period of the forcing is the same as the natural period of the spring, we are in trouble. We'll deal with this case later. But first look what happens for different values of w. Here are some graphs of the solution that satisfies $y(0) = y'(0) = 0$. First, when $w = 1.5$, we see a periodic motion, both in the y-v plane and as the graph of $y(t)$.

Figure 10.2

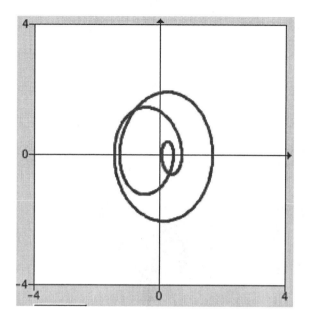

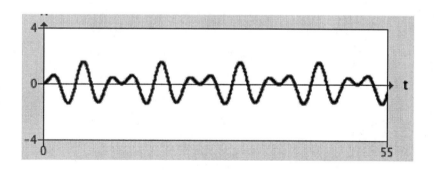

But when $w = \sqrt{2}$, the solution is no longer periodic.

Figure 10.3

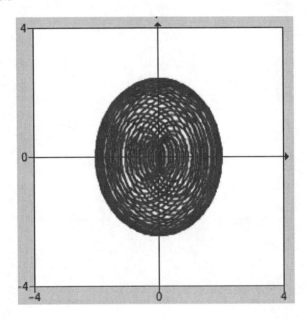

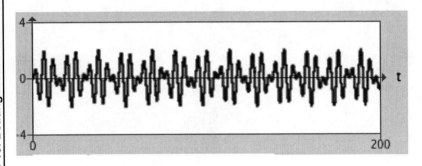

And when *w* gets very close to 1, say *w* = 1.06, we get **beating modes**.

Figure 10.4

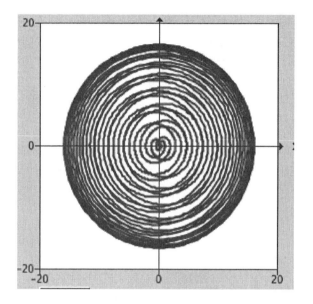

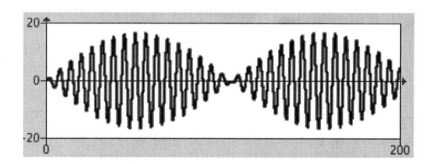

Note that the y-coordinate runs here from -20 to 20. So the mass sometimes undergoes large oscillations and sometimes does not. Why? Since w is very close to 1, the natural period and the forcing period are close together. This means that for a long period we are forcing the spring exactly the way it naturally wants to go, so the oscillations become larger and larger. But eventually, $\cos(t)$ and $\cos(wt)$ are relatively far apart, and this means that we are forcing in the direction opposite to that the spring wants to go, resulting in smaller and smaller oscillations.

Now let's go back to the case $w = 1$, that is, the differential equation

$y'' + y = \cos(t)$.

We no longer have our old solution

$$\frac{1}{1-w^2}\cos(wt) + k_1 \cos(t) + k_2 \sin(t)$$

since w is now equal to 1. Indeed, we cannot have a solution of the form $A\cos(t) + B\sin(t)$ since that expression solves the homogeneous equation. So what to do? How about making the guess $y(t) = At\cos(t) + Bt\sin(t)$? Plugging this into the differential equation then yields

$y'' + y = -2A\sin(t) + 2B\cos(t)$.

Setting this equal to $\cos(t)$ tells us that $B = 1/2$ while $A = 0$. So our general solution now becomes

$$\tfrac{1}{2}t\sin(t) + k_1 \cos(t) + k_2 \sin(t).$$

Therefore the solution that satisfies $y(0) = y'(0) = 0$ is just

$y(t) = \frac{1}{2} t \sin(t)$.

Note that this expression then oscillates back and forth with increasing amplitude; that is, our spring keeps moving alternately higher and lower. Note that in the graph below, the y-values extend from -100 to 100.

Figure 10.5

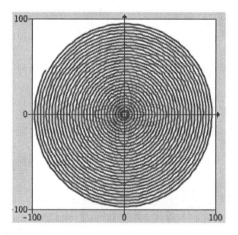

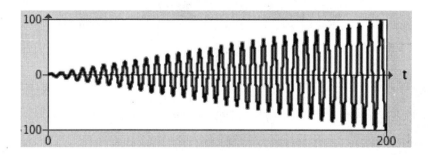

This is the phenomenon called **resonance**. We keep pushing and pulling our mass-spring system in just the way it wants to go, so the oscillations keep getting larger and larger, eventually leading to disaster.

Some experts blame resonance for the collapse of the Tacoma Bridge (a.k.a. "Galloping Gertie") back in 1940, just 4 months after the bridge was opened to traffic. A similar phenomenon led to the closing of London's Millennium Bridge in June 2000, just 2 days after it was opened to pedestrian traffic. Indeed, when crossing a bridge, soldiers no longer march in step for fear that their steps will coincide with the resonant frequency of the bridge.

Important Terms

beating modes: The type of solutions of periodically forced and undamped mass-spring systems that periodically have small oscillations followed by large oscillations.

resonance: The kind of solutions of periodically forced and undamped mass-spring systems that have larger and larger amplitudes as time goes on. This occurs when the natural frequency of the system is the same as the forcing frequency.

steady-state solution: A periodic solution to which all solutions of a periodically forced and damped mass-spring system tend.

Suggested Reading

Blanchard, Devaney, and Hall, *Differential Equations*, chaps. 4.3 and 4.5.

Hirsch, Smale, and Devaney, *Differential Equations*, chap. 6.2.

Roberts, *Ordinary Differential Equations*, chap. 6.1.

Strogatz, *Nonlinear Dynamics and Chaos*, chap. 7.6.

Relevant Software

Blanchard, Devaney, and Hall, *DE Tools*, Beats and Resonance, Mass Spring.

Problems

1. **a.** Solve the periodically forced first-order equation

    $$y' + y = \sin(t).$$

    b. What happens to all solutions of this equation?

    c. Sketch the graphs of some of these solutions in the region $0 < t < 20$.

    d. What happens to solutions if we change this equation to

    $$y' - y = \sin(t)?$$

    e. What happens to these solutions as time moves backward?

2. Find the steady-state solution of the forced mass-spring system with spring constant 2, damping constant 1, and forcing term $\sin(t)$.

3. What is the general solution of the forced mass-spring system given by $y'' + y = e^{-t}$?

4. Consider the mass-spring system $y'' + 3y = \sin(wt)$. For which value of w is this system in resonance?

5. For which values of w is the function $\cos(wt) + \cos(t)$ a periodic function?

6. Consider the undamped and forced mass-spring system $y'' + y = \cos(wt)$. Using the results of the previous problem, determine the values of w for which the solutions of this system are periodic functions.

Exploration

Consider the case of a pair of undamped harmonic oscillators.

$$y_1'' = -w_1^2 y_1$$
$$y_2'' = -w_2^2 y_2$$

These equations are decoupled, so we can easily solve both. But let's view these solutions a little differently. The quantity w_2/w_1 is called the frequency ratio. First show that the pair of solutions of these 2 equations is periodic if and only if the frequency ratio is a rational number. Plot a solution $(y_1(t), y_2(t))$ in the plane when, say, $w_1 = 2$ and $w_2 = 5$. What if $w_1 = 1$ and $w_2 = 2^{1/2}$? Change these equations to polar coordinates (r_j, θ_j) for $j = 1, 2$. What is the new system of differential equations? Do you see that $r_j' = 0$? So we can now think of solutions as residing on a torus (i.e., the surface of a doughnut). What would these solutions look like on the torus?

Beating Modes and Resonance of Oscillators
Lecture 10—Transcript

Welcome back. Over the last two lectures, we've been dealing with second-order differential equations, specifically the mass-spring or the harmonic oscillator differential equation. Remember that was $y'' + by'$, b was the damping constant, plus $ky = 0$. We saw a lot of different behaviors.

This time I'd like to look at the periodically forced harmonic oscillator. What that means is we'll look at the same differential equation, but now there'll be a nonhomogeneous term, a term $G(t)$, which we'll take to be a periodic function. What I'm thinking of doing is instead of attaching the spring to the roof, what I'll do is I'll hold the spring in my hand and move it up and down periodically. Then the question is what happens to the corresponding mass?

In order to understand that, we have to recall what happens when we don't have any periodic forcing, so let me go back and go over the different cases we saw then. Here is the harmonic oscillator equation. Remember that the first case was the overdamped case. That's when our damping constant was relatively large, and what we saw was that all solutions simply gravitated right down to equilibrium without crossing the equilibrium point, gliding right down to rest.

We also saw that there was a bifurcation point, the so-called critically damped case. As we lowered the damping, eventually we reached a point where we're just about to have the underdamped case. That was the critically damped case. Still, we saw all solutions gravitated right down to rest.

Then if we kept lowering the damping constant, we moved into a regime where we saw standard behavior for the mass-spring system. Now our mass-spring system oscillated around the equilibrium point. You don't see it that much right there because the damping is still large. On the other hand, if I lower the damping considerably, almost down to 0, then you do see oscillations. Now the spring oscillates down to rest. This is the underdamped case; this is the familiar case.

Then the final case we saw was when there was no damping at all, when our damping constant was equal to 0. There, solutions behaved periodically. There, the mass went up and down periodically without ever coming down to rest. No matter where we started this mass-spring system, we saw the same periodic behavior. That's the undamped case.

Let's turn to the periodically forced harmonic oscillator. Remember that's going to be $y'' + by' + ky =$ some forcing function, $G(t)$. Remember what we did in the first order case, when we had a first order linear homogeneous equation, $y' + ky = G(t)$. We solved that in a two-step process. The first step was to look at the homogeneous equation. That was $y' + ky = 0$, no forcing term on the right, or $y' = -ky$, and we know how to solve that. One solution is e^{-kt}, so our general solution of the homogeneous was some constant times e^{-kt}. That was step one.

Step two was to guess some solution of the nonhomogeneous case. You guessed some solution depending on what the right hand side was, then our general solution of the first-order linear nonhomogeneous equation was the sum of our guess and the general solution of the homogeneous problem.

Back to the periodically forced harmonic oscillator, let's take, as I said earlier, the forcing term to be a periodic function. For simplicity, I'll let $G(t)$ be some constant times $\cos(\omega t)$. I'll usually pick that constant, the so-called amplitude to be 1, so $G(t)$ will be $\cos(\omega t)$. It's a periodic function with period $2\pi/\omega$, and ω is now going to be a parameter for us.

First case, let's assume that we have nonzero damping. Our damping constant, b, is nonzero. Let's do a specific example of that. We'll take the damping constant to be 3, the spring constant to be 2, and for simplicity here, let me not include ω. Let's take the forcing term to be $\cos(t)$. Our differential equation for this periodically forced mass-spring system is $y'' + 3y' + 2y = \cos(t)$. We'll solve this exactly the same way we did with the first-order case; we'll first solve the homogeneous equation.

What is that? The homogeneous equation is $y'' + 3y' + 2y = 0$. We know to form the characteristic equation, the characteristic equation was gotten by guessing e^{st}, plugging it into the left, and we would find equation e^{st}, factor of the characteristic equation, $s^2 + 3s + 2$, and that's got to be equal to 0. The only way that will be equal to 0 is if $s^2 + 3s + 2$ vanishes, but of course, $s^2 + 3s + 2$ factors into $(s + 2)(s + 1)$.

We know some solutions. When $s = -2$, or $s = -1$, we've got an exponential solution, and as we saw over the last two lectures, it then follows that any constant times e^{-2t} plus any other constant times e^{-t}, is the general solution of the homogeneous. We have that; that's step one.

Step two, let me guess one solution of the nonhomogeneous equation. What are we going to guess here? Our right hand side is $\cos(t)$, so could we guess $A \cos(t)$? Not quite, if we plugged $A \cos(t)$ into our differential equation, remember we're going to have a $3y'$ term there. That would give us a sine term, so that wouldn't work. How about guessing a constant times $\sin(t)$? No again, same thing, the y'' and the $2y$ term will give us sines, and we want cosines out on the right. So what to guess? The best guess is probably a constant times sine, say $A \sin(t)$ plus another constant, say B times $\cos(t)$.

Let's make that guess. I'll plug $A \sin(t) + B \cos(t)$ into the differential equation. When I do, plug in $A \sin(t) + B \cos(t)$, I've got to take the second derivative, and I get $-A \sin(t) - B \cos(t)$. Then I've got to take 3 times the derivative of $A \sin(t) + B \cos(t)$. I get $-3B \sin(t) + 3A \cos(t)$, and then finally, I have to add 2 times my guess, and I get $2A \sin(t) + 2B \cos(t)$. So there's the left hand side. All of that has to be equal to, on the right, $\cos(t)$. Altogether, I've got on the left $(-A - 3B + 2A) \sin(t) + (-B + 3A + 2B) \cos(t) = \cos(t)$. The only way that left hand side is going to be equal to the cosine is if the coefficient of $\sin(t)$ is 0, and the coefficient of $\cos(t)$ is 1. We need $A - 3B$ to be 0 and $3A + B$ to be 1, so the first equation says that A must be $3B$, and the second equation then says, plugging that in, $10B = 1$, so $B = 1/10$, and then $A = 3B$, so $A = 3/10$. We have it; A must be $3/10$, and B must be $1/10$.

We have our solution to the nonhomogeneous equation. It is $(3/10) \sin(t) + (1/10) \cos(t)$, and altogether, that gives us the general solution of this periodically forced and damped harmonic oscillator. Our general solution is the general solution of the homogeneous constant times e^{-2t} plus another constant times e^{-t} plus that one solution of the nonhomogeneous equation we just found.

Let's go to the computer and see what these solutions look like. Here is the situation. We have cos(*t*) as our periodically forcing term, we have a spring constant of 2, and a damping constant of 3. When we start out at any initial condition, what we see is our solution does something to start, but then it settles into a periodic motion. Start out over here. Again, our solution does something to start, but eventually settles into a periodic motion. No matter where you start your solution, you eventually end up with a periodic solution. This is what we call the steady state solution.

Every solution tends to the steady state eventually. Why is that? Think about it; the general solution of the homogeneous equation is a bunch of decaying exponential terms. They're going to 0 very quickly. What's left, then, is our one solution of the nonhomogeneous equation, that (3/10) sin(*t*) + (1/10) cos(*t*); that's a periodic solution. That's the only thing that's left after a short time. All of our solutions tend to that steady state solution.

In this case, you saw that they tended to the steady state solution very quickly, but if our damping is relatively small, let me let the damping be around 0.2, then what happens is our solution runs around for a while sort of aimlessly, but eventually settles into the steady state solution. No matter where I start this solution, it runs around, crazily at first, but eventually becomes a steady state solution. Of course, since we have damping, our solution is, our general solution of the homogeneous equation is disappearing, and all we're left with is our one particular solution of the nonhomogeneous equation.

Let's go to the second case, the case where we have no damping. That means our damping term, *B*, is equal to 0, and so our equation is $y'' + ky$, let me pick *k* to be 1, equals cos(ωt). Again, omega is a parameter that we're going to vary in this undamped case.

To solve it, we do the same two-step process. First, solve the homogeneous equation. That homogeneous equation is $y'' + y = 0$, and we've seen that over and over again before. We know that the general solution of the homogeneous equation is $k_1 \cos(t) + k_2 \sin(t)$. Ah-ha, here's the difference from the damped case. This term is not going to disappear as time goes on. Our general solution of the homogeneous equation is going to stay put. That's going to complicate things.

Next, we've got to move to the nonhomogeneous equation, so $y'' + y = \cos(\omega t)$. We've got to guess one solution of the nonhomogeneous equation. What would you guess? Before, we had y'' plus a constant times y' plus another constant times y, and that forced us to guess a combination of sines and cosines, but now we don't have that y' term. We have $y'' + y = \cos(\omega t)$, so why not guess some constant times $\cos(\omega t)$.

Let's do that, guess $B \cos(t)$, then y' is $-B\omega \sin(\omega t)$, and then y'' is $-B$, pull out another omega, $\omega^2 \cos(\omega t)$. So when we guess $B \cos(\omega t)$, our differential equation, $y'' + y = \cos(\omega t)$ now reads $B(1 - \omega^2) \cos(\omega t) = \cos(\omega t)$. So we must have $B = 1/(1 - \omega^2)$.

One solution of our nonhomogeneous equation is $\dfrac{1}{1-\omega^2} \cos(\omega t)$.

For the general solution, we've got to add to that the general solution of the homogeneous, and that's $k_1 \cos(t) + k_2 \sin(t)$. This is fine as long as ω is not equal to 1. Remember, we've got that term $1/(1 - \omega^2)$. If ω is equal to 1, we've got a 0 in the denominator that gives us problems. This is fine as long as the forcing period, remember that's $2\pi/\omega$, is not the same as the natural period. The natural period is 2π. Remember our spring has spring constant 1. The solution's $\cos(\omega t)$. As long as the forcing period is different from the natural period, we've got our solution.

Let's turn now to the computer, and see what happens in this case here. Here we go, we have $\cos(\omega t)$. I've first set ω to be 2, and here's what happens to the solutions, look, periodic. Here's another solution. Notice it wraps around twice. Here's another solution, periodic; it sort of wraps around the origin twice. Here's another solution, periodic, but it is sort of making two little wiggles. All solutions here are periodic; ω is rationally related to our natural frequency, that's 1, well it is 2π. So we have 2π, and ω is 4π; they are rationally related.

What if I let ω be 1.5? We see a periodic solution, but this periodic solution wraps around a little bit more often. Another solution, again, behaves periodically. Another solution, again, periodic. Now ω is 1.5. Our natural

189

frequency, the multiple of cos(*t*) and sin(*t*), is 1. They are rationally related. We see that our sum of cos(ω*t*) and cos(*t*) and sin(*t*) give you a periodic solution.

Let me let ω be 1.25, and again we see a periodic solution. It takes a while for it to become periodic, but it is periodic. Another solution runs around for a while, but turns out to be periodic.

If omega is rationally related to 1, in this case, it's a rational number, then we always get periodic solutions. No matter where you start your initial condition, very close to 0, you always end up behaving periodically.

If I were to put in an irrational number, what would happen then? It is kind of hard to put irrational numbers into this computer, but let me put in a rational number that's very close to, but rationally unrelated to 1. We put in something around 0.2, and now what you see is the solution's running around for a while, more, and what's happening here. It seems to be going crazy.

Wait a minute. Look at the spring on the left. It seems to be coasting back down. There in the phase plane we see our solution coming back, and if this were actually an irrational number, this would not be periodic. It would start behaving kind of crazily. Now, it's starting to go back out, so if you have an irrational ω, you don't get periodic solutions; cos(ω*t*) and sin(*t*) and cos(*t*) are not rationally related.

One more example, let me let ω be very close to 2. It looks periodic, but in fact, you see some little wobbles in there. It's kind of getting a little thick. It's not periodic. If we were to take an irrational ω, this thing would never behave periodically. It's close to periodic motion, but not exactly there.

One other thing that I'd like to bring up, suppose ω is very close to 1. What does that mean? If ω is very close to 1, that means that we're pushing our spring up and down very much related to the way the spring naturally wants to go up and down. Remember, we're in the undamped case, so our spring would oscillate back and forth with a certain period, in fact, period 2π. Now we're pushing it almost at the same rate. As the spring wants to go down, we're pushing it down. As the spring wants to come up, we're pulling it up, down and up, down and up. That's what's going to cause beating modes.

Let me show you that. Let me again let ω be very close to 1, and what we saw, just as in the previous example, is we're pushing our spring up and down the way it wants to go for a while, but then we're not quite at the same thing. Then we're sort of pushing our spring in the opposite direction as the spring wants to go, so that damps down the up and down motion. The motion, the periodic motion, gets lower and lower. Now again, we start pushing in the same direction that the spring wants to go, so the oscillations get larger and larger. If I let ω be even closer to 1, those oscillations get even larger. We're pushing for a long time in the same direction as the spring wants to go. Oscillations are getting bigger and bigger, they keep getting bigger, but eventually, we go out of phase. Eventually, we're pushing against the spring, the direction it wants to go, and so now, these oscillations die down a bit. This is what's called beating modes. For a long time, our spring is oscillating very lowly, but then, as time goes on, it starts to oscillate with higher and higher amplitude.

In fact, let me show you this a different way. Let me plot the solutions for these beating modes with time going up to 200. Here's the same differential equation, only you're not seeing the mass-spring anymore, and now here're the graphs of $x(t)$ and $y(t)$. What you see is, for a while, the amplitude is very small; as time goes on, the amplitude increases, then it comes back and is very small, increases, comes back, and is very small. That's the so-called beating modes. No matter where I start the solution, you see the same sort of behavior. That's what happens when ω is very close to 1.

One last case to do, remember when we solved the nonhomogeneous equation, our equation did not work when ω was equal 1, so let's go to the case $y'' + y = \cos(t)$. You cannot have a solution of the form $A \cos(t) + B \sin(t)$ for our nonhomogeneous equation because that was a solution of the homogeneous equation.

What are we going to do to get that nonhomogeneous solution? We saw this before. When your first guess for the nonhomogeneous equation didn't work, the next best guess was to guess t times your previous guess.

Let's do that. Let's guess $At\cos(t) + Bt\sin(t)$. There's our guess, so plug that into the differential equation. We've got to compute y', but the derivative, by the product rule is $A\cos(t)$ and then $-At\sin(t)$. There's the derivative of $At\cos(t)$. Then secondly, the derivative of $Bt\sin(t)$ is, differentiate the t, you get $B\sin(t)$ and then plus Bt, differentiate the $\sin(t)$, you get a $\cos(t)$. There's the first derivative.

Now take the second derivative. The second derivative is $-A\sin(t)$. Differentiate the $-At\sin(t)$, you get $-A\sin(t)$ and $-At\cos(t)$. Now differentiate the second two terms. We'll get $B\cos(t)$, and then the derivative of $Bt\cos(t) = B\cos(t) - Bt\sin(t)$.

There it is. If you aggregate all of those terms, plug into the differential equation, what you'll find is $y'' + y = -2A\sin(t) + 2B\cos(t)$, and all of that has to be equal on the right to $\cos(t)$. So that says $-2A\sin(t) + 2B\cos(t) = \cos(t)$. We need A to be 0 and B to be 1/2. We've got it. We've got the solution, $(1/2)t\sin(t)$ plus our general solution of the homogeneous, that was a constant times $\cos(t)$ plus another constant times $\sin(t)$.

Think of what this solution is. Think of a solution that solves the initial condition $y(0) = 0$ and $y'(0) = 0$. That solution is our solution of the nonhomogeneous equation. That's $(1/2)t\sin(t)$. What happens to that solution? You know $(1/2)\sin(t)$ behaves periodically as time goes on, but now we've got t times that solution. What happens to that as t gets large? This sine is oscillating back and forth between ± 1. Multiplying it by t makes it oscillate back and forth between larger and larger numbers. These oscillations keep increasing in amplitude as time increases.

Start out at time 0, right near the origin, we see our solutions have increasing amplitude in the phase plane. They're moving away from the origin. The mass is moving up more and more. Oh, it is still moving up. Look what's happening up here. Oh, no. The mass is going through the roof. It's still increasing. Higher and higher, and now down into the cellar.

What's happening? Notice the mass is increasing down in amplitude even further. It's going way into the cellar. This is what's called resonance. This solution is going to keep oscillating further and further and your spring is going to break. There is no way that you can have a real life spring that behaves, in this case, when you're forcing period is exactly the same as your natural period. One other way to see that, let me look at the graphs of x and of y as a function of t going from 0 to 200, and there's the solution. Notice, by the way, that this runs from, I think, -100 to $+100$ in the y direction. No matter where we start this solution, we get this resonance behavior. The mass keeps going up and down.

There's an application of this. If you've got a mass-spring system that's in resonance, the natural period and the forcing period are the same, then disasters can happen. As an application of all of this, a number of years ago, back in the 1940s, there was the Tacoma Narrows bridge accident where the bridge started oscillating, oscillating, oscillating, and it oscillated too much and completely collapsed. The bridge was known as Galloping Gertie.

Even a few years ago, back in the year 2000, the same thing happened with London's Millennium Bridge. The bridge started oscillating quite a bit. Some people have said that this was caused by resonance, the same thing that we've just been looking at. Other people disagree. It's not exactly clear. On the other hand, the oscillations did start increasing as time went on, and that's what led to the collapse of the bridge.

That concludes our investigation into the mass-spring systems. Next time, we'll come back and take on more general linear systems of differential equations.

Linear Systems of Differential Equations
Lecture 11

We now move on to linear systems of differential equations. Much of the theory surrounding linear systems of differential equations involves tools from the area of mathematics known as linear algebra, including matrix arithmetic, determinants, eigenvalues, and eigenvectors.

Let's consider 2-dimensional, autonomous linear systems. These are systems of differential equations of the form

$x' = ax + by$

$y' = cx + dy,$

where a, b, c, and d are all parameters. We will often write this in matrix form $Y' = AY$, where Y is the 2-dimensional vector

$$Y = \begin{pmatrix} x \\ y \end{pmatrix}$$

and A is the 2 × 2 matrix

$$A = \begin{pmatrix} a & b \\ c & d \end{pmatrix}.$$

The product of the matrix A and the vector Y is the new vector whose first entry is the dot product of the first row of A with Y and whose second entry is the dot product of the second row of A with the vector Y—that is, the vector

$$AY = \begin{pmatrix} ax + by \\ cx + dy \end{pmatrix}.$$

A solution of such a linear system is then a vector depending on t (a curve in the plane) given by

$$Y(t) = \begin{pmatrix} x(t) \\ y(t) \end{pmatrix},$$

where $x(t)$ and $y(t)$ are the solutions of the given system. So $x'(t) = ax(t) + by(t)$, and $y'(t) = cx(t) + dy(t)$.

For example, consider the linear system

$$Y' = \begin{pmatrix} 1 & 1 \\ 3 & -1 \end{pmatrix} Y = \begin{pmatrix} 1 & 1 \\ 3 & -1 \end{pmatrix} \begin{pmatrix} x \\ y \end{pmatrix}.$$

We can easily check that one solution is $x(t) = e^{2t}$, $y(t) = e^{2t}$. We write this solution in vector form as

$$Y(t) = \begin{pmatrix} e^{2t} \\ e^{2t} \end{pmatrix} = e^{2t} \begin{pmatrix} 1 \\ 1 \end{pmatrix}.$$

Note that this is a straight line solution of the system; in the phase plane, this solution lies along the straight line $y = x$. It tends to the origin as time goes to $-\infty$ and to ∞ as time goes to ∞.

Another solution is $x(t) = e^{-2t}$, $y(t) = 3e^{-2t}$, or in vector form,

$$Y(t) = \begin{pmatrix} e^{-2t} \\ -3e^{-2t} \end{pmatrix} = e^{-2t} \begin{pmatrix} 1 \\ -3 \end{pmatrix}.$$

Again, this is a straight line solution, this time lying on the line $y = -3x$ in the phase plane and tending to the origin as time increases. These solutions in the phase plane look as follows.

Figure 11.1

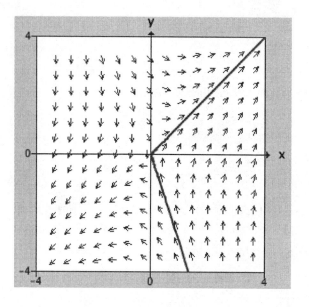

As earlier, any expression of the form

$$Y(t) = c_1 e^{2t} \begin{pmatrix} 1 \\ 1 \end{pmatrix} + c_2 e^{-2t} \begin{pmatrix} 1 \\ -3 \end{pmatrix}$$

for any values of c_1 and c_2 is also a solution. Here is a collection of such solutions in the phase plane.

Figure 11.2

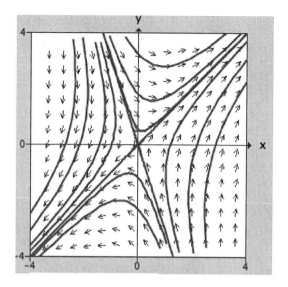

We say that the equilibrium point at the origin in this phase plane is a **saddle point**.

In fact,

$$Y(t) = c_1 e^{2t} \begin{pmatrix} 1 \\ 1 \end{pmatrix} + c_2 e^{-2t} \begin{pmatrix} 1 \\ -3 \end{pmatrix}$$

is the general solution to this differential equation. In order to see why this is true, we must be able to determine values of c_1 and c_2 that solve any initial value problem of the form

$$Y(0) = \begin{pmatrix} x_0 \\ y_0 \end{pmatrix}.$$

That is, we must be able to find c_1 and c_2 so that

$$Y(0) = c_1 \begin{pmatrix} 1 \\ 1 \end{pmatrix} + c_2 \begin{pmatrix} 1 \\ -3 \end{pmatrix} = \begin{pmatrix} x_0 \\ y_0 \end{pmatrix}.$$

Notice that the 2 vectors (1, 1) and (1, −3) are linearly independent. That is, they do not lie along the same straight line passing through the origin. This means that by the parallelogram rule, we can always stretch or contract these vectors (by multiplying by c_1 and c_2) so that the resulting vectors add up to (x_0, y_0).

Figure 11.3

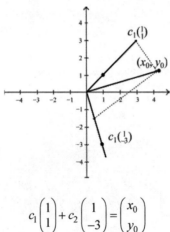

$$c_1 \begin{pmatrix} 1 \\ 1 \end{pmatrix} + c_2 \begin{pmatrix} 1 \\ -3 \end{pmatrix} = \begin{pmatrix} x_0 \\ y_0 \end{pmatrix}$$

Whenever we have 2 such solutions whose initial vectors are linearly independent, a similar combination of them gives us the general solution.

As another example, consider

$$Y' = \begin{pmatrix} -1 & 0 \\ 1 & -2 \end{pmatrix} Y.$$

We can easily confirm that the only equilibrium solution is at the origin. To find the other solutions, note that the first equation reads $x' = -x$. So we know that the general solution here is $k_1 e^{-t}$. Then the second equation is $y' = -2y + k_1 e^{-t}$, which is a first-order linear and nonhomogeneous equation. We know how to solve this. The solution is $y(t) = k_2 e^{-2t} + k_1 e^{-t}$. So our solutions to the linear system are

$x(t) = k_1 e^{-t}$

$y(t) = k_2 e^{-2t} + k_1 e^{-t}.$

In vector form, this solution is as below.

$$Y(t) = k_1 e^{-t} \begin{pmatrix} 1 \\ 1 \end{pmatrix} + k_2 e^{-2t} \begin{pmatrix} 0 \\ 1 \end{pmatrix}$$

When $k_1 = 0$ and $k_2 = 1$, we get the straight line solution $x(t) = 0$ and $y(t) = e^{-2t}$ lying along the x-axis. And when $k_1 = 1$ and $k_2 = 0$, we get another straight line solution $x(t) = e^{-t}$, $y(t) = e^{-t}$, this time lying along the line $y = x$. It is the same as before, but now both straight line solutions tend inward to the origin. We see that the phase plane is a sink.

Figure 11.4

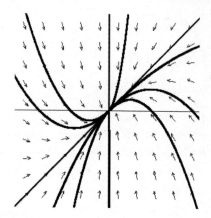

Thus we see again 2 straight line solutions for this system. This will be the way we proceed to solve linear systems; we'll first hunt down some straight line solutions and then use them to put together the general solution.

This straight line method does not always produce straight line solutions, however. There are many other possible phase planes for linear systems, which we produce with the same method.

For example, for the system

$$Y' = \begin{pmatrix} 0 & -1 \\ 1 & 0 \end{pmatrix} Y,$$

we have $x'(t) = -y(t)$ while $y'(t) = x(t)$. So some solutions are $x(t) = c\cos(t)$, $y(t) = c\sin(t)$ for any constant c. These solutions all lie along circles in the phase plane, and the equilibrium point at the origin is now called a **center**.

Figure 11.5

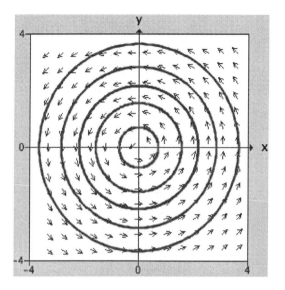

Let's look at a different example.

$$Y' = \begin{pmatrix} -0.1 & 1 \\ -1 & -0.1 \end{pmatrix} Y$$

We easily determine that some solutions are of the form

$$Y(t) = ce^{-0.1t} \begin{pmatrix} \cos(t) \\ -\sin(t) \end{pmatrix}.$$

Without the exponential term, these solutions would all lie along circles as above. But, since we have the exponential of $-0.1t$, these solutions must now spiral in toward the origin as time increases. Our phase plane now looks as follows, and this equilibrium solution is called a **spiral sink**.

Figure 11.6

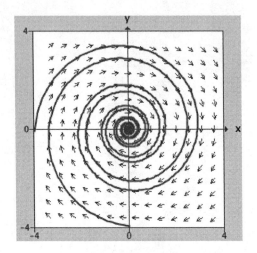

The first question concerning linear systems is what are the equilibrium points. Certainly the origin, (0, 0), is always an equilibrium point, but are there others? If so, they would be solutions to the system of linear algebraic equations

$ax + by = 0$

$cx + dy = 0.$

The first equation says that $x = -(b/a)y$ (as long as a is nonzero). Substituting this into the second equation, we find

$(ad - bc)y = 0.$

So there are only 2 possibilities here, either $y = 0$ or $ad - bc = 0$. In the first case, we then have that x is also 0, so we get the equilibrium point at the origin we already know about. In the other case, $ad - bc = 0$, we then have that any y value works; so $x = -(b/a)y$. So this yields a straight line of equilibrium solutions given by $x = -(b/a)y$.

A similar argument works if $a = 0$, at least as long as one of the coefficients in the matrix is nonzero. If all the entries in the matrix equal 0, we have a pretty simple system—namely, $x' = 0$ and $y' = 0$, so all points (x, y) are equilibrium points. We'll neglect this simple case in the future.

So, to summarize, for the linear system of ODEs $Y' = AY$, we have

1. A straight line of equilibrium points if $ad - bc = 0$

2. A single equilibrium point at the origin if $ad - bc \neq 0$.

The quantity $ad - bc$ will reappear often in the sequel; it is called the **determinant** of the matrix. We abbreviate it $\det A = ad - bc$.

Notice also that a linear system of algebraic equations

$ax + by = 0$

$cx + dy = 0$

has nonzero solutions if and only if $\det A = 0$.

Important Terms

center: An equilibrium point for a system of differential equations for which all nearby solutions are periodic.

determinant: The determinant of a 2-by-2 matrix is given by $ad - bc$, where a and d are the terms on the diagonal of the matrix, while b and c are the off-diagonal terms. The determinant of a matrix A (det$A$, for short) tells us when the product of matrix A with vector Y to give zero ($A\ Y = 0$) has only one solution (namely the 0 vector, which occurs when the determinant is non-zero) or infinitely many solutions (which occurs when the determinant equals 0).

saddle point: An equilibrium point that features one curve of solutions moving away from it and one other curve of solutions tending toward it.

spiral sink: An equilibrium solution of a system of differential equations for which all nearby solutions spiral in toward it.

Suggested Reading

Blanchard, Devaney, and Hall, *Differential Equations*, chap. 3.1.

Hirsch, Smale, and Devaney, *Differential Equations*, chap. 2.4.

Kolman and Hill, *Elementary Linear Algebra*, chap. 3.

Lay, *Linear Algebra*, chap. 3.

Roberts, *Ordinary Differential Equations*, chap. 7.

Strang, *Linear Algebra and Its Applications*, chap. 4.

Strogatz, *Nonlinear Dynamics and Chaos*, chap. 5.1.

Relevant Software

Blanchard, Devaney, and Hall, *DE Tools*, Linear Phase Portraits.

Problems

1. What is the product of the matrix $\begin{pmatrix} 1 & 3 \\ 2 & 1 \end{pmatrix}$ and the vector $\begin{pmatrix} 2 \\ 2 \end{pmatrix}$?

2. **a.** Write the system of equations

   $x' = y$

   $y' = -x$

   in matrix form.

   b. What are the equilibrium points for the above system?

3. Are the vectors $(1, 2)$ and $(-4, -8)$ linearly independent?

4. **a.** Is

   $\begin{pmatrix} e^{7t} + e^{t} \\ e^{2t} \end{pmatrix}$

   a solution of the system

   $x' = x + y$

   $y' = 2y$?

b. Is $\begin{pmatrix} e^{2t} \\ 0 \end{pmatrix}$ a solution of the above equation?

5. Consider the linear system below.

$$x' = x$$
$$y' = x + 2y$$

a. Write this system in vector form.

b. Since the system can be decoupled, first find the solution of $x' = x$ and then use that to find the solution of $y' = x + 2y$.

c. Write this solution in vector form.

d. Do the 2 above solutions generate the general solution?

e. Sketch the above solutions in the phase plane.

Exploration

We have seen that the determinant of a 2×2 matrix allows us to decide when a system of 2 linear equations has a nonzero solution. What about the case of 3 linear equations, such as the below?

$$a_{11}x + a_{12}y + a_{13}z = 0$$
$$a_{21}x + a_{22}y + a_{23}z = 0$$
$$a_{31}x + a_{32}y + a_{33}z = 0$$

Can you find a formula (depending on the a_{ij}) that tells you when this system of equations has nonzero solutions?

Linear Systems of Differential Equations
Lecture 11—Transcript

Here we go again. Over the last few lectures, we have been dealing specifically with the mass-spring system. We thought of it as a second-order linear homogeneous differential equation. We could have thought of it as a linear system, and in fact, we plotted the phase plane. We were doing exactly that, looking at a linear system. In the next few lectures, we're going to deal with more general, linear systems of differential equation. These will be very important when we go to nonlinear systems; linearization will be one of the main tools that we'll use.

In this part of the course, we'll primarily be concerned with two-dimensional autonomous linear systems, in fact, two-dimensional constant coefficient linear systems. That is, linear systems of the form $x' = ax + by$ and $y' = cx + dy$. Here, a, b, c, and d are all constants.

To handle these linear systems, we're going to have to make an excursion into the field of mathematics known as linear algebra. Most people who've seen linear algebra regard that as a field that is concerned with solving linear equations, not at all with differential equations. The fact is that when you see the linear algebra come up in differential equations, you really see, geometrically, why this field is so important.

Many of my students who've had a little bit of linear algebra before they take the differential equations class, come up and say wow, yeah, now I understand why this topic or that topic is important. As an aside note, you don't need to know any linear algebra to understand this part of the course. I'll delve deeply into anything we need from linear algebra as we go along.

Let's begin with say the matrix notation. We're going to need the concept of matrices and vectors, so let me review or tell you what I mean by matrix notation. We're going to take the system, $x' = ax + by$ and $y' = cx + dy$, and abbreviate it using vector notation. I'll write the vector capital Y as the column vector little x and little y, our dependent variables. I'll also introduce the coefficient matrix, and I'll call it capital A. The matrix will be a 2 by 2 matrix that consists of a square with 4 numbers in it. The first row is the coefficients of my first differential equation, so a and b, and the second row of the matrix is the coefficients of the second equation, c and d. Our 2 by 2 matrix, A, is given by $A = (a, b, c, d)$.

That's the way I'll enunciate a matrix. Now we're going to write our differential equation as the vector Y' is the matrix A times the vector Y. What do I mean by matrix A times the vector Y? I mean to take the product of the 2 by 2 matrix, (a, b, c, d) and the vector little x little y, what we do is take the dot product of the first row of the matrix with our vector. The dot product means I multiply the first term, a, by x and add to it the product of the second term, b with y. So (a, b) dot product with (x, y) gives me $ax + by$, the right hand side of my first differential equation. Then secondly take the dot product of the second row with the vector (x, y), that gives me $cx + dy$ as the second term of the vector.

Altogether what you see is $Y' = AY$ is the vector that's the right hand side of the differential equation, namely $ax + by$, and $cx + dy$, and Y' means the vector little x' little y'. That's vector notation for linear systems. As an example, take the matrix $(1, 2, 3, 4)$. What's the product of that matrix with the vector $(1, -1)$? We take the dot product of the first row, $(1, 2)$, with our vector $(1, -1)$, that gives us $(1 \times 1) + (2 \times -1)$ or -1. Secondly, take the dot product of $(3, 4)$ with $(1, -1)$, and that gives me $3 - 4$, or -1.

Let's turn to an example of a linear system of differential equations. Let me look at, for example, $x' = x + y$ and $y' = 3x - y$. In matrix form, this is the linear system. Vector Y' is matrix $(1, 1, 3, -1)$ times the vector Y.

Where to begin? As always, we look for equilibrium points. What's an equilibrium point for this system of differential equations? Just look at it. Clearly $x = 0$ and $y = 0$ is one equilibrium point. Both of the right hand sides of that system of differential equations vanish at the origin. Are there others? In order to have equation 1 vanish, we need $x = -y$, but then in equation 2, this says that $0 = 3x$, that's $-y$, so $-3y - y = 0$. Therefore, y must be 0, but then from our first equation, $x = 0$. What we have for this linear system is only one equilibrium point at the origin.

As a guide to what's coming later, notice that for any linear system, $x' = ax + by$ or $cx + dy$, that linear system always has an equilibrium point at the origin. We always have an origin equilibrium point. The question that'll come up later is, are there other equilibrium solutions? In this particular case, the answer is no. The origin is our only equilibrium point.

What about other solutions for this system of differential equations? I claim that the solution $Y(t)$ given by the vector (e^{2t}, e^{2t}) is a solution. I will actually write that by factoring out the e^{2t} from the vector giving me e^{2t} times the vector $(1, 1)$. By that, I mean multiply each term by e^{2t}.

I claim that e^{2t} times the vector $(1, 1)$ is a solution. We have to check that. Is that the case? Let's see. Is it true that that satisfies? Take the derivative of (e^{2t}, e^{2t}). Does that equal matrix $(1, 1, 3, -1)$ times e^{2t} times $(1,1)$? First term, differentiate e^{2t}, and that's our $x(t)$ term. We get $2(e^{2t})$. Does that equal $x + y$? Yes, both x and y are e^{2t}, so x' does equal $x + y$. Second term, that's e^{2t}. That's our $y(t)$ term. Does that satisfy $y' = 3x - y$? The answer is yes. The derivative of e^{2t} is $2(e^{2t})$, and $3x - y = 3(e^{2t}) - e^{2t}$, again, e^{2t}.

We have a solution. This is going to be a very important solution. This is what we call a straight-line solution. Look at this solution. It's $(e^{2t} \times 1, e^{2t} \times 1)$. The x and y components of this vector are the same. That is, for each t, this solution lies along the straight line, $y = x$ in the plane. It's a straight-line solution.

Notice that x and y are always positive; e^{2t} is always positive, and as t gets large, e^{2t} gets very large, so this solution runs out to infinity along the line $y = x$. On the other hand, when t is negative, e^{2t} is small, but positive, and in fact as t goes to minus infinity, e^{2t} goes to 0. This is a straight-line solution that scoots along the line $y = x$ out to infinity as time goes on, and down to the origin, our equilibrium point, as time goes backwards.

Then it's easy to check that, as we've seen for other linear differential equations, any constant times that solution is also a solution, so Ce^{2t} times $(1,1)$ is also a solution to this differential equation.

How about other solutions? Are there other solutions for this differential equation? Here's another one. Let's try the solution $Y(t)$ is given by e^{-2t} and $-3e^{2t}$, As I did earlier, let me pull out e^{-2t} from that vector and I'm left with e^{-2t} times $(1, -3)$. Is that a solution? You have to check it. Quickly, does the first term, $x(t)e^{-2t}$ satisfy the differential equation? The derivative of e^{-2t} is $-2e^{-2t}$. Does that equal $x + y$? Yes, $x = e^{-2t}$, $y = -3e^{-2t}$, so $x + y = -2e^{-2t}$, and that equals x'. Similarly, you can check that $-3e^{-2t}$, our $y(t)$ term, also satisfies that differential equation.

Here we have a second solution of this differential equation. This solution of the differential equation lies along a different straight line. If we look at the y coordinate, the y coordinate is always -3 times the x coordinate, the y coordinate is $-3e^{-2t}$. The x coordinate is e^{-2t}. This is a straight-line solution lying along the line $y = -3x$, the line through the origin, with slope -3.

We have an e^{-2t} term here, what happens to this solution? The x coordinate is always positive, e^{-2t}. The y coordinate $-3e^{-2t}$ is always negative, so what happens as time goes on? We're on the negative part of the line $y = -3x$. As time goes on, our exponential tends to 0, so this time the solution tends to the equilibrium point at the origin as time goes on, and as time goes backward, it shoots off to infinity, a straight-line solution like the previous example, only it's going in the opposite direction. We have that any constant times that solution is also a solution to the differential equation, so any constant times e^{-2t} times the vector $(1, -3)$ is solution to this differential equation.

Let me go and look at what these solutions look like in the phase plane. Here is our matrix. The matrix is $(1, 1, 3, -1)$, and here are our two straight-line solutions. We have one straight-line solution lying along the line $y = x$. You look at the vector field, and you see that the vector field is pointing away from the origin along that straight line. Solutions here are running off to infinity as time goes on. As time goes backwards, they're receding into the origin.

Similarly, we saw any constant times that is a solution. If I multiplied by -1, I would get the solution that lies on the opposite side of the line $y = x$. This solution, as time goes on, runs off to infinity. As time goes backwards, it tends into the origin. We also have another solution, this solution lying along the line $y = -3x$. This solution now tends into the origin as time goes on, and tends away from the origin, off to infinity, as time goes backwards. In this particular case, we have two straight-line solutions, one leaving the origin, the other coming into the origin. We're going to see, as we delve into linear algebra, that those are the solutions that give us everything.

Before going much further, let me stop and talk about the linearity principle. Given two solutions, Y_1 and Y_2, of a linear system, then it's a fact that a constant times Y_1 plus another constant times Y_2 is also solution of the linear system. We've already seen that if you have one solution, a constant times it is also a solution. It's easy to check that the sum of two solutions is also a solution, so the linearity principle says that if you have two solutions, then you've really got infinitely many solutions. Any constant times the first plus a constant times the second is a solution to this system of differential equations.

That brings up two questions. First, what do these other solutions look like, and then secondly, when is this combination of two solutions the general solution? In the example we just saw, I gave you two solutions. Can I use those two to manufacture the general solution?

First question is, what do these solutions look like? For example, let's do the first solution. Remember, that was e^{2t} times vector $(1, 1)$. The second one was e^{-2t} times vector $(1, -3)$. What does the solution that's the combination of these two solutions look like? What does any constant times the first plus any other constant times the second look like?

What happens as time goes to minus infinity? As time goes to minus infinity, that solution with an e^{2t} times $(1, 1)$ gets smaller and smaller. It goes to 0. That says that the first term constant, e^{2t}, $(1, 1)$, sort of vanishes as time goes backwards leaving us only with our second solution, e^{-2t} times vector $(1, -3)$.

As time goes forward, what happens? As time goes forward, the exact opposite thing happens. As time goes forward, the solution involving e^{-2t} now tends to 0. That leaves us essentially with only the first solution, a constant times e^{2t} times $(1, 1)$. As time goes forward, our solutions look like one of the solutions we found. As time goes backward, they look like the other solution we found.

Let's look and see what these other solutions look like in the phase plane. Here we see our two straight-line solutions, one running away from the origin, the other one tending into the origin. If I take any other solution, notice what happens. In backward times, it looks like the straight-line solution running along the line $y = -3x$. In forward time, it looks like the other straight-line solution. All solutions have that property. They look like one of the two straight-line solutions in forward time, and the other straight-line solution in backward time.

Here is the full phase plane picture for this linear system of differential equations. two very special solutions are straight-line solutions, and then all other solutions looking like the straight-line solutions in the far forward time or the far backward time. This is one of the many phase planes that we're going to see for linear systems of differential equations. This picture is what's known as a saddle phase plane. The origin is called a saddle point.

The question is, is the solution, $c_1 e^{2t}(1, 1) + c_2 e^{-2t}(1, -3)$, our general solution? That is, can we always find values of c_1 and c_2 that allow us to find the solution corresponding to any initial value problem, say the initial value problem $y(0)$ is the given vector (x_0, y_0)? Plug 0 into that solution, we get $c_1 e^0 (1, 1) + c_2 e^0 (1, -3)$. Do those two vectors add up to our given initial condition, (x_0, y_0)? That is, can we solve the equations $c_1 + c_2 = x_0$? That's the first component of our equation. The second is $c_1 - 3c_2 = y_0$. Here x_0 and y_0 are some given constants, any given real numbers. Can we find c_1 and c_2 that satisfy those equations?

Look at the first equation. The first equation says that $c_1 = x_0 - c_2$, and then the second equation says that $c_1 = 3c_2 + y_0$. We must have $x_0 - c_2 = 3c_2 + y_0$. That says that $x_0 - y_0 = 4c_2$ or $c_2 = (x_0 - y_0)/4$. Yes, we can find a c_2 that works; c_2 is the quantity $(x_0 - y_0)$ divided by 4. Then plugging that into the second equation, we see that c_1 must be equal to $3c_2 + y_0$, or $c_1 = (3/4)x_0 - y_0 + y_0$, or $c_1 = (3/4)x_0 + (1/4) y_0$. We have it. We can find c_1; c_1 is given by that expression in terms of x_0 and y_0, and c_2 is given by the other expression. We have the general solution to this system of differential equations.

We don't want to be doing that algebra all the time. There is actually an easier way of seeing that that's the general solution. Remember at time 0, one of our solutions is at (1, 1), whereas the other one is at (1, −3). Think of those as two vectors in the plane. We have to find a pair of constants that multiply the vector (1, 1,) and the vector (1, −3) and add up to any vector in the plane, say, (x_0, y_0). Can you do that geometrically?

Yes, look at that vector (1, 1) and (1, −3). If I stretch (1, 1) out and I squeeze (1, −3) a little bit, I can arrange so the those squeezed vectors and stretched vectors form parts of a parallelogram whose diagonal is our given point (x_0, y_0). That is, if I lengthen the vector (x_0, y_0), and compress the vector (1, −3) just the right amount, I can arrange so that those vectors add up to (x_0, y_0). Geometrically, we see that, yes, we can find a c_1 and c_2 that work. We do have the general solution.

That's a notion that's going to come up over and over again. I'll say that two vectors are linearly independent in the plane if they don't point along the same straight line through the origin. Two vectors that point at some angle that's not 180 or 0 degrees are linearly independent, whereas two vectors that point in the same direction, they're not linearly independent. They are dependent. Two vectors that point in the opposite directions, they're not linearly independent. The fact is that if 2 vectors are linearly independent, then given any point whatsoever in the plane, say (x, y), or call it the vector Z, we can always find a c_1 and c_2 that work. That is, we can always find a c_1 and c_2 such that c_1 times the first plus c_2 times the second adds up to the vector (x, y). That's pretty clear if our vector, V, is the vector (1, 0), and our vector, W, is the vector (0, 1) because then if the vector Z is the point (x, y) in the plane, all we have to do is multiply the vector, V, by the number x and multiply the vector W by the number y. We end up with the sum $xv + yw = Z$.

That is true as long as V and W are linearly independent. You can always stretch or compress or flip the vectors to the other side so that the corresponding vectors will add up exactly to your point, (x, y). If V points off in the northeast direction, W points of in the west direction, and (x, y) is somewhere off in the northeast, we can, for example in this picture, shrink V and flip W to the other side and manipulate it so that the sum of those two vectors, by the parallelogram rule, adds up exactly to (x, y). That gives us a geometric technique for deciding when we have the general solution.

One more example, let's look at the linear system, $x' = -x$, and $y' = x - 2y$. Matrix form, its vector Y' is $(-1, 0, 1, -2)$. Equilibrium solutions, the first equation says that x must be equal to 0, but then the second equation $y' = x - 2y$, says that y must be 0. We only have an equilibrium point at the origin. Are there other solutions?

Look at that first equation. The first equation is $x' = -x$. We know the solution to that. It's any constant times e^{-t}. Then what's our second equation? Our second equation is $y' = x$, but we know what x is. It is e^{-t} or $k_1 e^{-t} - 2y$. Our second equation is a first order linear nonhomogeneous equation, $y' - 2y + k_1 e^{-t}$, and we know how to solve that.

The solution to the homogeneous equation is $y(t)$ equals some other constant times e^{-2t} and then our nonhomogeneous solution to the nonhomogeneous equation is $y' = -2y + k_1 e^{-t}$. We'll have to guess something of the form a constant times e^{-t}, not k_1 or k_2, some other constant.

If you do the math, you'll see that constant, given by k_1 that we had before, actually works. The solution is $x(t)$ is $k_1 e^{-t}$ and $y(t) = k_2 e^{-2t} + k_1 e^{-t}$. There is one solution to the differential equation. Then we already have another solution, so our combination of solutions by the linearity principle the first solution, $k_1 e^{-t} (1, 1)$ plus the second solution $k_2 e^{-2t} (0, 1)$.

Is that the general solution to this differential equation? Yes, at time 0, we're at $(1, 1)$, and $(0, 1)$. In each case, those vectors are linearly independent. We have the general solution. What do these solutions look like? Take the first solution, $k_1 = 0$ and $k_2 = 1$. If $k_2 = 1$, we have the solution $e^{-2t}(0, 1)$. That's the solution. It's a straight-line solution lying along the y-axis; x is always 0, and y is e^{-2t}, another straight-line solution.

What about the second solution? Let's take the second solution to be $k_1 = 1$, and $k_2 = 0$, then we get $x(t) = e^{-t}$ and $y(t)$ is e^{-t} also. Again, x and y are the same, another straight-line solution this time lying along the line $y = x$, a straight-line solution coming into the origin along that straight line.

Let me go look at what this system looks like. Here are our two straight-line solutions. This straight-line solution comes into the origin, and this straight-line solution also comes into the origin, so in this case, all of our solutions tend into the origin. We have decaying exponentials in our general solution. All solutions tend into the origin, a very different phase plane from what we've seen before. All solutions here tend to the origin.

This now sort of opens up the question of how do we proceed? To find the general solution of systems of differential equations, what we'll do in the future is try to find those straight-line solutions. When we find those straight-line solutions, we'll be done. In order to do that, we need some more tools from linear algebra. That's what we'll investigate in the next lecture.

An Excursion into Linear Algebra
Lecture 12

Recall the linear system that we solved last time:

$$Y' = \begin{pmatrix} 1 & 1 \\ 3 & -1 \end{pmatrix} Y.$$

There were 2 straight line solutions given by

$$Y_1(t) = e^{2t} \begin{pmatrix} 1 \\ 1 \end{pmatrix} \text{ and } Y_2(t) = e^{-2t} \begin{pmatrix} 1 \\ -3 \end{pmatrix}.$$

$Y_1(t)$ lies along the diagonal line $y = x$, and $Y_2(t)$ lies along the line $y = -3x$. Note that

$$\begin{pmatrix} 1 & 1 \\ 3 & -1 \end{pmatrix} \begin{pmatrix} 1 \\ 1 \end{pmatrix} = \begin{pmatrix} 2 \\ 2 \end{pmatrix} = 2 \begin{pmatrix} 1 \\ 1 \end{pmatrix} \text{ and } \begin{pmatrix} 1 & 1 \\ 3 & -1 \end{pmatrix} \begin{pmatrix} 1 \\ -3 \end{pmatrix} = \begin{pmatrix} -2 \\ 6 \end{pmatrix} = -2 \begin{pmatrix} 1 \\ -3 \end{pmatrix}.$$

We'll see these equations again.

How do we solve the general linear system? As above, we have often seen that there are solutions of a system that run along a straight line. How do we find these solutions? Well, we would need the vector field to either point straight in toward the origin or point directly away from the origin. That is, we would need the right-hand side of our system, namely AY, to be some multiple of Y, say λY. That is, we would need to find a vector Y such that $AY = \lambda Y$ or $AY - \lambda Y = 0$, where 0 is the zero vector—that is, (0, 0). Note that this is precisely what is happening in our example above.

We usually write this latter equation as $(A-\lambda I)Y = 0$, where I is the identity matrix

$$I = \begin{pmatrix} 1 & 0 \\ 0 & 1 \end{pmatrix}.$$

Written out, these equations become

$$(a-\lambda)x + by = 0$$
$$cx + (d-\lambda)y = 0.$$

So to find places where the vector field points toward or away from the origin, we must find a value of λ for which the above equations have a nonzero solution. Such a λ-value is called an **eigenvalue** for our system, and any nonzero solution of the equation $(A-\lambda I)Y = 0$ is called an **eigenvector** corresponding to λ.

So how do we determine the eigenvalues? We need to find a value of λ for which the above system of algebraic equations has nonzero solutions. But we saw earlier that that happens when the determinant of the corresponding matrix is zero. So λ_0 is an eigenvalue if it is a root of the equation $\det(A-\lambda I) = 0$. Let's write that out explicitly. We find that

$$\det(A-\lambda I) = (a-\lambda)(d-\lambda) - bc = \lambda^2 - (a+d)\lambda + ad - bc = 0.$$

This equation is also called the characteristic equation. Note that it is a quadratic equation, so there are usually 2 distinct roots for this equation. These roots are the eigenvalues for the matrix. Then we can find the corresponding eigenvector by simply finding a nonzero vector solution to the equation

$$(A - \lambda_0 I) Y = 0,$$

where λ_0 is the given eigenvalue.

By the way, note that in the characteristic equation

$$\lambda^2 - (a+d)\lambda + ad - bc = 0,$$

we see the term $ad - bc$. That's what we called the determinant of the matrix A, or det$A$. There is also the term $a + d$ (i.e., the sum of the main diagonal terms in the matrix). This expression is called the **trace** of matrix A, which we write as TrA. So the characteristic equation can be written

$$\lambda^2 - (Tr\ A)\lambda + \det A = 0.$$

The important observation is that if λ is an eigenvalue for the matrix A and Y_0 is its corresponding eigenvector, then we have a solution to the system given by

$$Y(t) = e^{\lambda t} Y_0,$$

which is easily checked.

For example, consider the linear system

$$Y' = \begin{pmatrix} 2 & 1 \\ 0 & -1 \end{pmatrix} Y.$$

We first find the eigenvalues: They are the roots of the characteristic equation

$$\det\begin{pmatrix} 2-\lambda & 1 \\ 0 & -1-\lambda \end{pmatrix} = 0,$$

which simplifies to $(2-\lambda)(-1-\lambda) = 0$. Clearly these roots are 2 and -1. So these are our eigenvalues. Next we find the eigenvector corresponding to the eigenvalue 2. To do this, we must solve the system of algebraic equations

$$(2-2)x + y = 0$$
$$0x + (-1-2)y = 0$$

that reduces to $y = 0$ and $-3y = 0$. Notice that these 2 equations are redundant; this is always the case since we know that we must find a nonzero solution to the eigenvector equation. Our solution is any (nonzero) vector with y-coordinate equal to 0. So for example, one eigenvector corresponding to the eigenvalue 2 is the vector

$$Y = \begin{pmatrix} 1 \\ 0 \end{pmatrix}$$

(we could have chosen any nonzero x-component).

So we have one solution to this system, namely

$$Y_1(t) = e^{2t} \begin{pmatrix} 1 \\ 0 \end{pmatrix}$$

or $x(t) = e^{2t}$, $y(t) = 0$. Note that this is a straight line solution lying on the x-axis. It is easy to check that any constant times this solution is also a solution to our system, so we find other straight line solutions lying on the x-axis—some on the positive side and some on the negative side. But all tend to (plus or minus) infinity as time goes on and to 0 as time goes backward.

For the eigenvalue -1, we similarly find a corresponding eigenvector

$$Y = \begin{pmatrix} 1 \\ -3 \end{pmatrix}$$

and thereby get the solution

$$Y_1(t) = e^{-t} \begin{pmatrix} 1 \\ -3 \end{pmatrix}.$$

This is also a straight line solution lying along the line $y = -3x$, which tends to the origin as time goes on and away from the origin (toward infinity) as time goes backward.

So we get a whole family of solutions of the form

$$Y(t) = k_1 e^{2t} \begin{pmatrix} 1 \\ 0 \end{pmatrix} + k_2 e^{-t} \begin{pmatrix} 1 \\ -3 \end{pmatrix}.$$

Indeed, this turns out to be the general solution. Below are the straight line solutions together with some other solutions in the phase plane. This is an example of a phase plane that has a saddle equilibrium point at the origin.

Figure 12.1

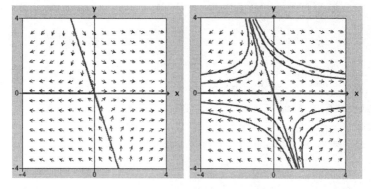

There are 2 other possible things that can happen when we have 2 real, distinct, and nonzero eigenvalues. Either both can be positive or both can be negative. In the former case, our equilibrium point is a (real) source, since our general solution will be of the form

$$k_1 e^{\lambda_1 t} Y_1 + k_2 e^{\lambda_2 t} Y_2,$$

where both λ_1 and λ_2 are positive. Thus all nonzero solutions will move away from the origin. In the other case, we have a (real) sink since all solutions will now tend to the origin.

For example, if our matrix is

$$\begin{pmatrix} 0 & 1 \\ -1 & -3 \end{pmatrix},$$

then the eigenvalues are the roots of $\lambda^2 + 3\lambda + 1 = 0$, which by the quadratic formula are $\dfrac{-3 \pm \sqrt{5}}{2}$. Both of these are negative, so we get a real sink at the origin.

Figure 12.2

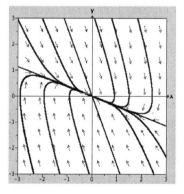

But if our matrix is

$$\begin{pmatrix} 0 & 1 \\ -1 & 3 \end{pmatrix},$$

the eigenvalues are $\dfrac{3 \pm \sqrt{5}}{2}$, which are now both positive, so we get a real source.

Figure 12.3

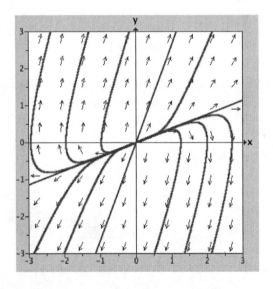

Then there are a number of other possibilities for the eigenvalues. These roots of the characteristic equation could be complex; real, nonzero, but not distinct (i.e., repeated roots); or zero. As we shall see in upcoming lectures, each of these cases gives very different phase planes.

Important Terms

eigenvalue: A real or complex number usually represented by λ (lambda) for which a vector Y times matrix A yields non-zero vector λY. In general, an $n \times n$ matrix will have n eigenvalues. Such values (which have also been called proper values and characteristic values) are roots of the corresponding characteristic equation. In the special case of a triangular matrix, the eigenvalues can be read directly from the diagonal, but for other matrices, eigenvalues are computed by subtracting λ from values on the diagonal, setting the determinant of that resulting matrix equal to zero, and solving that equation.

eigenvector: Given a matrix A, an eigenvector is a non-zero vector that, when multiplied by A, yields a single number λ (lambda) times that vector: $AY = \lambda Y$. The number λ is the corresponding eigenvalue. So, when λ is real, AY scales the vector Y by a factor of lambda so that AY stretches or contracts vector Y (or $-Y$ if lambda is negative) without departing from the line that contains vector Y. But λ may also be complex, and in that case, the eigenvectors may also be complex vectors.

trace: The sum of the diagonal terms of a matrix from upper left to lower right.

Suggested Reading

Blanchard, Devaney, and Hall, *Differential Equations*, chaps. 3.2–3.3.

Hirsch, Smale, and Devaney, *Differential Equations*, chap. 3.1.

Kolman and Hill, *Elementary Linear Algebra*, chap. 7.

Lay, *Linear Algebra*, chap. 5.

Roberts, *Ordinary Differential Equations*, chap. 7.

Strang, *Linear Algebra and Its Applications*, chap. 5.

Strogatz, *Nonlinear Dynamics and Chaos*, chaps. 5.1–5.2.

Relevant Software

Blanchard, Devaney, and Hall, *DE Tools*, Linear Phase Portraits.

Problems

1. Compute the determinant of the matrix $\begin{pmatrix} 1 & 3 \\ 2 & 1 \end{pmatrix}$.

2. What do you know about the equilibrium points for the system

   $$Y' = \begin{pmatrix} 1 & 3 \\ 2 & 1 \end{pmatrix} Y \, ?$$

3. Compute the trace and the determinant of the matrix

   $$\begin{pmatrix} 2 & 1 \\ 1 & 2 \end{pmatrix}.$$

4. **a.** Sketch the direction field for the linear system

$$Y' = \begin{pmatrix} 2 & 0 \\ 0 & -1 \end{pmatrix} Y.$$

b. Do you see any straight line solutions for this system?

5. What are the eigenvalues of the matrix

$$\begin{pmatrix} 2 & 1 \\ 1 & 2 \end{pmatrix}?$$

6. What are the eigenvalues of an upper triangular matrix

$$\begin{pmatrix} a & * \\ 0 & b \end{pmatrix},$$

where * can be any number?

7. What are the eigenvalues and eigenvectors for the very special matrix

$$\begin{pmatrix} 0 & 0 \\ 0 & 0 \end{pmatrix}?$$

8. Find the eigenvalues and eigenvectors of the matrix

$$\begin{pmatrix} 1 & 3 \\ \sqrt{2} & 3\sqrt{2} \end{pmatrix}.$$

9. Find the general solution of

$$Y' = \begin{pmatrix} 3 & 1 \\ 1 & 3 \end{pmatrix} Y$$

and sketch the phase plane.

10. Find the solution to the previous system satisfying the initial condition $Y(0) = (1, 0)$.

Explorations

1. Consider the linear system

$$Y' = \begin{pmatrix} -1 & 1 \\ 0 & -2 \end{pmatrix} Y.$$

First find both straight line solutions. Then determine how all other solutions tend to the origin as time goes on.

2. In the last exploration of linear algebra, we saw that that the determinant of a 3 × 3 matrix allows us to decide when a system of 3 linear equations has a nonzero solution. This should allow us to expand the notion of eigenvalue and eigenvector to this case. So find the eigenvalues and eigenvectors of the matrix

$$\begin{pmatrix} 3 & 0 & 1 \\ 0 & 2 & 0 \\ 0 & 1 & 1 \end{pmatrix}.$$

What would be the behavior of the solutions of the corresponding linear system of differential equations?

An Excursion into Linear Algebra
Lecture 12—Transcript

Hello again, our goal today is to figure out how to solve linear systems of differential equations. Solving these systems is going to involve an excursion into the field of mathematics known as linear algebra, and what you'll see is these solution techniques involve a lot of computational ideas, a lot of different ideas from linear algebra. I'll go through them over the next couple of lectures, and hopefully, by then everything will be familiar to you.

Let's get going. We're going to start with a linear system, $x' = ax + by$ and $y' = cx + dy$. Remember, last time we wrote that in matrix form. We introduced the vector, y is vector (x, y) and matrix A is the coefficient matrix whose entries are (a, b, c, d) as in matrix form, our differential equation is vector y' is matrix AY.

Remember, we went through an example last time. In that case, the matrix was $(1, 1, 3, -1)$, or the differential equation was $x' = x + y$, $y' = 3x - y$, and we found three solutions. One of those solutions was the vector solution (e^{2t}, e^{2t}), or the way I wrote it was $e^{2t}(1, 1)$. A second solution was e^{-2t} times the constant vector $(1, -3)$, and then thirdly, we saw that the only equilibrium solution was at the origin.

Let me go back and look at the phase plane in this case. Here is our phase plane and we found those two very special solutions one of which lies along the line $y = x$ and the other of which lies along the line $y = -3x$. Those were our straight-line solutions. They'll become very important as we enter linear algebra. Then we actually saw that all other solutions just veered off from one solution toward the other as time goes on. Finally, we saw that there was only one equilibrium point, here, at the origin.

How do we solve the general linear system? It's going to be a two-step process right now. First, we have to find all the equilibrium solutions, and then, akin to that example we did, we have to find those other straight-line solutions. Those other straight-line solutions had a special form. One of them was $e^{2t}(1, 1)$, and that number, 2, is going to be called an eigenvalue. The vector $(1, 1)$ is its corresponding eigenvector.

We're going to have to find an eigenvalue and an eigenvector. Similarly, in that other straight-line solution, e^{-2t} (1, −3), that −2 is the other eigenvalue, and its corresponding eigenvector is (1, −3). We'll get to that in a moment, but let's answer the first question. Where does this linear system have equilibrium points?

If $x' = ax + by$, and $y' = cx + dy$, certainly when x and y are 0, we have an equilibrium point, but are there others? Let's see. To find the others, we have to solve the system of algebraic equations, $ax + by = 0$, and $cx + dy = 0$. It's not a differential equation anymore; that's an algebraic equation.

That's easy to solve. Equation number one says that x must be equal to $(-b/a)y$, assuming a is nonzero. If it is 0, it is easy to handle. Then equation two says plug $x = (-b/a)y$ into the second equation, then multiply through by a, and the second equation says that $(ad - bc)y$ must be 0. That quantity $(ad - bc)y$ must be 0. That means there're two possible cases for other equilibrium solutions.

The first case is y could be 0. If $y = 0$, from our first equation, then x must be 0, and we already know that (0, 0) is a solution. We want to know if there are other solutions. The only way we can have other equilibrium solutions is if that quantity $ad - bc = 0$. If that's the case, then our expression $(ad - bc)y = 0$ says that any y value works; any y value whatsoever satisfies that equation. Then the first equation says that x must be equal to $-b/a$ times that any y value. That's an equation of a straight line, so if $ad - bc = 0$, we get a straight line of equilibrium points, many more equilibrium points.

This constant, $ad - bc$, is going to come up all the time in the future. It has a name; it's what we call the determinant. Given a matrix, the determinant is the product of the main diagonal terms, ad, then subtract off the product of what we call the off diagonal terms, bc.

To summarize, for the linear system vector $y' = Ay$, we have two cases. Case one, there's a single equilibrium point at the origin if the determinant of A is nonzero. Case two, we have a straight line of equilibrium points if the determinant, $ad - bc$, is equal to 0. The determinant determines the number of equilibrium points.

One little problem here is one special case. What happens if your matrix was (0, 0, 0, 0)? Then the determinant is 0, but now all points are equilibrium points. Every point is an equilibrium solution, so that's a very special case. It will never come up, usually.

For later use, notice what we just did. We saw that if you took the linear system of algebraic equations, $ax + by = 0$ and $cx + dy = 0$, then that system of equations has nonzero solutions if and only if the determinant of the matrix, A, was equal to 0.

For example, if you take a linear system of equations, $2x + 3y = 0$, $4x + 6y = 0$, what are the solutions there? The determinant is $(2)6 - (3)4$. That's $12 - 12$, or 0, so we must have multiple solutions. Of course, look at those two equations. The second equation is twice the first equation, so anything that solves $2x + 3y = 0$ is a solution. That's the straight line, $y = (-2/3)x$.

Another example, take $2x + 3y = 0$, first equation, and $3x + 2y = 0$. The determinant is $(2)2 - (3)3$. That's $4 - 9$, not equal to 0, and only one solution of that system of equations, namely, the origin.

That takes care of the equilibrium solutions portion. Let's move to the straight-line solutions. Here's where linear algebra really comes in. How are we going to find a straight-line solution to this differential equation? We need to find a line in the plane, going through the origin, where the vector field is everywhere tangent to it.

That is, we have to find vectors that are pointing either directly into the origin or directly away from the origin where those vectors are given by our vector field. What does it mean for a vector to point in or away from the origin at some point? Given the point V, that vector must be either a constant times V that's positive, or a constant times V that's negative.

What do we need? We need our vector field to have that property, but our vector field is A times some vector. It's AV. We need AV then to be equal to some multiple of V. Let's call that multiple λ. We need the vector AV, our vector field, to be equal to some constant, λ, times the position V.

Let me show you what that means geometrically. Here I've got a linear system. The matrix is (2, 1, 1, −1), here's my vector V, and what I'm plotting at V is the right hand side of the differential equation, the vector field. As I move V around, you see the vector field changes, but every once in a while, like right there, the vector field is pointing in the same direction as my initial point, V. The vector field is some multiple of the initial point V, $AV = \lambda V$.

As I run around, most often, the vector field does not point in the same direction as my point V, but right about there, it does. At that point, V, the vector field is pointing directly into the origin, so again, AV, the vector field, is equal to a constant times V, this time a negative value of V. We see there are certain places where the vector field points in the same direction as our initial point, like right there. Those are vectors that solve the equation $AV = \lambda V$.

That is our springboard to finding straight-line solutions because if we find a vector such that $AV = \lambda V$, then we have a solution. Our solution is e to that number, λ, times T times the vector we found, $e^{\lambda t}$ times a constant vector. That's a straight-line solution.

Why is $e^{\lambda t}V$ a solution to this differential equation? We know how to do it. Let's plug it into the differential equation. Plug $e^{\lambda t}$ times that constant vector, V, into $y' = Ay$, and we first differentiate y'. But, y is $e^{\lambda t}V$, the only t occurs in that exponential term, its derivative is $\lambda e^{\lambda t}$ times our constant vector, V, and that's y'.

Is that equal to the matrix, Ay? Take A, multiply it by our purported solution, $e^{\lambda t}V$. A times vector $e^{\lambda t}V$, in each term of that vector, there's an exponential term, $e^{\lambda t}$, so we can pull that out. We get either the λt times matrix A times vector V, but remember, AV was that special vector which was λV. When we plug our solution into Ay, we get $e^{\lambda t}\lambda V$, exactly what we got when we differentiated y.

Yes, $e^{\lambda t}V$ is a solution to the differential equation, so that is a straight-line solution to the differential equation. To summarize, to find these straight-line solutions, we need to find a vector, V, that satisfies $AV = \lambda V$, or bringing the λV over to the other side, we need to find a vector that satisfies $AV - \lambda V = (0,$

0). Think of V as being the vector (x, y). Let me write this in a slightly funny way, but a traditional way. Let me take what we call the identity matrix, that's the matrix whose diagonal entries are 1 and off diagonal entries are 0, so (1, 0, 0, 1). Let me multiply that matrix by λ. I get λs down the diagonal. Let me take that matrix with λs down the diagonal and subtract it off from the matrix A. What do I get? I subtract each individual entry and end up with the matrix whose entries are $(a - \lambda, b, c, d - \lambda)$. That's a new matrix. That's the matrix we call $a - \lambda$ times the identity.

What do we need? We need to find a vector, (x, y) such that that matrix, $(A - \lambda V, c, d - \lambda)(x, y) = (0, 0)$. Written out in longhand, we have to find a value of λ such that the system of linear equations, $A - \lambda x + by = 0$, and $cx + d - \lambda y = 0$. We need to find a λ value for which these equations have a nonzero solution. Such a λ value has a name, it's the eigenvalue, and then that nonzero solution is an eigenvector associated to that eigenvalue.

Think of where we were just a bit ago. We saw that a system like that has a nonzero solution if the corresponding determinant is equal to 0. We have to take the determinant of the matrix $A - \lambda i$, and if that determinant is equal to 0, we know we have nonzero solutions.

Let's do it. What's the determinant of $A - \lambda i$? Down the diagonal, we have $(a - \lambda)(d - \lambda)$. Off of the diagonal, we have bc, so all together, our equation is $(a - \lambda)(d - \lambda) - bc = 0$. We have to find λ values that satisfy that equation, but if you multiply that out, what you see is the equation reads $\lambda^2 - a\lambda - D\lambda$. Then we combine that to $-(a + d)\lambda$, and then finally, there's an ad, $a - bc$ term, and all of that's equal to 0, so $\lambda^2 - (a+d)\lambda + ad - bc = 0$.

This is what we call the characteristic equation. That's a quadratic equation. Any root of that quadratic equation is a λ value that works. Any root is a λ value that's an eigenvalue that'll generate a straight-line solution for this.

In this equation, we see the term $ad - bc$, that's what we call the determinant. That shows up in the characteristic equation. We also see the term $a + d$. That's also going to appear later in this course very often. That's what's called the trace of the matrix. The trace of the matrix is the sum of the diagonal entries, and the determinant is $ad - bc$. Our characteristic equation written more succinctly is λ^2 minus the trace of the matrix times λ plus the determinant of the matrix equals 0.

What we can do is, given the characteristic equation, we can write that down immediately, solve the quadratic equation, get the eigenvalues, and then use the eigenvalues to determine that nonzero eigenvector, the solution of that algebraic equation that gives us the second part of our solution. It's kind of complicated, but let's go through an example.

Let's take the matrix $(2, 1, 0, -1)$. Our linear system is $y' = (2, 1, 0, -1)y$. Our first thing to do is to determine the eigenvalues, those λ values. To do that, we need the characteristic equation. The characteristic equation says, take your matrix, subtract off λs down the diagonal, and set it equal to 0. Take the determinant of that matrix, $(2 - \lambda, 1, 0, -1 - \lambda)$ and set the determinant equal to 0.

What's the determinant here? It's $(2 - \lambda)(-1 - \lambda) - (0)1$, so the determinant here is $(2 - \lambda)(-1 - \lambda)$. You've got to set that equal to 0. That's pretty good. We don't even have to use the quadratic formula here. We know the roots are 2 and -1. It's kind of a free gift from the math department. Look at your matrix. It was what we call an upper triangular matrix, that lower entry on the left was 0. When that happens, then the eigenvalues are displayed right along the diagonal. You see 2 and -1 are your eigenvalues. Similarly, if they were 0 in the upper right and upper triangular matrix, again you'd see the eigenvalues displayed along the diagonal.

We have our eigenvalues, 2 and -1. We have to get the eigenvectors corresponding to 2 and -1. First, let's find the eigenvector corresponding to 2. How do we get that? We take our matrix, we subtract off the eigenvalues, 2, down the diagonal, and then take the corresponding equation, set it equal to 0, and solve. We know we can. We know we can get nonzero solutions.

In this case, our matrix was $(2, 1, 0 -1)$. Start with the eigenvalue 2, subtract it off the diagonal, and you get $(2 - 2)x + y = 0$. That's the first equation. The second equation reads $0x + (-1 - 2)y = 0$, or your equation is reduced to $y = 0$ and $-3y = 0$. They are redundant. You can find nonzero solutions to that system of algebraic equations. Any vector with $y = 0$ works.

For example, I could choose the vector (1, 0) or anything else 0 as long as it's not 0. Any such vector is an eigenvector corresponding to the eigenvalue 2. There it is. By what we did earlier, we have a solution. The solution is e to the eigenvalue t, e^{2t}, times its eigenvector (1, 0), and notice that's the solution $e^{2t}0$. That's a straight-line solution moving along the x-axis away from the origin.

Good, we've found one straight-line solution. Let's go to the other eigenvalue. We saw that was -1. We have to do the same computation. Subtract off the eigenvalue down the diagonal of your matrix, then multiply that matrix by (x, y) and set it equal to 0. When you subtract off -1 down the diagonal, you end up with a system of equations, $(2 - -1)x + y = 0$, and $0x + (-1 + 1)y = 0$, or $3x + y = 0$ and the second equation is $0 = 0$. We have nonzero solutions. Any nonzero vector that satisfies $3x + y = 0$ works.

For example, I could let x be 1, and then y has to be -3. There's our eigenvector for the eigenvalue -1. More importantly, there is another straight-line solution, the solution e^{-t}, e to the eigenvalue t, times its eigenvector, $(1, -3)$, is a straight-line solution for the system of differential equations. Here, written out, our solution is e^{-t}, the x component, $-3e^{-t}$ in the y component, so our straight-line solution lies along the line $y = -3x$ in the phase plane.

From what we did last time, we know the eigenvectors, (1, 0) and (1 -3). They're linearly independent, so we're done. We have the general solution. Any constant times our first solution plus any other constant times our second solution is a solution to the differential equation, and from that, we can solve any initial value problem.

Here's our matrix, matrix (2, 1, 0, -1). Here are the two straight-line solutions we just found. One corresponded to eigenvalue 2 along the x-axis; we see a straight-line solution running away along the x-axis. The other one corresponded to the eigenvalue, -1 with eigenvector (1, -3), that's a straight-line solution that runs directly into the origin, right down here starting where y is negative. You see a straight-line solution running into the origin.

With those straight-line solutions, we can generate all possible solutions by taking any combinations of the first and the second. As before, since one of our eigenvalues is negative and the other eigenvalue is positive, we see that all other solutions look like they tend to one straight-line solution in forward time and tend to the other one in backward time, again, an example of a saddle phase plane.

There are going to be all kinds of different things that come up when we try to find these general solutions, try to find these eigenvalues and eigenvectors. For example, we could find a case where, unlike the example we just did, both of our eigenvalues were positive. That means our general solution would be of the form $k_1 e^t$ times its eigenvector, plus k_2 times another e^t times the eigenvector. That means we have increasing exponentials everywhere, so all of our solutions would run away from the origin. That's what we call a real source.

On the other hand, the eigenvalues could both be negative, and those exponential terms $e^{\lambda t}$ would be decaying exponentials, so now all solutions would come into the origin. We'd have what we call a real sink.

For example, let's look at the matrix $(0, 1, -1, -3)$. What are the eigenvalues here? Step one, subtract off λs down the diagonal, $-\lambda 1, -1, -3, -\lambda$. Step two, take the determinant of that, set it equal to 0, but the determinant, $-\lambda(-3 - \lambda) + 1 = 0$, or if you work that out, it's $\lambda^2 + 3\lambda + 1 = 0$. Find the roots of that characteristic equation. What you'll see is you get the roots, $(-3 \pm \sqrt{5})/2$. When you have minus sign up there, you've got a negative eigenvalue, $-3 + \sqrt{5}$, that's also negative. You have two negative eigenvalues here. You should get a sink; all solutions should come into the origin.

Here is our matrix, $(0, 1, -1\ -3)$. There are our two straight-line solutions. I didn't compute the eigenvectors, but you can go ahead and do that. Notice that all other solutions come into the origin. All other solutions tend to 0. The equilibrium point at 0 is a sink.

Were I to change this coefficient -3 to 3, then what would happen is both of your eigenvalues would now become positive. You would find your eigenvectors get two straight-line solutions. Solutions would all be tending away from the origin along those straight-line solutions, and all other solutions would be tending off to infinity. All other solutions would leave the origin and tend off to infinity, an example of a real source.

I didn't compute the eigenvectors here. That's because in the next couple of lectures, we'll do some more computation of eigenvectors, but it's the same as the process that I just carried out. One question comes up. In these pictures, why, when we had a sink, did almost all solutions come in tangentially to one of those straight-line solutions? When we have a source, why did all of them leave tangentially to one of those straight-line solutions?

That usually happens when you have different eigenvalues. For example, look at the simple case, $x' = -x$, $y' = -2y$. In matrix form that's $y' = (-1, 0, 0, -2)$. That's both an upper and lower triangular matrix. It's what we call a diagonal matrix, so the eigenvalues are displayed right along the diagonal. They are -1 and -2.

We could go and compute the eigenvectors here, but we know what the solution to this differential equation is. It's $x' = -x$, and $y' = -2y$. We know some solutions. For $x' = -x$, our solution is e^{-t}. For $y' = -2y$, we know our solution is e^{-2t}.

What do these solutions look like? If $x = e^{-t}$, $y = e^{-2t}$, notice $y = x^2$. That means it should lay long a parabola. It turns out we have two straight-line solutions on the axes, and all other solutions come in tangential to that straight-line solution along the axis.

There's another reason why this is true. Look at our solutions. They're of the form e^{-t} and y is of the form e^{-2t}. What happens to y as time goes on? It goes to 0, but it goes to 0 much faster than x; e^{-2t} goes to 0 much faster than e^{-t}, and that's why that y term disappears very quickly leaving you with an x term toward the end. That's why all solutions, except that vertical straight-line solution, come in tangential to the origin.

To summarize, given the linear system, $y' = Ay$, our method of solving will be first, find the eigenvalues. That's take the roots of the characteristic equation, determinant of $(A - \lambda I) = 0$. Second, given the eigenvalues, you find the corresponding eigenvectors. You take the matrix A, subtract off your eigenvalue down the diagonal, and multiply that by vector xy. Set it equal to 0 in linear algebra equations. Third, you then get the straight-line solutions $e^{\lambda t}$ times your given eigenvector, and then finally, we get the general solution.

One quick thing, notice if you make a mistake computing your eigenvalues, then you're never going to be able to find an eigenvector because if you don't have an eigenvalue, that linear system has only 0 solutions. That's a bonus. You can't go wrong if you make a mistake computing your eigenvalues.

Finally, it's clear that there's going to be many more possibilities. The roots of that characteristic equation, that quadratic equation, could be complex, could be real but repeated, and could be 0. In each case, we get very different phase planes. These are the things that we'll begin to discuss in the next lecture.

Visualizing Complex and Zero Eigenvalues
Lecture 13

Recall that to solve the linear system of differential equations given by

$$Y' = \begin{pmatrix} a & b \\ c & d \end{pmatrix} Y,$$

we followed a 4-step process. First we wrote down the characteristic equation

$$\lambda^2 - (a+d)\lambda + (ad-bc) = \lambda^2 - T\lambda + D = 0,$$

where T is the trace of the matrix and D is the determinant. Second, we solved this quadratic equation to find the eigenvalues

$$\lambda_\pm = \frac{T \pm \sqrt{T^2 - 4D}}{2}.$$

Third, for each eigenvalue $\lambda_\pm$, we found the corresponding eigenvectors $V_\pm$ by solving the system of algebraic equations below.

$$(a-\lambda)x + by = 0$$
$$cx + (d-\lambda)y = 0$$

Finally, this gave us the general solution below.

$$k_1 e^{\lambda_- t} V_- + k_2 e^{\lambda_+ t} V_+$$

But the roots of the characteristic equation may be complex—so what happens in that case?

Let's begin with the simple example

$$Y' = \begin{pmatrix} 0 & 1 \\ -1 & 0 \end{pmatrix} Y.$$

We actually know how to solve this system since the system is just $x' = y$ and $y' = -x$. So we really have $x'' = -x$, and we know the solution of this second-order equation. The general solution is $x(t) = k_1\cos(t) + k_2\sin(t)$. Therefore $y(t) = x'(t) = -k_1\sin(t) + k_2\cos(t)$.

Altogether, this yields

$$Y(t) = k_1 \begin{pmatrix} \cos(t) \\ -\sin(t) \end{pmatrix} + k_2 \begin{pmatrix} \sin(t) \\ \cos(t) \end{pmatrix}.$$

Let's see how this solution arises out of the eigenvalue/eigenvector process. Our characteristic equation here is $\lambda^2 + 1 = 0$, so the eigenvalues are the imaginary numbers i and $-i$. To find the eigenvector corresponding to i, we must solve the system of equations

$-ix + y = 0$

$-x - iy = 0.$

These 2 equations are redundant, since multiplying the second equation by i gives the first equation. So the eigenvector corresponding to the eigenvalue i is any nonzero vector that satisfies $y = ix$. So for example, one eigenvector is

$$\begin{pmatrix} 1 \\ i \end{pmatrix}.$$

We therefore get the solution

$$Y(t) = e^{it} \begin{pmatrix} 1 \\ i \end{pmatrix}.$$

But there is a problem here. We started with a system of real differential equations but ended up with a complex solution. Let's remedy this using Euler's formula. Our solution can be expanded via Euler to

$$Y(t) = (\cos(t) + i\sin(t))\begin{pmatrix} 1 \\ i \end{pmatrix} = \begin{pmatrix} \cos(t) \\ -\sin(t) \end{pmatrix} + i\begin{pmatrix} \sin(t) \\ \cos(t) \end{pmatrix}.$$

This is still a complex solution to a real differential equation. But were we to plug this expression into the equation $Y' = AY$, the real part would remain real and the imaginary part would remain imaginary. That means that the real part of this complex solution,

$$Y_{real}(t) = \begin{pmatrix} \cos(t) \\ -\sin(t) \end{pmatrix},$$

and the imaginary part,

$$Y_{imag}(t) = \begin{pmatrix} \sin(t) \\ \cos(t) \end{pmatrix},$$

are both solutions of this differential equation. And notice that the real part starts out at the point $(1, 0)$ whereas the imaginary part begins at $(0, -1)$. So these are independent solutions. It is interesting that using just one of the eigenvalues gives us a pair of solutions that then generate all possible solutions (below) exactly as we saw earlier.

$$Y(t) = k_1 \begin{pmatrix} \cos(t) \\ -\sin(t) \end{pmatrix} + k_2 \begin{pmatrix} \sin(t) \\ \cos(t) \end{pmatrix}$$

Clearly, when $k_1 = 0$, our solutions all lie on circles centered at the origin. In fact, that's true for any values of k_1 and k_2 (as long as they are both not equal to zero). So our phase plane here is seen in Figure 13.1.

Figure 13.1

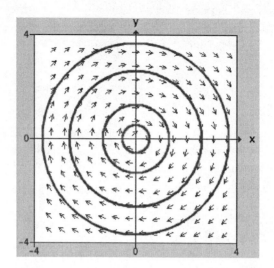

This type of equilibrium is center. In general, if the eigenvalues are $\pm ia$, the solutions may lie on circles or ellipses.

When we compute the roots of a characteristic equation, we may also find complex roots of the form $a + ib$ and $a - ib$ where, unlike the previous case, a may be nonzero. In this case, the exact same procedure as above yields a complex eigenvector V for one of the eigenvalues. This then yields, by Euler's formula, solutions of the form

$$e^{(a+ib)t}V = e^{at}(\cos(bt) + i\sin(bt))\,V.$$

If we then break this vector solution into its real and imaginary parts, we find 2 real solutions to the linear system, and it turns out that these solutions are independent. So we find the general solution just as before.

The difference here is that our new solutions always involve the exponential term $e^{(at)}$. If $a > 0$, then this term tends to infinity as time goes on, whereas if $a < 0$, this term goes to zero as time goes on. As the remaining terms in the solution involve $\sin(bt)$ and $\cos(bt)$, it follows that if $a > 0$, solutions spiral away from the origin (the equilibrium is a **spiral source**), but if $a < 0$, solutions spiral in toward the origin (a spiral sink).

For example, for the linear system

$$Y' = \begin{pmatrix} -0.1 & 1 \\ -1 & -0.1 \end{pmatrix} Y,$$

we can easily check that the eigenvalues are $-0.1 + i$ and $-0.1 - i$. Similar computations to the above then yield the general solution

$$Y(t) = k_1 e^{-0.1t} \begin{pmatrix} \cos(t) \\ -\sin(t) \end{pmatrix} + k_2 e^{-0.1t} \begin{pmatrix} \sin(t) \\ \cos(t) \end{pmatrix}.$$

The corresponding phase plane is the below.

Figure 13.2

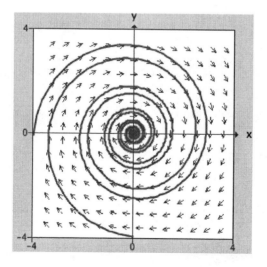

If we consider instead the linear system

$$Y' = \begin{pmatrix} 0.1 & 1 \\ -1 & 0.1 \end{pmatrix} Y,$$

the eigenvalues are $0.1 + i$ and $0.1 - i$, and the general solution becomes

$$Y(t) = k_1 e^{0.1t} \begin{pmatrix} \cos(t) \\ -\sin(t) \end{pmatrix} + k_2 e^{0.1t} \begin{pmatrix} \sin(t) \\ \cos(t) \end{pmatrix}.$$

Solutions now spiral away from the origin, as seen below.

Figure 13.3

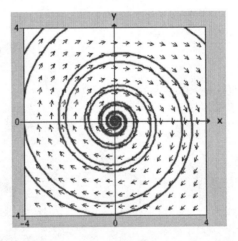

So far we have figured out how to handle the major types of linear systems, sinks, saddles, sources, and centers. But there are some other special cases, such as when we have zero eigenvalues and real and repeated eigenvalues.

If we have just one eigenvalue that is equal to zero (and so the other is either positive or negative), we know more or less how to solve the system. We find our eigenvector V corresponding to the eigenvalue 0, so we then have a solution of the form $Y(t) = k_1 e^{0t} V$. Our solution is just the constant

$k_1 V$. So all points that lie on the straight line passing through the origin and the point V in the plane are constant or equilibrium solutions. Then the sign of the other eigenvalue a determines whether the remaining solutions tend toward or away from these equilibria.

As an example, consider the system

$$Y' = \begin{pmatrix} 0 & 1 \\ 0 & -1 \end{pmatrix} Y.$$

Check on your own that the eigenvalues are 0 and -1 and the general solution is

$$Y(t) = k_1 e^{-t} \begin{pmatrix} -1 \\ 1 \end{pmatrix} + k_2 \begin{pmatrix} 1 \\ 0 \end{pmatrix}.$$

So we have a line of equilibrium points along the x-axis, and all other solutions tend toward these equilibria. The phase plane is the below.

Figure 13.4

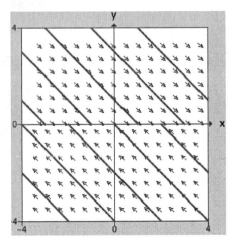

As we'll see in the next lecture when we describe the trace-determinant plane, linear systems often undergo bifurcations when there is a family of such systems that has either a zero eigenvalue or a center. For example, consider the family of differential equations

$$Y' = \begin{pmatrix} -1 & 0 \\ 1 & A \end{pmatrix} Y$$

The eigenvalues here are -1 and A. So, whenever $A < 0$, we have a sink; whenever $A > 0$, we have a saddle. A bifurcation occurs when $A = 0$, and the system has a zero eigenvalue.

Important Term

spiral source: An equilibrium solution of a system of differential equations for which all nearby solutions spiral away from it.

Suggested Reading

Blanchard, Devaney, and Hall, *Differential Equations*, chaps. 3.2–3.3.

Hirsch, Smale, and Devaney, *Differential Equations*, chap. 3.1.

Kolman and Hill, *Elementary Linear Algebra*, chap. 7.

Lay, *Linear Algebra*, chap. 5.

Roberts, *Ordinary Differential Equations*, chap. 7.

Strang, *Linear Algebra and Its Applications*, chap. 5.

Strogatz, *Nonlinear Dynamics and Chaos*, chaps. 5.1–5.2.

Relevant Software

Blanchard, Devaney, and Hall, *DE Tools*, Linear Phase Portraits.

Problems

1. a. For the system of differential equations

    $$Y' = \begin{pmatrix} 2 & 5 \\ -1 & -2 \end{pmatrix} Y,$$

    first compute the characteristic equation.

   b. What are the eigenvalues of this matrix?

   c. Compute the asssociated eigenvectors.

2. a. Compute the eigenvalues associated to

    $$Y' = \begin{pmatrix} 0 & 1 \\ 0 & -1 \end{pmatrix} Y.$$

   b. Find the associated eigenvectors for this system.

3. Rewrite the second-order differential equation for the mass-spring system as a linear system and find the eigenvalues. Do you see any connection with our earlier method of solving this system?

4. **a.** What are the eigenvalues and eigenvectors of the matrix

$$\begin{pmatrix} a & b \\ -b & a \end{pmatrix}?$$

b. Determine the types of the equilibrium points for the linear system corresponding to the above matrix, and sketch the regions in the a-b plane where this matrix has different types of eigenvalues.

5. **a.** Find the general solution of the system

$$Y' = \begin{pmatrix} 1 & 3 \\ \sqrt{2} & 3\sqrt{2} \end{pmatrix} Y$$

and sketch the phase plane.

b. Find the solution to this system satisfying the initial condition $Y(0) = (1, 0)$.

Exploration

Consider all possible 2×2 matrices of the form

$$A = \begin{pmatrix} a & b \\ c & d \end{pmatrix}.$$

Determine the set of a-, b-, c-, and d-values for which this matrix has either a zero eigenvalue or pure imaginary eigenvalues (i.e., eigenvalues of the form $\pm \mu$ for some $\mu \neq 0$). Express these sets in terms of the trace and the determinant of A.

Visualizing Complex and Zero Eigenvalues
Lecture 13—Transcript

Good, glad to see you came back. That last lecture was quite full. We saw all sorts of linear algebra, trace, determinant, eigenvalues, and eigenvectors. That went pretty quickly. Don't worry; we have a lot more cases to do. We'll be going over all of these concepts in a variety of settings over the next two lectures.

Let me summarize what we do to solve linear systems of differential equations. Take the linear system y' is matrix (a, b, c, d) times vector y. Step number one is to write down the characteristic equation. That means subtract off λs down the diagonal, take the determinant of that matrix, the determinant of the matrix $A - \lambda I$ and set it equal to 0. You get a quadratic equation. As we saw, it's the equation $\lambda^2 - (a + d)\lambda + (ad - bc) = 0$, or in front of λ, we have minus the trace of our matrix. Then the constant term, $ad - bc$ is the determinant of our matrix, so the characteristic equation is λ^2 minus the trace times λ plus the determinant. Then we solve that characteristic equation. The roots of the characteristic equation are what we call the eigenvalues. They were λ, $\pm$ call them, $\dfrac{T \pm \sqrt{T^2 - 4D}}{2}$ where T is the trace of our matrix and D is the determinant. Look at that. We have eigenvalues that depend only on T and D. Our matrix depended on 4 parameters, but really, the eigenvalues depend on just two numbers, the trace and the determinant. That will come back a lot later.

That was step two. Step three was, for each eigenvalue, then we find the corresponding eigenvector. What is that? That's gotten by solving the following linear system of algebraic equations. Subtract off λ, your eigenvalue down the diagonal, multiply that matrix by xy, set it equal to 0, and you know you can find nonzero solutions. Find that nonzero solution, and it's an eigenvector.

Then you've got one solution to your differential equation, e to the eigenvalue times t times the corresponding eigenvector is a solution. If you use the other eigenvalue to generate another solution, that then gives you the general solution.

The question is, since we have a quadratic equation, those roots could be complex. What happens then? Let's start with a simple example. Start with matrix $y' = (0, 1, -1, 0)y$, or written out in longhand, $x' = y$, $y' = -x$. We've seen that simple differential equation before, $x'' = y' = -x$, $x'' = -x$. We know solutions; the solutions are sines and cosines. We saw earlier that the general solution of the linear, second-order homogeneous equation $y'' + y$ was any constant times $\cos(t)$, plus any constant times $\sin(t)$, and then if you take the derivative of that, you get the same constant times minus sine plus $k_2 \sin(t)$. We have the general solution of this linear system.

The general solution is $k_1 \begin{pmatrix} \cos(t) \\ -\sin(t) \end{pmatrix} + k_2 \begin{pmatrix} \sin(t) \\ \cos(t) \end{pmatrix}$.

How does that come up in the context of eigenvalues and eigenvectors? Let's repeat that process. Here's the matrix $(0, 1, -1, 0)$. Find the eigenvalues. Subtract off λs down the diagonal, and take the determinant. The determinant, a characteristic equation, then becomes $\lambda^2 + 1 = 0$. We see where they're coming from. The roots here, the eigenvalues are $\pm i$, the imaginary number i.

Let's look first at the eigenvalue i. What is the corresponding eigenvector? Remember what we have to do. We have to subtract off our eigenvalue, i, down the diagonal. We get the matrix $(-i, 1, -1, -i)$. Then we have to multiply that matrix by xy and set it equal to 0. We get the equation $-ix + y = 0$. That's the first equation; the second equation is $-x -iy = 0$, and we have to solve those equations.

Look at those equations. It's $-ix + y = 0$. That says that y is ix, and the second equation is redundant. Any vector, nonzero, that satisfies $y = ix$ is an eigenvector. For example, let x be 1, then y must be i. There's our eigenvector $(1, i)$ and now we have one solution. Our one solution is e^{it}, e to the eigenvalue t, times its eigenvector, $(1, i)$.

Look at this. We have a complex solution to a real differential equation. How to get back to the real world? Here comes Euler again. Remember Euler's formula, $e^{it} = \cos(t) + i\sin(t)$, so we can write our solution as $\cos(t) + i\sin(t)$ $(1, i)$. If you break that vector up into its real and imaginary part, you get a sum of two vectors. If we multiply $\cos(t) + i\sin(t)(1)$ we get $\cos(t) + i\sin(t)$. If we multiply $\cos(t) + i\sin(t)(i)$, we get $-\sin(t) + i\cos(t)$.

Let's take the part of that solution that has no Is in it, the real part. That's the vector $(\cos(t), -\sin(t))$, and let's take the vector that has Is it, that's the vector $(\sin(t), \cos(t))$, and so our solution can be written in complex form as $(\cos(t), -\sin(t)) + i(\sin(t), \cos(t))$. I'm going to call the first vector, $(\cos(t) -\sin(t))$, the real part of our solution, and the other vector, $(\sin(t), \cos(t))$ the imaginary part. No is in the imaginary part. I'm factoring out that i, so the imaginary part is $(\sin(t), \cos(t))$.

We've seen that both of those are already solutions to that known differential equation. Both of those are actual real solutions to our differential equation. Look what's happened. We started with a complex eigenvalue, got a complex eigenvector, generated a complex solution, but then it turns out that the real part and the imaginary part of that complex solution are both solutions. Even better, at time 0, our real part of the solution is $\cos(0) - \sin(0)$, that's at $(1, 0)$. Whereas the imaginary part, $(\sin(t), \cos(t))$ at time 0 is at $(0, 1)$. The vector $(1, 0)$ and $(0, 1)$ are linearly independent, so we have the general solution.

That's another free gift from the math department. It says that when you've got a complex eigenvalue, you find the eigenvectors, you write out the complex solution, and what just happened here always happens. The real part of your solution and your imaginary part of the solution are both solutions to the differential equation, and more importantly, they're linearly independent. With just one eigenvalue, you can generate the general solution.

Let's go and see what's happening in this equation on the phase plane. There's our differential equation, $x' = y$, $y' = -x$, and what happens to our solutions? Our solutions run around in circles, of course, $\cos(t)$ and $-t$ are just solutions. These solutions, if you look at the direction field are going around the origin in the clockwise direction. Solutions could also go in the counterclockwise direction. For example, if I take the negative of that vector field, that is, the system $x' = -y$, $y' = x$, now solutions go in the opposite direction. They run around in the counterclockwise direction.

More generally, solutions don't have to lie on circles when you've got imaginary eigenvalues as we just saw. For example, if you take the system $x' = y$ and $y' = -4x$, then you can check that the eigenvalues are imaginary. They're $\pm 2i$, but now in this case, solutions run around not in circles, but ellipses.

If you were to go to another linear system, take the linear system y' is matrix $(1, 1, -5, -1)$, so $x' = x + y$, $y' = -5x - y$. You can check that the system also has eigenvalues $\pm 2i$, pure imaginary, but now the solutions lie on ellipses, but these ellipses aren't centered on the x or the y-axis. That's what happens in general. When you've got a linear system of differential equations whose eigenvalues are pure imaginary, most often your solutions will be ellipses that lie out in the plane. In any event, your solutions are periodic.

What else goes on? When we solved that equation, we found pure imaginary eigenvalues. It's just as easy to get eigenvalues that are complex numbers of the form a $\pm$ ib, so you could find eigenvalues that have a real part. Then what would you do? Then what you do is go and find the corresponding eigenvector. The eigenvector to one of those eigenvalues would be a complex vector. Your solution would then be e^{at} times $e^{(ib)t}$ times that complex eigenvector, but of course, again, by Euler, the $e^{(ib)t}$ breaks up into cosines and sines. You can write that as $\cos(bt) + i\sin(bt)$.

What you would do in this more general case is then multiply that all out. You'd get a horrible-looking complex vector, break it into its real part, the stuff that doesn't involve i, and the imaginary part, the stuff that involves i, and just as before, you would get a pair of solutions to that linear system. In fact, as before, that pair of solutions would be independent. You would get the general solution.

The one difference here is these solutions are going to involve not only $\cos(bt)$ and $\sin(bt)$, but they'd also involve e^{at}. When a was 0, when you're in the pure imaginary case, eigenvalues $\pm ib$, we got periodic solutions, but now if a is nonzero, we have some spiraling going on. For example, if a is greater than 0, then our solutions must spiral away from the origin, whereas if a is less than 0, we have an e^{at}, and that's a decaying exponential. We're multiplying our periodic solution by something that's getting smaller, and solutions would spiral toward the origin.

That gives us three different cases in the case of complex eigenvalues. You get an eigenvalue, $a + ib$, if a is greater than 0 you've got solutions that spiral away. We call that phase plane a spiral source. If a is less than 0, solutions spiral in, you've got a spiral sink, and the intermediate case that we just saw when a is 0 and solutions are periodic, we call that a center, three different cases, spiral sources, spiral sinks, and centers.

Let me give one quick example of this. Let's take the matrix (−0.1, 1, −1, −0.1). What are the eigenvalues? The trace is −0.1 + −0.1, that's −0.2, and the determinant is (−0.1)(−0.1), that's 0.01 plus 1. We get determinant is 1.01. The characteristic equation is $\lambda^2 + 0.2\lambda + 1.01 = 0$. We've seen that characteristic equation back in the mass-spring system. I like it. Our eigenvalues are given by −0.2, the trace, plus or minus the trace squared, minus 4 times the determinant, that's 0.04 − 4.04, and that's −4, all over 2. We have our eigenvalues. Get rid of the 2s, you see the eigenvalues are $-0.1 \pm i$. We have complex eigenvalues with a real part that's negative. We should have a spiral sink.

Let me not go through the laborious computations of the eigenvectors. If you go ahead and do that you'll find that the real and imaginary parts of your solution break up into the exponentials, but the real part is (cos(t), − sin(t)) and the imaginary part is (sin(t), cos(t)). That would give us the general solution because at time 0, those two vectors, as we saw earlier, are independent.

Here's our matrix. It's (−0.1, 1, −1, −0.1). We saw that we should have a spiral sink, and in fact, that's exactly what happens. All solutions spiral into the origin. That's the phase plane, that's a spiral sink, reminiscent of what we saw when we did underdamped harmonic oscillators.

If I were to change this matrix, not to be −0.1, but to be + 0.1, along the diagonal, so + 0.1 along the diagonal, then what you see is the real parts of our eigenvalues would turn out to be + 0.1, and then the imaginary parts would be ±i. We have real parts that are positive. We, then, have a spiral source, and there is the solution to this system of equations. Solutions spiral away from the origin no matter where you start because you've got an increasing exponential, solutions go far away in forward time, and back into the origin in backward time.

We've seen lots of different phase planes for linear systems so far. We've seen saddles, real sinks and sources, spiral sinks and sources, and centers, but there are many other special cases. For example, there are 0 eigenvalues and there may be repeated eigenvalues.

Let's delve into one of those cases there. Suppose you happen to have an eigenvalue that's 0. Then you can still find an eigenvector that corresponds to that eigenvalue. You get one solution, e to the eigenvalue t times V, or $e^{0t}V$. That's the constant vector, V. If you multiply that by a constant, you can get any constant, times (V, y) and get a whole straight line of solutions that are constant. You get a whole collection of solutions that lie along that straight line.

Let's do one example of that. Let's take the matrix (0, 1, 0, −1). The trace of this matrix is −1, add up the diagonal entries, −1. The determinant is (0 × −1) − (0 × 1). The determinant is 0. Your characteristic equation is $\lambda^2 + \lambda = 0$. Eigenvalues are −1 and 0. Of course, we didn't have to do that. That's an upper triangular matrix. We have a 0 on the lower left, so you can read the eigenvalues off from the diagonal. The eigenvalues are 0 and −1.

What's the eigenvector for the eigenvalue 0? Subtract off 0s down the diagonal. It doesn't do much, leaves you with the same matrix. Multiply that by xy, and solve. The first equation gives $0x + y = 0$. The second equation gives $0x − y = 0$. So $y = 0$ is our solution. Any vector with $y = 0$ is an eigenvector. For example, one eigenvector corresponding to the eigenvalue 0 is, say, the vector (1, 0). We get a straight line of equilibrium points that lie along the x-axis. Any constant, 0, is an eigenvector for the eigenvalue 0, so they are all equilibrium points.

What about the other eigenvalues? Our other eigenvalue is -1. We could find the eigenvector, but let's do it a slightly different way. Our system of differential equations was $x' = y$, and $y' = -y$. We know how to solve that second equation, $y' = -y$ has solution e^{-t}, so any constant times e^{-t} is our $y(t)$ solution, so we have the $y(t)$ solution.

What is x'? Well, $x' = y$, $x' = Ce^{-t}$. What is x? If x' is Ce^{-t}, we can integrate both sides to get x. If we integrate both sides, we get that $x(t)$ is the integral of e^{-t} is $-e^t$, so it's C - e^{-t})plus some arbitrary constant, call it D. We have x; $x = -Ce^{-t} + D$. Think about it. What is y in relationship to x? So y was Ce^{-t}, x was $-Ce^{-t} + D$, so $y = -x + D$. All of our other solutions lie along the straight line with slope -1 and having y intercept some constant, D. All of our other solutions are straight-line solutions, but they all involve an e^{-t} so they're decaying. They're tending down to one of the equilibrium points lying on the x-axis.

There is another way of seeing that. We had the matrix $(0, 1, 0, -1)$, and one of the eigenvalues was -1. You could go and solve for the eigenvector for -1, and what you get is the vector $(1, -1)$. The vector $(1, -1)$ is the eigenvector corresponding to the eigenvalue, -1. One way to check that is multiply your matrix times the vector, and see if you get -1, the eigenvalue, times that vector. Take the matrix $(0, 1, 0, -1)$. Multiply it by the vector $(1, -1)$. You get the vector $(-1, 1)$. You get the vector that's basically the negative of the vector you started with, so yes, our matrix times $(1, -1)$ is -1 times the vector $(1, -1)$. That says that $(1, -1)$ is an eigenvector.

Let's see what these solutions look like in the phase plane. Here's our matrix, $(0, 1, 0, -1)$. Here're the straight-line solutions. We have a straight-line solution along the x-axis. That corresponds to the eigenvector corresponding to the eigenvalue 0, and we saw that they were all equilibrium points. Notice the vectors above point in, and the vectors down below point in as well.

Then secondly, we saw that the vector $(1, -1)$, right here, was an eigenvector corresponding to the eigenvalue -1, and notice we have a straight-line solution which is pointing into the origin, another straight-line solution pointing into the origin. As we just saw, all other solutions lie along the $y = -x + D$. All other solutions lie on a straight line coming into one of the equilibria on the *x*-axis. All other solutions are straight-line solutions not tending to 0, but tending to our equilibrium point, a very different kind of phase plane compared to what we've seen. When you have 0 eigenvalues, things are a little different.

We saw a straight line of equilibrium points with all solutions coming into it. It's entirely possible to have a straight line of equilibrium points with all solutions leaving it. For example, look at the matrix (1, 0, 1, 0). What are the eigenvalues here? This is a triangular matrix. The upper right hand corner is 0, so our eigenvalues are displayed along the diagonal. The eigenvalues are 1 and 0. We have an eigenvalue 0. We should get a straight line of equilibrium points corresponding to that eigenvalue, and then we have another eigenvalue that's 1, so we should have another straight-line solution that's leaving the origin.

Let's look and see what happens in that phase plane. Here are the straight-line solutions. We see a straight line leaving along the line $y = x$. That is because the vector (1, 1) is our eigenvector corresponding to the eigenvalue 1. Check that. Take your matrix (1, 0, 1, 0), multiply it by (1, 1), and guess what. You get (1, 1), so matrix times the vector (1, 1) is 1 times the vector (1, 1). That's our eigenvector.

If you check, the eigenvector corresponding to the eigenvalue 0 is anything that's nonzero and lying along the *y*-axis, like 0 and 1, and now, because our second eigenvalue, 1, is positive, solutions all run away on straight lines from one of those equilibrium points. When you have 0 eigenvalues, it's possible to have the other eigenvalue come in or the other eigenvalue go away, in which case, you've got solutions that do the same thing. The third possibility is when you have repeated 0 eigenvalues or repeated real eigenvalues.

That's what's going to come up next time, but also next time, we're going to see lots of interesting bifurcations. As a prelude, let me look and see what happens when we consider a family of systems of differential equations. For example, let's look at the linear system given by the matrix, which I've displayed right here, let's say −1, 0, 1, and let's call it *A*, so our matrix is (−1, 0, 1, *A*). *A* is a parameter. What's happening here? As we change *A*, we're going to get a different family of linear systems. What happens in each member of those families?

What are our eigenvalues? Our eigenvalues are -1 and A because this is a lower triangular matrix. Eigenvalues are displayed along the diagonal. Eigenvalues are -1 and A. If A is positive, we have one eigenvalue that's negative, and another eigenvalue that's positive. We have a saddle. If $A = 0$, we have a 0 eigenvalue situation, and a second eigenvalue, -1, and if A is negative, we have two negative eigenvalues, so we have a sink.

You see a bifurcation in this family as A passes through 0. For any A that's positive, a saddle, for any A that's negative, a sink, and the way you get there is by going to the 0 eigenvalue line. That's a bifurcation line. Let's look at that.

For example, if A is positive, we saw we had a saddle. Here I've chosen A to be 1, and you see a saddle phase plane. For any A that's positive, we always get a saddle phase plane, for example, if A is 0.1, then again a saddle phase plane. Because A is very small, these other solutions are coming in very close to our solution that's leaving the origin. When $A = 0$, that's where we get the bifurcation. These solutions actually come into the equilibrium points that lie along the 0 eigenvector, the y-axis. Finally, when A becomes negative, say -0.2, we undergo a bifurcation. All solutions come into the origin. All of the sudden, all of our solutions tend to 0. We have negative eigenvalues; we have a sink.

This is what we're going to try to do when we summarize all of linear systems. We want to see where these bifurcations occur and how they occur. That will be what we call the trace determinant plane. That's the summary of all linear systems which comes up next.

Summarizing All Possible Linear Solutions
Lecture 14

We have solved almost every type of linear system. The only remaining case is that of repeated eigenvalues. The easy case in this situation is when our system assumes the form

$$Y' = \begin{pmatrix} a & 0 \\ 0 & a \end{pmatrix} Y.$$

We see immediately that the eigenvalues are both given by a, and any nonzero vector is an eigenvector since

$$\begin{pmatrix} a & 0 \\ 0 & a \end{pmatrix} \begin{pmatrix} x \\ y \end{pmatrix} = a \begin{pmatrix} x \\ y \end{pmatrix}.$$

Therefore all solutions are straight line solutions (assuming a is nonzero) either tending to the origin (if $a < 0$) or tending away (if $a > 0$).

Here is the phase plane when $a = 1$.

Figure 14.1

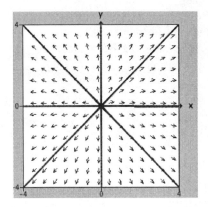

This case, however, is unusual. Most often we find that the eigenvectors lie along a single straight line. To find the other solutions, consider the system given by

$$Y' = \begin{pmatrix} 0 & 1 \\ -1 & -2 \end{pmatrix} Y.$$

It is easy to check that the eigenvalues here are -1 (repeated), and there is a single eigenvector given by $(1, -1)$. So we have one straight line solution given by

$$Y_1(t) = k_1 e^{-t} \begin{pmatrix} 1 \\ -1 \end{pmatrix}.$$

To get the other solutions, recall that we have actually seen this example before; it is the linear system corresponding to the mass-spring system determined by the equation $y'' + 2y' + y = 0$. We saw earlier that this was the critically damped harmonic oscillator whose general solution was given by $y(t) = k_1 e^{-t} + k_2 t e^{-t}$. As a system, our solution is then given by

$$Y(t) = \begin{pmatrix} y(t) \\ v(t) \end{pmatrix} = \begin{pmatrix} k_1 e^{-t} + k_2 t e^{-t} \\ -k_1 e^{-t} - k_2 t e^{-t} + k_2 e^{-t} \end{pmatrix},$$

which we can rewrite as

$$Y(t) = k_1 e^{-t} \begin{pmatrix} 1 \\ -1 \end{pmatrix} + k_2 t e^{-t} \begin{pmatrix} 1 \\ -1 \end{pmatrix} + k_2 e^{-t} \begin{pmatrix} 0 \\ 1 \end{pmatrix}.$$

Note that the eigenvector $(1, -1)$ appears twice in this expression. But there is another term involving

$$k_2 e^{-t} \begin{pmatrix} 0 \\ 1 \end{pmatrix}.$$

We easily check that the vector (0, 1) is actually the solution of the equation

$$(A - \lambda I)\begin{pmatrix} x \\ y \end{pmatrix} = \begin{pmatrix} 1 \\ -1 \end{pmatrix},$$

where λ is our known repeated eigenvalue. In fact, this is the prescription for finding the general solution for a linear system with repeated eigenvalue λ and a single eigenvector. So the general solution is

$$Y(t) = k_1 e^{\lambda t} V + k_2 t e^{\lambda t} V + k_2 e^{\lambda t} W$$

where we have

$$(A - \lambda I)W = V.$$

The phase plane for the above system now has only one straight line solution, while all other solutions also tend to the origin since the terms e^{-t} and te^{-t} both tend to zero as time goes on.

Figure 14.2

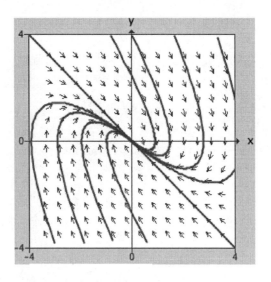

Indeed, as seen above, all other solutions tend to the origin in a direction tangential to the straight line solutions. The reason for this is that each of the 3 terms in the solution involves either e^{-t} or te^{-t}. By calculus, as t tends to infinity, e^{-t} tends to zero much more quickly than te^{-t}, forcing all solutions to come into the origin tangent to the line containing the eigenvector.

Now let's summarize the entire situation for linear systems by painting the picture of the trace-determinant plane. For a linear system of the form

$$Y' = \begin{pmatrix} a & b \\ c & d \end{pmatrix} Y,$$

we have seen that the characteristic equation is given by

$$\lambda^2 - T\lambda + D\lambda = 0.$$

Here the trace T is given by $a + d$, and the determinant D is $ad - bc$. Then the eigenvalues are given by

$$\frac{T \pm \sqrt{T^2 - 4D}}{2}.$$

Using this formula, we can summarize the situation for linear systems in writing as follows.

1. The system has complex eigenvalues if $T^2 - 4D < 0$. So we have

    a. a spiral sink if $T < 0$,

    b. a spiral source if $T > 0$, and

    c. a center if $T = 0$.

2. The system has real, distinct eigenvalues if $T^2 - 4D > 0$. So we have

   a. a saddle if $D < 0$,

   b. a real sink if $D > 0$ and $T > 0$, and

   c. a real source if $D > 0$ and $T < 0$.

3. The system has a single zero eigenvalue if $D = 0$ and $T \neq 0$.

4. The system has repeated eigenvalues if $T^2 - 4D = 0$.

We can also summarize this pictorially by drawing the trace-determinant plane. In this figure, the horizontal axis is the trace axis and the vertical axis is the determinant axis. Each matrix then corresponds to a point in this plane given by (T, D). In the figure below, the different regions then correspond to places where our linear system has different phase planes. The parabola given by $T^2 - 4D = 0$ represents the matrices that have repeated eigenvalues; the trace axis $D = 0$ is where we have a zero eigenvalue; and the positive determinant axis is where we have a center. The regions between these 3 curves are where the other types of phase planes occur.

Figure 14.3

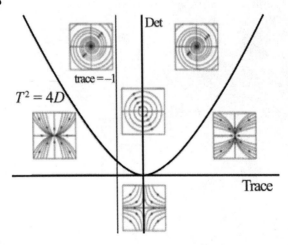

This figure also represents the bifurcation diagram for linear systems. For if we have a family of linear systems that depends on a parameter k, then the corresponding curve that this family produces in the trace-determinant plane shows how and when bifurcations occur. For example, consider the family of linear systems given by

$$Y' = \begin{pmatrix} 0 & k \\ -1 & -1 \end{pmatrix} Y.$$

The trace here is -1, and the determinant is k. So this family of systems lies along the vertical line given by trace $= -1$ and the determinant increasing. We see that this family starts out as a saddle, and then crosses the zero eigenvalue line and becomes a real sink. We eventually find repeated eigenvalues followed by a spiral sink.

Suggested Reading

Blanchard, Devaney, and Hall, *Differential Equations*, chaps 3.5 and 3.7.

Hirsch, Smale, and Devaney, *Differential Equations*, chaps. 3.3 and 4.1.

Strang, *Linear Algebra and Its Applications*, chap. 5.

Strogatz, *Nonlinear Dynamics and Chaos*, chap. 5.2.

Relevant Software

Blanchard, Devaney, and Hall, *DE Tools*, Linear Phase Portraits.

Problems

1. **a.** Sketch the direction field for the system

$$Y' = \begin{pmatrix} 2 & 0 \\ 0 & 2 \end{pmatrix} Y.$$

   b. What are the eigenvalues for this matrix?

   c. Find all possible eigenvectors for this matrix.

   d. What is the general solution of this system?

   e. How do typical solutions of this system behave?

2. Find the general solution of the linear system

$$Y' = \begin{pmatrix} -1 & 0 \\ 1 & -1 \end{pmatrix} Y$$

   and sketch the phase plane.

3. Find the general solution of the linear system

$$Y' = \begin{pmatrix} 0 & 0 \\ 1 & 0 \end{pmatrix} Y$$

   and sketch the phase plane.

4. **a.** Sketch the curve in the trace-determinant plane that the following family of linear systems moves along as a changes.

$$Y' = \begin{pmatrix} a & a \\ 1 & 0 \end{pmatrix} Y$$

b. Find all a-values where this system undergoes a bifurcation (i.e., the type of the equilibrium point at the origin changes).

5. Repeat problems 4a and 4b for the system

$$Y' = \begin{pmatrix} a & 2 \\ -2 & 0 \end{pmatrix} Y.$$

Exploration

Consider the 3-parameter family of linear systems given by

$$Y' = \begin{pmatrix} a & b \\ c & 0 \end{pmatrix} Y.$$

First fix $a > 0$ and describe the analogue of the trace determinant plane (i.e., sketch the regions in the b-c planes where this system has different types of equilibria. Then repeat this question for $a = 0$ and $a < 0$. Put all of this information together to create a 3-dimensional model that describes all possible behaviors of this system.

Summarizing All Possible Linear Solutions
Lecture 14—Transcript

Welcome back. I hope you're surviving all this linear algebra, one more section on linear systems to go. What I'd like to do today is get the big picture of what's happening for linear systems, namely the trace-determinant plane. We'll put everything we've seen together in this one picture.

Before we do that, we've solved almost all linear systems. There's one kind of linear system that we haven't solved yet. That's the case of repeated eigenvalues. That's what we'll do first. It turns out there are two cases of repeated eigenvalues, the easy case and the difficult case. Let me start with the easy case.

Let's take the differential equation y' is matrix $(b, 0, 0, b)y$. Clearly, the eigenvalues are displayed down the diagonal. We have a diagonal matrix, upper and lower triangular. We have repeated eigenvalues, b. What are the eigenvectors? Look at this. Take that matrix, $(b, 0, 0, b)$, and multiply it by any vector whatsoever. What do you get? If the vector is (x, y), multiply by $(b, 0, 0, b)$, you get (bx, by). You get b times that vector.

What does that say? That says that any nonzero vector is an eigenvector corresponding to this repeated eigenvalue, b. That means that all solutions lie on straight lines, not just those two special ones that we saw in the case of real sink sources and saddles. In particular, if b is negative, then we know our eigenvalue is negative. We have an e^{bt}. Solutions should run into the origin along that straight line, and if b is greater than 0, solutions should run away.

Let's look at the phase plane in this particularly simple case. Here is the case where I've chosen b to be 1. My matrix is $(1, 0, 0, 1)$, the identity matrix, and now every solution runs away from the origin along a straight line, a very special case. Every vector is an eigenvector. Had I chosen b to be negative, say -1, then we have eigenvalues that are repeated, -1. It's a diagonal matrix, so all solutions run into the origin along those straight lines. There's a very simple case of repeated eigenvalues.

Let's go into the more complicated case. Let me take the system $y' = (b, 1, 0, b)y$. I've replaced that 0 in the upper right hand corner with a 1. Nonetheless, our matrix is still upper triangular. The eigenvalues, b, are repeated. We have to find eigenvectors for that eigenvalue, b.

First, notice that one eigenvector is $(1, 0)$. If you take that matrix, $(b, 1, 0, b)$, and multiply it by the vector $(1, 0)$, you get the vector $(b, 0)$. That is, you get the vector b times the vector $(1, 0)$. That says that $(1, 0)$ is an eigenvector for this repeated eigenvalue, but it turns out that that's the only eigenvector that you can find. What to do?

How to find the rest of the solutions? Let's go back to our original equation, write it out in longhand, it's $x' = bx + y$, and $y' = by$. Look at that second equation, $y' = by$. We know how to solve that; $y(t)$ must be some constant times e^{bt}. We have $y(t)$. That says that our first equation is now $x' + bx + y$, but y is $k_1 e^{bt}$. Our first equation is first order, linear nonhomogeneous once again. It's $x' = bx + k_1 e^{bt}$.

We know how to solve these first-order linear nonhomogeneous equations. We first get the general solution of the homogeneous. That is $k_2 e^{bt}$. We have the general solution of the homogeneous is $k_2 e^{bt}$, and we need one solution of the nonhomogeneous. The term to the right involves e^{bt}. The usual situation is to guess a solution, and your first guess would be, let me guess a constant times e^{bt}, say Ee^{bt}. That isn't going to work because we know that's a solution of the homogeneous equation. We've seen this situation before. What we've seen is when your natural guess for a solution of the nonhomogeneous doesn't work, then the next best guess is to guess t times your guess.

What we'll do is guess Ate^{bt}, that's our guess for the solution of the nonhomogeneous equation, plug it into the equation $x' = bx + k_1 e^{bt}$, and after a little bit of algebra, and a bunch of product rules, you'll find that $a = k_1$. We've done exactly this when we looked at the critically damped harmonic oscillator several lectures ago.

We've determined one solution of the nonhomogeneous. It's $k_1 t e^{bt}$, so we have our general solution of the equation. Our general solution is $x(t) = k_2 e^{bt}$, that's the general solution of the homogeneous, plus our now one particular solution of the nonhomogeneous, $k_1 t e^{bt}$. That's our $x(t)$ solution, and remember our $y(t)$ solution was $k_1 e^{bt}$. There's our solution. You can check pretty easily that this is the general solution of this differential equation.

What do we do in general? Look at this solution to our repeated eigenvalue differential equation. Let me break it up into a bunch of different little pieces. Our first equation is $x(t) = k_2 e^{bt}$. Let me write that first piece as $k_2 e^{bt}$ times the vector, $(1, 0)$. That vector, $(1, 0)$, was our eigenvector. There's a second piece up there. It's $k_1 t e^{bt}$. Let me write that as $k_1 t e^{bt} (1, 0)$, our eigenvector again. Then we have a third term down in the y term, that's $k_1 e^{bt}$. Let me write that as $k_1 e^{bt}$ times the vector, $(0, 1)$. In matrix form, using vectors, we see that our solution is $k_2 e^{bt}$ times the eigenvector, plus $k_1 t e^{bt}$ times this vector, plus $k_1 e^{bt}$ times this vector, $(0, 1)$.

What is that vector $(0, 1)$? It turns out that that vector $(0, 1)$ is what's sometimes called a generalized eigenvector. The vector $(0, 1)$ is a solution, let me call it x, of the equation matrix A minus eigenvalue i times x equals, not $(0, 0)$ like we did in the case of finding eigenvectors, but now equals our old eigenvector, namely $(1, 0)$. That vector, $(0, 1)$ is a solution to a somewhat different linear algebraic system of equations. It's $(A - \lambda i)xy = (1, 0)$.

How do we get that? We have to solve that equation. Our matrix was $(b, 1, 0, b)$. We subtract off our eigenvalues, b, down the diagonal. That leaves us with matrix $(0, 1, 0, 0)$, so the linear system of algebraic equations we have to solve is $(b - b)x + y = 1$ and $(b - b)y = 0$. The first equation says that $y = 1$, and the second equation says that, as you well know, $0 = 0$. Any vector with $y = 1$ is a solution to that special system of linear algebraic equations.

That, in fact, is the general method of solving linear systems when you've got repeated eigenvalues, b. Your general solution will be of the form $y(t)$ is some constant, k_2, times e^{bt} times the eigenvector that you can find, then plus $k_1 t e^{bt}$ times that same eigenvector, and then plus $k_1 e^{bt}$ times the vector, say W, where W is the solution of the linear algebraic equation $(A - \lambda i)W$ equals the eigenvector, V. Let me not go through the details there. They are very similar to what we've seen many times in the past.

Instead, let me go to the computer and see what the solutions look like for this case, the special case, of repeated eigenvalues. Here is our matrix. I've chosen b to be 1, so our matrix is (1, 1, 0, 1). We get only one eigenvector here. That eigenvector lies along the x-axis, say (1, 0), and what we see is we have a solution that's running away from the origin corresponding to the eigenvalue 1 whose eigenvector lies on the x-axis.

What happens to all other solutions? All other solutions involve either an e^{1t} or a $t(e^{1t})$. All of those other solutions must also tend away from the origin, te^{bt} goes off to infinity as time goes on, and in fact, that's what we see. All other solutions leave the origin and tend off to infinity. In fact, what happens, is all these other solutions, as in the case of a real source, leave the origin tangentially to the direction of our one eigenvector.

If I were to choose the matrix, not (1, 1, 0, 1), but let me put in −1, and −1 so my matrix is (−1, 1, 0, −1). Eigenvalues are displayed along the diagonal. They're both −1, so we have repeated eigenvalues with one eigenvector. However, solutions all come into the origin. We have an e^{-t} term, e to the eigenvalue t. That's e^{-t}. We've also got a te^{-t}. As time goes on, e^{-t} goes to 0 and t goes off to infinity, but remember we worked with that on l'Hopital's rule a while before. We see that te^{-t} also goes to 0. All solutions actually come into the origin in the case of this repeated eigenvalue −1.

That finishes off solving all possible linear systems of differential equations. What I'd like to do now is try to summarize the entire situation. Given the linear system, $y' = (a, b, c, d)y$, what does the dictionary of all possible solutions look like? We know that that matrix depends on 4 parameters, so you might think that the parameter space is a 4-dimensional space, but it's really only two dimensional because remember our characteristic equation depended on the trace and the determinant. Our characteristic equation was $\lambda^2 - T\lambda + D = 0$. Recall the trace is the quantity $a + d$, and the determinant is the quantity $ad - bc$. From that characteristic equation, we found both of our eigenvalues. Our eigenvalues depended only on the trace and the determinant. They were T, that's trace, $\pm\sqrt{T^2 - 4D}$, D is the determinant, all divided by 2. Let me summarize this situation two ways.

267

First, in words, and then as a picture. We have our eigenvalues, $(T \pm \sqrt{T^2 - 4D})/2$. What's the phase plane for various values of T and D? First, we have the square root of $T^2 - 4D$. If that $T^2 - 4D$ is negative, we have complex eigenvalues, and we saw three special cases. If $T^2 - 4D$ is negative, we have a complex eigenvalue whose real part is $T/2$. If T is less than 0, if the trace is negative, we have a spiral sink. If the trace is positive, we have a spiral source. If T is equal to 0, our eigenvalues are pure imaginary. We have a center. That's case 1, when $T^2 - 4D$ is negative.

What about when $T^2 - 4D$ is positive? If $T^2 - 4D$ is positive, in the numerator, we have $T \pm$ the square root of $T^2 - 4D$. We have two real eigenvalues. What's the phase plane? Assuming $T^2 - 4D$ is positive, if D is negative, then we have a saddle. Why is that? Look inside the square root sign. You've got $T^2 - 4D$, remember D is negative, so we have T^2 plus something positive. What's inside the square root sign is bigger than T^2, so our eigenvalues are T plus something bigger than T, and then T minus something bigger than T. The first eigenvalue, T plus the square root is positive, but the second eigenvalue T minus it, is negative. We have a positive and a negative eigenvalue. In that case, we have a saddle.

If the determinant is less than 0, and $T^2 - 4D$ is positive, saddle. What if the determinant is positive? If the determinant is positive, inside the square root sign we have $T^2 - 4D$ minus something. We have T^2 minus something that is going to make that inside the square root smaller. What do we have? Our eigenvalues are T plus the square root, that's positive, and now we've also got T minus something that's smaller than T, so that's also positive if T is positive. In this case, we have two real and positive eigenvalues. We have a real source. Then the exact opposite is true if D is positive and T is negative, you've got two real negative eigenvalues. You've got a real sink.

What other cases come up? Notice that if the determinant equals 0, and the trace is not 0, then one of our eigenvalues equals 0. If the determinant equals 0, then under the square root, we have the square root of T^2, that's T, so we have $T \pm T$. One of them is going to cancel out and give us a 0 eigenvalue. When the determinant is 0 and the trace is not equal to 0, we have a 0 eigenvalue. The final case, the one we just did, what happens if $T^2 - 4D = 0$? If $T^2 - 4D = 0$, our eigenvalues are $T/2$. They're repeated.

There's a verbal summary of what's happening for linear systems of differential equations. What I want to do now is give you a global picture of what's going on for solutions of linear systems of differential equations. This is what we call the trace-determinant plane. I'm going to plot the trace horizontally, and the determinant vertically. Then, given the matrix, I'll compute its trace and its determinant. That will give me a point in the trace-determinant plane. What I'll do now is draw a picture of what we have talked about, all possible situations for different systems of differential equations.

For example, we know that if $T^2 - 4D = 0$, there we have repeated eigenvalues. In a trace-determinant plane, $T^2 - 4D = 0$, that's $T^2 = 4D$, $4D = T^2/4$. Think of D as going vertically, as being y, T as going horizontally, that's x. That would say in y, x terms y is $x^2/4$. That is, the curve $T^2 - 4D = 0$ is actually a parabola. What I'll first do is draw that parabola in the trace-determinant plane. If my matrix, if it puts me on the parabola, we have repeated eigenvalues.

The next special case was when the determinant equals 0. If the determinant equals 0, then we've seen we have 0 eigenvalues. Determinant equals 0 is the trace access, so that gives us another curve where we have special eigenvalues. The parabola $T^2 - 4D = 0$ repeated eigenvalues. The trace access determinant equals 0 gives us 0 eigenvalues.

All those other regions in between are places where those different types of phase planes arose, so let's go and look at this picture. Here is a picture of the trace-determinant plane on the left. Right here, you see a parabola running down. That's the parabola $T^2 - 4D = 0$. Horizontally, you see the trace axis. That's running right through the middle of the picture. That's the trace axis where the determinant equals 0, and vertically, right in the middle of the picture, is the determinant axis. All of these other regions are regions where we have specific kinds of linear systems of differential equations.

For example, in our previous summary, we saw that if $T^2 - 4D$ were negative, then we would have complex eigenvalues. Here's the curve $T^2 - 4D = 0$. If I'm above that curve, then $T^2 - 4D$ is negative, along the determinant axis. The determinant up here would be positive, T would be 0, and

$T^2 - 4D$ would be negative. The entire region above this parabola is where we have complex eigenvalues.

We saw that if the trace were negative, we have a spiral sink. If I choose any matrix that gives me a trace-determinant that lies in this orange region here, we know we have a spiral sink. For example, if I choose a matrix, let me choose the matrix (0, 1), and then these two points. That lies in the region where we have a spiral sink, and here is the phase plane. Solutions all spiral in no matter where we choose this matrix, we see solutions spiraling into the origin. Sometimes they spiral in very quickly so that you don't see it. It doesn't look like that's spiraling, but if I were to magnify that, you'd see, in fact, that it is spiraling.

Another region, if $T^2 - 4D$ is negative, so we're above the parabola but T is positive, then we know we have a real source. In this region here, any matrix that puts us there gives us a spiral source. For example, if I choose that matrix, the matrix (0, 1), and then these two values, you can check that the trace is positive. $T^2 - 4D$ is negative. We get a spiral source, all solutions spiral away from the origin. All solutions leave the origin spiraling as they do. The region in between, that's the determinant axis; that's where we have a center. This is a little funny in this program. The picture comes out a little bit wrong for reasons that I won't get into. What we get are solutions that circle around the origin periodically. All matrices drawn along the determinant axis have centers for their phase plane.

What happens if $T^2 - 4D$ is positive? Three different regions here, we saw that if $T^2 - 4D$ is positive, and the determinant was less than 0, that means we're down in this blue region here. There we have a saddle. Let me choose a matrix there, and now this matrix, with its given trace and determinant is a saddle. You see solutions looking like one solution in one direction, like another solution in the other direction, and in fact, the eigenvectors are right there. There is a typical saddle picture. We always get a saddle when, in the trace-determinant plane, we're below the trace axis where the determinant is negative.

We finally saw that if $T^2 - 4D$ is positive, but trace is negative, then we're over here in this region here, and in this region, what we have is a real sink. Notice that all solutions now come into the origin. That region to the left is a region where we always have a real sink. Then finally, when $T^2 - 4D$ is positive but T is also positive, we're in this region here. We have, you guessed it, a real source. That green region is the region where we see real sources.

There are a couple of special cases left. That's when we're on the parabola $T^2 - 4D = 0$ or on the trace axis. That's where our special kinds of differential equations arise. For example, let me see if I can get it on that parabola. Yes, there I've chosen a point, a matrix, that puts me on that parabola. We see that we have one straight line of solutions leaving the origin, and all other solutions leave tangential to it. This is the case of repeated eigenvalues.

Were I to go over to this region here and choose something on the repeated eigenvalue curve, now since above us we have a spiral sink and below us we have a real sink, this is where we have repeated eigenvalues that are negative. All solutions come in to the origin.

Finally, right down here on the trace axis is where we saw the determinant was 0. There we have a 0 eigenvalue. The other eigenvalue is negative. All solutions come into one of those equilibrium solutions lying along the eigenvector corresponding to 0. Same thing to the right, I can find a point on that trace axis, then I get a 0 eigenvalue where the other one is positive, so solutions run away from the eigenvector corresponding to the eigenvalue 0.

Not only is this a summary of everything that's going on in the set of linear systems of differential equations, it also shows us where bifurcations occur. Let's take a little trip around this region. Let me follow this rectangular path in the trace-determinant plane, and then plot the corresponding phase planes. When I do, we move pretty quickly, but you're seeing a representative collection of all the different possible phase planes.

You might see spiral sinks and sources to begin with, now a saddle, and then back to where we started. Let's do that a little bit more slowly. We start out on the repeated eigenvalue curve and then move into the orange region where we have a spiral sink. Then suddenly, boom, we hit the determinant axis. We have a change, a bifurcation. We have a center. Then keep going and immediately transition to spiral source, and then eventually hit the repeated eigenvalues line. Notice there's one straight-line solution here. Keep going, now descend, and we're into the real source region. Then we hit the 0 eigenvalue line, the trace axis, and then finally we're in the region where we have a saddle. Come back up to the trace axis, and then into the region where we have a real sink. This is a nice record of not only all possible phase planes, but also what's going on, what's the bifurcations that are happening.

Let me end with a little quiz. Here's a little animation. What trail are we following in the trace-determinant plane with this sequence of phase planes? It is a little hard to see. Let me do it again, a saddle there, and looks like we end up with a spiral sink. We start out with a saddle, keep going, saddles, then boom, we've hit the 0 eigenvalue line. It looks like all solutions are coming into the origin in real sink, and now we're in the spiral sink region. Where have we gone? It looks like we started out with a saddle, moved vertically up through the real sink region, and into the spiral sink region.

One last one for you to contemplate, what happens here? What trek am I taking now? We started with a real sink, then spiral sink, then spiral source, and then real source. Unlike the previous example where we went vertically, this time we're going horizontally above the trace axis. The point is, in the trace-determinant plane, when you draw any curve, you're going to see bifurcations exactly when you cross those special lines, $T^2 - 4D = 0$ and the trace axis. That's a summary of the bifurcations in linear systems, another trek back to bifurcations.

Next time, we're done with linear systems, we'll move on to nonlinear systems of differential equations.

Nonlinear Systems Viewed Globally—Nullclines
Lecture 15

We now move on to nonlinear systems of ordinary differential equations. In this lecture we introduce one of the main techniques that we use to understand the behavior of solutions of nonlinear systems. This tool, called nullclines, allows us to partition the phase plane into certain regions where we know the approximate direction of the vector field.

Given a nonlinear system $x' = F(x, y)$ and $y' = G(x, y)$, the x-nullcline will be the set of points at which $x' = 0$—that is, $F(x, y) = 0$. At these points, the vector field points vertically, either up or down. Similarly, the y-nullcline is the set of points where $y' = 0$, so $G(x, y)$ vanishes. At these points, the vector field points horizontally, either to the left or the right. Note that the equilibrium points are then given by the points of intersection of the x- and y-nullclines.

Most importantly, in the regions between the nullclines, assuming our vector field is continuous, the vector field must point in 1 of 4 directions: northeast, northwest, southeast, or southwest. This very often allows us to get a good handle on the qualitative behavior of nonlinear systems.

Let's start with a simple nonlinear system.

$x' = y - x^2$
$y' = 1 - y$

The x-nullcline is the parabola $y = x^2$. The vector field is vertical along this parabola. It points to the right above the parabola since $x' > 0$ in this region, and it points to the left below the parabola. The y-nullcline is the horizontal line $y = 1$, so we immediately see equilibrium points at $(-1, 1)$ and $(1, 1)$. Moreover, the vector field is tangent to this nullcline. So we have solutions that move right along this line in the region above the parabola and to the left on the portions below the parabola. We also have that $y' < 0$ above the line $y = 1$ and that $y' > 0$ below this line. Therefore we know the direction of solutions in all the regions between the nullclines.

If a solution starts in the region above the parabola but below the line $y = 1$, it must proceed up and to the right. It cannot cross the line $y = 1$, since we have a solution there and solutions cannot cross by the **existence and uniqueness theorem**. It also cannot hit the parabola because to do so, it must turn and become vertical. But the vector field always points upward here, so this is not possible. So solutions in this region must tend to the equilibrium point at $(1, 1)$. The same thing happens in the region to the right of the parabola and above $y = 1$; all these solutions tend to the same equilibrium point. In the left region below the parabola but above $y = 1$, solutions must tend to infinity.

In the region above the parabola and $y = 1$, solutions have 4 choices: They can tend directly to $(1, 1)$. They can cross the right branch of the parabola but then enter the region where all solutions tend to $(1, 0)$. They can hit the left branch of the parabola, in which case they tend to infinity. Or they can tend to the equilibrium point at $(-1, 1)$. In the lower region, solutions similarly have 4 possible choices, but we know more or less what happens to all solutions. It appears that $(1, 1)$ is a sink and $(-1, 1)$ is a saddle. But the latter is not clear: Can we have more than a single curve tending to that equilibrium point? We'll see how to handle this in the next lecture.

Let's go back to the predator-prey system

$x' = x(1 - x) - xy$

$y' = -y + axy.$

Here we first choose $0 < a < 1$, so the rate of predation is small. Recall that the computer showed us that all solutions tended to the equilibrium point at $x = 1, y = 0$.

Figure 15.1

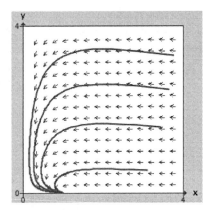

The x-nullcline will be the set of points in the phase plane where the x-component of the vector field vanishes. So the x-nullcline is the set of points where the vector field points vertically. In this example, the x-nullcline is given by $x(1-x) - xy = 0$, so that $xy = x(1-x)$. Thus the x-nullcline consists of the pair of straight lines $x = 0$ and $y = 1 - x$. Similarly, the y-nullcline is the set of points where $y' = 0$ (i.e., where the vector field points horizontally). In this case, the y-nullclines are the lines $y = 0$ and $x = 1/a$.

Figure 15.2

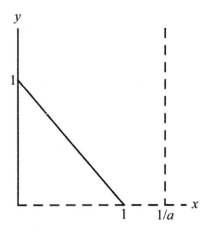

In this figure, the solid lines are the *x*-nullclines and the *y*-nullclines are dashed. As before, the places where an *x*- and a *y*-nullcline intersect are the equilibrium points for the system. So in this case, the equilibria are (0, 0) and (1, 0). Note also that the *y*-nullcline $x = 1/a$ and the *x*-nullcline $y = 1 - x$ meet at a point where the *y*-coordinate is negative. This means that this is not an equilibrium point for the predator-prey system, since both *x* and *y* are nonnegative.

Most important, at all points in one of the regions between the nullclines, the vector field always points in 1 of 4 directions: northeast, northwest, southeast, or southwest. And we can determine this direction by simply computing the vector field at a single point in the region. For example, at (1, 1), our vector field is given by $x' = -1$, $y' = -1 + a < 0$; so the vector field points down and to the left at all points between the lines $y = 1 - x$ and $x = 1/a$. In the triangle bounded by $y = 1 - x$ and the *x*- and *y*-axes, the vector field points down and to the right. And in the region to the right of $x = 1/a$, the vector field points up and to the left.

Figure 15.3

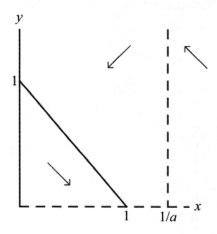

As a consequence, we completely understand what happens to all solutions of this system. Solutions that start to the right of $x = 1/a$ must move up and to the right until they hit the line $x = 1/a$. Then they move down and to the left. So there are only 2 possibilities for what happens to these solutions: Either the solution tends to the equilibrium point at $(1, 0)$ or it crosses the line $y = 1 - x$. (They can never cross the x- or y-axes by the uniqueness theorem.)

In the final case, in the triangular region, solutions move down and to the right. Since these solutions can never cross the x-axis, they must also tend to the equilibrium point at $(1, 0)$. We therefore know that all solutions of this predator-prey system necessarily tend to the equilibrium point at $(1, 0)$, assuming that our initial populations are both positive. Therefore we know that when $a < 1$, the predator population goes extinct.

Nullclines do not always give the complete picture of the phase plane. For example, if $a > 1$ in our predator-prey system, we get the following nullcline picture. Here the nullclines $x = 1/a$ and $y = 1 - x$ do cross at a new equilibrium point $(1/a, (a - 1)/a)$.

Figure 15.4

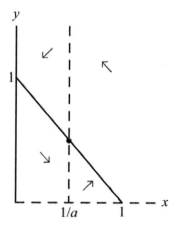

We do know that every solution starting at a point with $x, y > 0$ must eventually enter the region $x \leq 1/a$. But does this solution tend to the new equilibrium point? If so, does the solution spiral into this equilibrium point, or does it not spiral? The computer seems to indicate that both types of behavior can happen. These figures below show the phase plane when $a = 1.2$ (left) and when $a = 2$ (right).

Figure 15.5

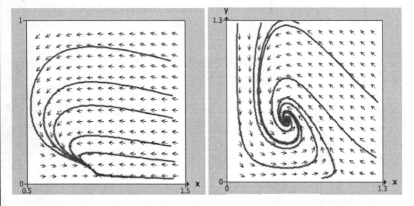

In addition, could the solution move around the equilibrium point in a periodic motion and never come to rest? This kind of local behavior near equilibria will be the subject of the next lecture.

As another example of nullcline analysis, consider the nonlinear system

$$y' = x - y^2.$$

The x-nullcline for this system is the parabola $y = x^2$, which opens upward; the y-nullcline is the parabola $x = y^2$, which opens to the right. These nullclines meet at the 2 equilibria: $(0, 0)$ and $(1, 1)$. In the 5 regions separated by the nullclines, we compute that the vector field points as follows.

Figure 15.6

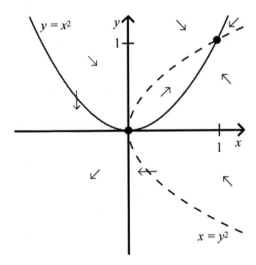

So (1, 1) appears to be a sink and (0, 0) a saddle. We don't know this for sure, but the next lecture will give us the local tool to make this assessment. The actual phase plane for this system is below.

Figure 15.7

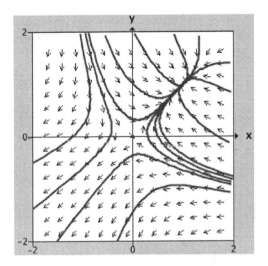

Important Term

existence and uniqueness theorem: This theorem says that, if the right-hand side of the differential equation is nice (basically, it is continuously differentiable in all variables), then we know we have a solution to that equation that passes through any given initial value, and, more importantly, that solution is unique. We cannot have two solutions that pass through the same point. This theorem also holds for systems of differential equations whenever the right side for all of those equations is nice.

Suggested Reading

Blanchard, Devaney, and Hall, *Differential Equations*, chap. 5.2.

Edelstein-Keshet, *Mathematical Models in Biology*, chap. 6.2.

Hirsch, Smale, and Devaney, *Differential Equations*, chap. 9.1.

Strogatz, *Nonlinear Dynamics and Chaos*, chap. 6.1.

Relevant Software

Blanchard, Devaney, and Hall, *DE Tools*, Linear Phase Portraits.

Problems

1. a. Find the x- and y-nullclines for the system

$$x' = y$$
$$y' = 2x.$$

b. Sketch the regions where the vector field points in different directions.

c. What are the equilibrium points for this system?

d. Sketch the direction field and phase plane for this system.

e. How many solutions tend to the equilibrium points?

2. Sketch the nullclines and phase plane in the region $x, y \geq 0$ for the following system.

$$x' = x(-x - 3y + 150)$$
$$y' = y(-2x - y + 100)$$

3. Sketch the nullclines and phase plane in the region $x, y \geq 0$ for the following system.

$$x' = x(100 - x - 2y)$$
$$y' = y(150 - x - 6y)$$

4. Using nullclines, determine the phase plane for the following system.

$$x' = x(1 - x)$$
$$y' = x - y^2$$

5. a. Consider the predator-prey system

$$x' = x(1-x) - Axy$$
$$y' = y(1-y) + xy,$$

where A is a positive parameter. Using nullclines, determine the values of A where different outcomes occur. In particular, determine at which parameters this system undergoes a bifurcation.

b. What changes will there be in the predator-prey system if we introduce a new parameter of B instead of A?

$$x' = x(1-x) - xy$$
$$y' = y(1-y) + Bxy$$

Exploration

Use the nullclines to investigate the phase plane for the following system that depends on a parameter A.

$$x' = x^2 - 1$$
$$y' = -xy + A(x^2 - 1)$$

At what values of A do you see a bifurcation? Explain what happens to solutions before and after the bifurcation.

Nonlinear Systems Viewed Globally—Nullclines
Lecture 15—Transcript

So, hello again. Now we move onto the next phase of this course, namely, nonlinear systems of differential equations. As I remarked earlier, most of the applications of differential equations arise in a setting of nonlinear systems. Unfortunately, these are the differential equations that are most often most difficult to solve. We'll see some techniques to at least approximate solutions as we go along, but first, what is a nonlinear system of differential equations? Well, basically, it's a collection of first order differential equations that have some nonlinear turns in them. For example, an x^2 or a sine y, an exponential sine of tangent x. These are all nonlinear terms, unlike the linear systems that we dealt with over the past few lectures, where only x and y standing alone appeared. Those were linear systems.

For simplicity in this part of the course, we will deal mostly with two dimensional nonlinear systems of differential equations. So that system will be x' at some function of x and y, and y' is some other function of x and y. Now, the systems could be non-autonomous. There could be say T terms in there, but we'll put that off for the most part.

As I said, we'll look at two tools. The first tool will be what we call the nullclines. The nullclines sort of give us the broad view of what is happening with systems of differential equations. In the next lecture, we will come back and look at some the narrow pieces where the nullclines don't work as well. That's the technique of linearization that harkens back to everything we did in the last couple of lectures.

So, let's begin with nullclines. First off, what is the x nullcline? Well, the x nullcline is the set of points in the plane where $x' = 0$, where our first differential equation vanishes. That means, since $x' = 0$, our vector field points vertically. It only has a y component. Secondly, what is the y nullcline? Well, the y nullcline is the set of points where $y' = 0$, where our second differential equation vanishes. Since there is no y component, this means now that the vector field points horizontally. It may point to the right. It may point to the left, whatever. There is no y component.

Now notice that the equilibrium points are the places where the nullclines meet, where both x' and y' vanish. More importantly, if you look at a region between any of these nullclines, then there the vector field can't point vertically, can't point horizontally, so it must point in one of 4 directions, either northeast, northwest, southeast or southwest. So, in all the regions between the nullclines, the vector field points essentially in one direction. That's the geometric technique that we'll use to get the broad picture of what's happening to the solutions of differential equations.

So, let me begin with a simple example. Let's say $x' = y - x^2$; $y' = 1 - y$. Nonlinear system because we have an x^2 term in the first equation. Let's begin with the x nullcline. The x nullcline is where x' vanishes, so where $y = x^2$. That is just a parabola in the xy plane. That is where $x' = 0$, so that's where the vector field points vertically. So, what is the x nullcline? Well, that's where x' vanishes. So that's where $y = x^2$. That's just the parabola in the xy plane. On that parabola, the vector field points vertically. We need to know now where the vector field points in the other regions in the plane, for example, above the parabola. Well, if we are above the parabola, say on the y-axis, then y positive and x is 0. So x' is positive in the entire region above that parabola. Similarly, x' is negative in the entire region below the parabola.

Now for the y nullcline. Well, the y nullcline is given by $1 - y = 0$, or just the horizontal line, $y = 1$. And immediately, we see our two equilibrium points where the x and y nullclines meet. They meet at the point 1,1 and at the point $-1,1$. So we have our equilibrium points.

Now, notice that on the y nullcline, y' is horizontal. So in this particular case, in fact, the vector field is tangent to our y nullcline. It points horizontally either left or right. Which way does it point? Well, above the x nullcline we know x' is positive, so the vector field along the y nullcline, between the two equilibrium points, must point to the right. Similarly, one off to the right or to the left of the x nullcline of the vector field must point to the left. So our solutions run off to the left in either of those directions.

Now again, we need to know the sine of y' off its nullcline. Well, if you're above the y nullcline, then you're above the line of $1 - y = 0$, $y = 1$. There, y'

is negative. If you're below the y nullcline, then y' is positive. So, we know which way the vector field points in all of these regions because we know where x' and y' are positive and negative. For example, if we take the region that lies below both of those nullclines, then we know that first x' must be negative and y' must be positive. So our vector field must point to the left and up everywhere below those two nullclines. On the other hand, if we look in the region between the two nullclines, then we have that y' is positive. The vector field points up, and x' is also positive. So the vector field points up and to the right in that region there. If we look at the region above both of these nullclines, the region inside the parabola and above the line $y = 1$, there we saw that y' is negative, but now, x' is positive. So our vector field points to the right and down. Then finally, in the remaining two regions of the upper regions, to the left and the right, there you can check that the vector field points to the left and down.

So we know which way the vector field points in all the regions between the nullclines. That then gives us a good idea of what happens to solutions. For example, what happens to any solution that starts in the Region A? That's the region between the two nullclines. Well, there we saw the vector field points up and to the right. That means solutions must increase. Now, they can never cross the y nullcline because the vector field is horizontal there, by existence and uniqueness, we cannot have solutions cross. That means our solution must continually go up and to the right. It can never cross the curve $y = x^2$. That's where the vector field is vertical. In order to do that, the solutions would have to turn around and go downhill, but I can't. In the Region A, it's always going up and to the right. Therefore, there is no other possibility but if you start with the solution in A, that solution must tend to the equilibrium point at 1,1.

What about solutions in Region B? Well, essentially the same thing. There we know solutions go down and to the left. Those solutions could never cross the line $y = 1$ because again, we know that the vector field is horizontal there. Again, the solutions could never cross the curve $y = x^2$. There we know the vector field is vertical there. In order to do that, the solutions would have to turn around and go uphill. That can't happen. So that means, in the Region B, again, the only thing solutions can do is tend downward and to the left and eventually end up tending to the equilibrium point at 1,1.

Another region, the Region C, off to the left, what happens there? There we know the vector field points down and to the left, so solutions must continually decrease in the y direction and tend off to the left in the x direction. The only possibility is solutions go to infinity.

Now what happens to solutions in Region D? Well, it turns out there are 4 possibilities for what happens to solutions there. For example, our solution could tend off to the right and hit the parabola $y = x^2$. There, the vector field is vertical, so the solution would go downhill into the other region, where we know that solutions all tend to the equilibrium point at 1,1. A second possibility is our solution could be going downhill and to the right in Region D, but it could end up just landing at the equilibrium point at 1,1. Or, our solution could still be descending and tend to the equilibrium point at $-1,1$. Or, the final possibility is our solution could go down and cross the parabola $y = x^2$ to the left, in which case we enter the Region C and off to infinity. Four possibilities, but we understand what happens in each of those cases.

The final case is what happens in Region E. There we know the vector field goes up and to the left. So again, there are 4 possibilities. Far to the right, our solutions could be going up and to the left and end up at the equilibrium point at 1,1. Or our solutions could increase and enter the Region A, in which case, we know what happens. Then the vector field turns to the right and again, tends to the equilibrium point at 1,1. A third possibility is our solutions could keep increasing and land at the equilibrium point at $-1,1$. Or, the fourth possibility is our solutions could miss that x nullcline and just keep tending off to the left, tending to infinity, as x goes to the left.

So now, let's look at our equilibrium points. It certainly looks like the equilibrium point at 1,1 is a sink. All solutions we saw that came near that equilibrium point tended to it. On the other hand, what about $-1,1$? Well, it kind of looks like $-1,1$ is a saddle, Solution C to come in close to it and then veer off in another direction. But that's a little problematic. That may not be true. Perhaps, we could have a whole collection of solutions tending into that equilibrium point either from above or from below. That means that we don't really have a complete grip on what's happening to solutions of differential equations near the equilibrium. We'll see that the next process, mainly linearization, will cure that for us.

First let me turn to the computer and see what happens to these solutions. Here is the direction field for this differential equation. What we see is solutions that start up high tend to go the equilibrium point at 1,1, a sink, just as we thought. On the other hand, as we move along, eventually these solutions go off in the other direction. Some solutions in the final Region E tend to infinity, others go to the sink at 1,1. If we start from above, sometimes solutions go off to infinity, and other times they go to the sink at 1,1. It sure looks like that equilibrium point at $-1,1$ is a saddle point. But we don't know that for sure. We'll find that out in the next lecture.

Now we can go to another example. Remember we talked for quite a bit about the Predator-Prey system. That was the system x' is given by $x(1-x) - xy$. That was our prey differential equation. The predator differential equation was y' is $-y$ plus, let's put in some parameter, a times xy. So there is the Predator-Prey differential equation that depends on the parameter a. Now remember, a few lectures ago, we saw that it seemed that if a was less than 1, all solutions tended to an equilibrium point at the point $x = 1$, $y = 0$. That is our species y went extinct, but our species x survived. Let me show you that again.

There is the direction field for the Predator-Prey system. Now we see that it appears that all solutions, no matter where you stop, tend to the equilibrium point down here at 1,0. How do we determine that? Well, let's turn to the nullclines. First, what's the x nullcline? Again, that is the set of points where $x' = 0$, so the vector field points vertically. So in this case, the neck of the x nullcline is given by $x(1 - x) - xy = 0$. Or $xy = x(1 - x)$. So that means we have a pair of nullclines in this case: $x = 0$—the y-axis is one nullcline— and the line $y = 1 - x$ is a second nullcline. That's where the vector field points vertically.

What about the y nullcline? Well, the y nullcline is where $y' = 0$. So $y' = -y + xya$. Let me for convenience pick a to be say ½. So the y nullcline is given by $-y + xy/2$. So the y nullclines are given by $y = 0$ (the x-axis), and the line $x = 2$. So we have a pair of nullclines again in this case. The places where the x and y nullclines meet are again the equilibrium points, so we see nullclines meeting at the origin and at that equilibrium point 1,0. The y nullcline given by $x = 2$, also meets the x nullcline, $y = 1 - x$, that straight line, at a point where $y = -1$, which is negative. So that's really not of interest in our population model.

The points now between these nullclines are places where the vector field points in one of those 4 directions, Northeast, Northwest, Southeast, or Southwest. We've got three regions here—the region off to the right where x is greater than two, the triangular region to the left near the origin, and then the region in between. What happens in those regions? For example, we could determine that by picking one point from each of these regions, determining the direction of the vector field at that point, and then we'd know that the vector field points in that same direction at all other points. For example, in that middle region, take that point 1,1 that lies in the middle region. There, $x' = -1$ and $y' = -1 + ½$. So, it's also negative. So x' and y' are negative. Our vector field points down and to the left. In fact, that happens at all points in this region.

What about in the triangular region bounded above by the line $y = 1 - x$ in the x and y-axis? Well, there we could take for example, x and y to be 0.01. Then you compute that at that point, $x' = 0.1(0.9) - 0.01$, which is positive. Similarly, you compute that $y' = -0.1 + 0.1(0.1)(½)$. That's 0.005. That's negative. So now our vector field x' is positive, y' is negative. Our vector field points down and to the right. Then in the final region to the right of say $x = 2$, then x' is given by $x(1 - x)$ for any x in that region. That quantity is negative. Similarly, $-xy$, that term is also negative. So x' is negative. y' is $-y + ½xy$. That's $y(-1 + ½x)$. So in that case, when x is greater than 2, we always have that y' is positive. So y' is positive; x' is negative; our vector field points up and to the left. Now we know what happens to all solutions. If you take a solution that starts to the right of $x = 2$, initially it goes up and to the left. It must eventually hit the line $x = 2$, where it becomes horizontal. It enters the next region where the vector field is pointing down and to the left. Now, that solution may come across and hit the line $y = 1 - x$. If that happens, it becomes vertical and then turns and heads to the right. As before, it could never cross the x-axis. There the solutions, we know, are running out to the equilibrium point, so our solution must then just end with x getting bigger and bigger to $x = 1$.

Another possibility is the solution could hit the line $x = 2$ and then enter the middle region, and then descend down and to the left, and also tend to the equilibrium point at 1,0. So that shows that no matter what initial condition you have, assuming both the initial populations are positive, then the solutions must go to the equilibrium point at 1,0.

So, the nullclines in this case, give us a complete picture of what is happening to solutions. This isn't always the case. Let me look at the case where say a is greater than 1. Let a be let's say 1.2. What we see now is it looks like solutions are tending to some equilibrium point that's not on the x-axis. But if I let a be larger, let's say a is 2. Now again, it looks like solutions are tending toward an equilibrium point that's not on the x-axis, but it looks like these solutions are spiraling in. So it looks like yes, all solutions tend to an equilibrium point, but do they spiral or do they not spiral? Again, we don't quite know everything that's happening for this differential equation using the nullclines. We don't know what's happening near the equilibrium points. We'll have to see. We'll have to use the process called linearization to determine exactly what's happening near these equilibrium points.

Now let's turn to one other example. Let me look at a system of differential equations that now depends on parameters a and c. First of all, the behavior of solutions, and secondly, whether we get any bifurcations. So here is my system. Look at $x' = y - x^2 - a$. Then $y' = x - y^2$. So we have this parameter a, which I'll vary. First, let me do the case where $a = 0$, so our differential equation for x is $x' = y - x^2$. First step is to find the nullclines. Well, the x nullcline is given by $y - x^2 = 0$, or again, the parabola $y = x^2$. Just as we saw in the first example, that's where the vector field is vertical. Above that parabola, x' is positive. Below that parabola, x' is negative. There is the behavior given by the x nullcline.

Our y nullcline is given by $x - y^2 = 0$, or $x = y^2$. That's a parabola opening to the right. If we're to the right of that parabola, then $x - y^2$ is positive, so y' is positive. On the other hand, if we go to the left of that parabola, then y' is negative. So we know where both x' and y' are positive and negative. We know the behavior of the vector field in all regions between the nullclines in this case.

For example, first off, well, let's look at the equilibrium points. We know that the equilibrium points are where these nullclines meet. So the parabola $y = x^2$ and $x = y^2$ meet each other at that point 1,1 and also at the origin. So we have our two equilibrium points. Now look and see which direction the vector field points in these regions between the nullclines. For example, in the region above the x nullcline, above the curve $y = x^2$. Well, there we know that x' is positive and y' is negative. So the vector field points down and to the right in that region.

What about the region between the two parabolas? There we know that y' is positive and x' is positive. So, in that region, the vector field points up and to the right, and so forth, with all the other regions. For example, to the right inside the parabola, $x = y^2$. We know y' is positive and x' is negative. So the vector field points up and to the left. Also, in the region to the left of both of the parabolas, there we know x' is negative and y' is negative, so the vector field points down and to the left. In the final region, in the upper right, we see that the vector field also points down and to the left. So we have a picture of what the vector field is doing in the regions between these nullclines.

So what happens to solutions? Well, for example, let's take solutions in the Region A. In the Region A, we know the solutions go up and to the right. They can never hit either the x or y nullcline because in order to do that, the solution would either to veer up and become vertical and hit the place where the vector field is horizontal. That can't happen. Or, it would have to go down and be vertical and hit the x nullcline. Again, that can't happen. The vector field is always pointing up and to the right. So the only thing that can happen for solutions in the Region A is those solutions will come up and tend to the equilibrium point at 1,1. Similarly, in the Region B, to the upper right, there solutions are pointing down and to the left. Again, they can never straighten out and hit either of the nullclines. Those solutions must again tend to the equilibrium point at 1,1.

What about in the Region C, to the left? There we saw the vector field was pointing down and to the left. So solutions must just go off to infinity. There's no other possibility. What about in the Region D, under the parabola $x = y^2$? There we saw the vector field pointed up and to the left. So now, we've got several possibilities. Our solution could go up and enter the

Region B, in which case, it turns horizontal and then goes downhill and tends to the equilibrium point at 1,1. Or, the solution could just keep going up and hit he equilibrium point at 1,1. Or, the solution could come up and enter the Region A, in which case, it takes a right hand turn and again goes to the equilibrium point 1,1. Or, the final possibility is includes the solutions going up and tending to the equilibrium point at the origin, or tending and hitting the lower part of the parabola $x = y^2$, in which case, the vector field gets horizontal. Then it enters the Region C and goes off to infinity. So, we have a good idea of what's happening to all solutions in the Region D. There are many possibilities, but we know all possible possibilities, whatever that means. Then finally, in the Region E, same thing, either our solutions could tend to the origin, could tend to 1,1, or it could go off to infinity by entering the Region C.

There is my direction field. Now what we see is wherever we start off to the right, we tend to the equilibrium point at 1,1, unless we go into that region to the left, in which case we go off to infinity. Similarly from above, either we go off to infinity or we tend to the equilibrium point at 1,1. Now again, at 0,0, it looks like we have a saddle point, but again, we don't know, maybe a whole bunch of solutions come into that equilibrium point. We'll use linearization to see that a little bit later.

So, there is the behavior of the solutions when our parameter $A = 0$. Remember, our first differential equation involved that parameter A. It was $x' = y - x^2 - A$. So in the more general case, our x nullcline is given by $y - x^2 - A = 0$, or $y = x^2 + A$. That is our x nullcline as the parabola opening upward: $y = x^2 + A$. So just by looking at the behavior of this parabola, as A changes, we see that we get a bifurcation. As we raise this parabola, $xy = x^2 + A$ up. Initially, when A is small, that parabola crosses the y nullcline at two points, just as in the case where $A = 0$, but eventually, there comes an A value where it touches at just a single point and then it goes above the parabola. So there is a bifurcation at an A value that I'll call A^*.

What happens for other values of A? Well, if A is small below A^*, then we basically have the same nullcline picture as we had before and we see the same behavior solutions. Either they go to the equilibrium point at 1,1, well now it's moved from 1,1 to some other point, or the other equilibrium point, or the solutions go off to infinity. But now, when A equals A^*, what happens? Well, those two equilibrium point merge and now we can compute again, just as we did before the directions of the vector field and we see something else happens. Those two equilibrium points have merged, but from the right, it looks like solutions are tending into that equilibrium point. It looks like a sink from the right, based on the configuration of the nullclines, but to the left, it looks like the solutions are tending off to infinity. So this equilibrium point looks like it's a sink from the right and a saddle from the left. We've got a change in the configuration of the equilibrium.

Now finally, what happens when A is greater than A^*? Well, now these two nullclines have separated. There are no equilibrium points. If we look at what happens to the vector field, just as before, we see that the configuration of the vector field now tells us everything. If solutions start inside the y nullcline, we must go up and to the left, across the y nullcline and then tend off to infinity. If solutions start inside the parabola given by the x nullcline, they go down and to the right, hit the nullcline and then again, go off to infinity. So we see that in this case, all solutions suddenly go to infinity.

To summarize, just by using the nullclines, what we see is if A is less than that bifurcation parameter A^*, lots of solutions go to the sink and other solutions seem to go to infinity. Maybe there are some that go to what looks like the saddle point. At the bifurcation point our two equilibrium points suddenly merge and then as soon as A is greater than A^*, they disappear. Now all solutions must go off to infinity. So, this is the process called nullclines that gives us the global point of view on what's happening to systems of differential equations. The questions that came up were local near the equilibrium points. That's what we'll deal with in the next lecture when we talk about linearization.

Nonlinear Systems near Equilibria—Linearization
Lecture 16

Recall that for the predator-prey system in the last lecture

$$x' = x(1-x) - xy$$
$$y' = -y + axy$$

we had an equilibrium point at $x = 1/a$, $y = (a-1)/a$ when $a > 1$. It appeared from the computer illustrations that sometimes solutions spiraled in to this point, but for other a-values this did not occur. To understand better what happens near equilibria in nonlinear systems, we use a procedure called linearization. This often allows us to use a linear system to figure out what is happening locally near a nonlinear equilibrium point.

To explain the process of linearization, let's first consider a first-order equation $y' = f(y)$, and let's assume we have an equilibrium point at $y = 0$. Recall from calculus that we can expand the function $f(y)$ as a **Taylor series** about $y = 0$. That is, we can write it as follows.

$$f(y) = f(0) + f'(0)y + \frac{f''(0)}{2!}y^2 + \frac{f'''(0)}{3!}y^3 + \cdots$$

The first term in this series is 0, since we have an equilibrium point at $y = 0$. Also, if y is very close to 0, the y^2 term (the third term in the series) is much smaller than the term involving just y (the second term). Similarly, the terms involving y^n for $n > 2$ are even smaller. This tells us that we can approximate the right-hand side of $y' = f(y)$ by just using the second term in the Taylor series, namely $f'(0)y$. Therefore, very close to $y = 0$, solutions of our nonlinear ODE $y' = f(y)$ should be close to solutions of the ODE $y' = f'(0)y$, which is just a linear ODE that we know how to solve.

If our equilibrium point is located at a nonzero value, say y_0, then a simple change of variables says that the same process is true: Solutions of the nonlinear ODE resemble the solutions of the linear equation $y' = f'(y_0)y$, at least in a very small neighborhood of $y = y_0$. There is one slight problem here. If $f'(y_0)$ happens to equal 0, then the linear system reduces to just $y' = 0$, for which all solutions are equilibria. This certainly will not be the case for the corresponding nonlinear system.

For a nonlinear system

$$x' = F(x, y)$$
$$y' = G(x, y)$$

with an equilibrium point at (x_0, y_0), we can perform a similar linearization, only now the corresponding linear approximation involves the partial derivatives of both F and G. That is, our nonlinear system has solutions near (x_0, y_0) that resemble the solutions of the following linear system.

$$x' = \frac{\partial F}{\partial x}(x_0, y_0) x + \frac{\partial F}{\partial y}(x_0, y_0) y$$

$$y' = \frac{\partial G}{\partial x}(x_0, y_0) x + \frac{\partial G}{\partial y}(x_0, y_0) y$$

Note that this is a linear system of ODEs. Again, what we have done here is just drop all the higher-order terms in the Taylor series expansions of both F and G. As above, solutions of our nonlinear system near the equilibrium point (x_0, y_0) should resemble the solutions of the linear system

$$Y' = JY,$$

where J is the matrix

$$\begin{pmatrix} \frac{\partial F}{\partial x}(x_0, y_0) & \frac{\partial F}{\partial y}(x_0, y_0) \\ \frac{\partial G}{\partial x}(x_0, y_0) & \frac{\partial G}{\partial y}(x_0, y_0) \end{pmatrix}.$$

The matrix J is called the **Jacobian matrix** for the nonlinear system.

As an example, consider the system below.

$$x' = x - xy^2$$
$$y' = -y + xy$$

There is no way to solve this system explicitly, but we certainly have an equilibrium point at the origin. The Jacobian matrix at an arbitrary point is

$$J = \begin{pmatrix} 1-y^2 & -2xy \\ y & -1+x \end{pmatrix}.$$

At the origin we get

$$J = \begin{pmatrix} 1 & 0 \\ 0 & -1 \end{pmatrix},$$

so our nonlinear system is close to $Y' = JY$. Since J has eigenvalues 1 and -1, our nonlinear system has a saddle at the origin. A glance at the phase plane shows that there is a lot more going on in this system.

Figure 16.1

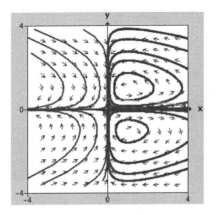

However, if we zoom in at the origin, we do see a part of the phase plane that looks like a saddle.

Figure 16.2

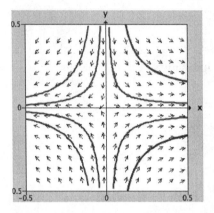

As in the first-order case, there are times when linearization does not work. In particular, if the Jacobian matrix associated to our equilibrium point has imaginary or zero eigenvalues, then the addition of the higher-order terms will usually greatly alter the linear phase plane. For example, the system

$$x' = y - x(x^2 + y^2)$$
$$y' = -x - y(x^2 + y^2)$$

has an equilibrium point at the origin. The corresponding Jacobian matrix is given by

$$J = \begin{pmatrix} 0 & 1 \\ -1 & 0 \end{pmatrix},$$

and the eigenvalues are easily seen to be the imaginary numbers i and $-i$. So the linearized system has a center at the origin. But this is not true of the nonlinear system (Figure 16.3), where we see a spiral sink.

Figure 16.3

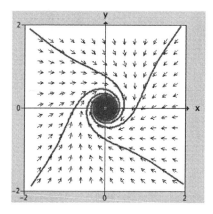

If we look instead at the system

$$x' = y + x(x^2 + y^2)$$
$$y' = -x + y(x^2 + y^2)$$

then again the linearized system has a center at the origin. But the same arguments as above show that this equilibrium point is now a source.

Let's return to the predator-prey system

$$x' = x(1 - x) - xy = F(x, y)$$
$$y' = -y + axy = G(x, y).$$

As we saw earlier, we have an equilibrium point at $(1/a, (a - 1)/a)$ when $a > 1$. The Jacobian matrix at an arbitrary point is

$$J = \begin{pmatrix} 1 - 2x + y & -x \\ ay & -1 + ax \end{pmatrix}.$$

At our equilibrium point, this matrix becomes

$$J = \begin{pmatrix} -1/a & -1/a \\ a-1 & 0 \end{pmatrix},$$

so the solutions near our equilibrium point should resemble those of the linearized system $Y' = JY$. The characteristic equation for the linearized system is

$$\lambda^2 + \lambda/a + (a-1)/a = 0,$$

so the eigenvalues are

$$\tfrac{1}{2a}\left(-1 \pm \sqrt{1 + 4a(1-a)}\right).$$

There are then 2 different cases. The first occurs when the term inside the square root is negative. In this case, the eigenvalues are complex with negative real parts, so the equilibrium is a spiral sink. If the term $1 + 4a(1 - a) > 0$, we certainly know that this term is less than 1. This is because $a > 1$, which makes $4a(1 - a)$ negative. Our eigenvalues are both real and negative, so in this case we have a real sink. An easy computation shows that the first case occurs when $a > 2 + 2\sqrt{2}$, and the second case occurs if $1 < a < 2 + 2\sqrt{2}$. This is what we saw in the previous lecture.

Important Terms

Jacobian matrix: A matrix made up of the various partial derivatives of the right-hand side of a system evaluated at an equilibrium point. The corresponding linear system given by this matrix often has solutions that resemble the solutions of the nonlinear system, at least close to the equilibrium point.

Taylor series: Method of expanding a function into an infinite series of increasingly higher-order derivatives of that function. This is used, for example, when we approximate a differential equation via the technique of linearization.

Suggested Reading

Blanchard, Devaney, and Hall, *Differential Equations*, chap. 5.1.

Edelstein-Keshet, *Mathematical Models in Biology*, chaps. 4.7–4.8.

Hirsch, Smale, and Devaney, *Differential Equations*, chaps. 8.1–8.3.

Strogatz, *Nonlinear Dynamics and Chaos*, chap. 6.3.

Relevant Software

Blanchard, Devaney, and Hall, *DE Tools*, HPG Linearizer, HPG System Solver, Vander Pol.

Problems

1. Compute the partial derivatives with respect to x and y of the function $F(x, y) = (x^2y + x^3 + y^3)$.

2. What is the Jacobian matrix for a linear system

   $x' = ax + by$
   $y' = cx + dy$?

3. **a.** Find all equilibrium points for the system

$$x' = x^2 + y$$
$$y' = x.$$

b. Compute the Jacobian matrix at each equilibrium point.

c. What is the type of these equilibrium points?

4. Find and determine the type of the equilibrium points for the Vander Pol equation given by

$$x' = y$$
$$y' = -x + (1 - x^2)y,$$

and then use the computer to see what else is going on for this system.

5. Give an example of a nonlinear system that has an equilibrium point for which the linearized system has a zero eigenvalue, but the nonlinear system has only one equilibrium point.

6. For the system

$$x' = y - 1$$
$$y' = y - x^2,$$

determine the types of each equilibrium point.

7. For the system

$$x' = \sin(x)$$
$$y' = \cos(y),$$

find all equilibrium points and use linearization together with the nullclines to sketch the phase plane.

8. For the system

$$x' = y - x^2 + A$$
$$y' = y,$$

determine the types of each equilibrium point and also the A-values where a bifurcation occurs.

Exploration

Consider the system of differential equations

$$x' = x/2 - y - (x^3 + y^2x)/2$$
$$y' = x + y/2 - (y^3 + x^2y)/2.$$

What is the type of the equilibrium point at the origin? Using the computer, find out what happens far away from the origin. To understand this behavior, change coordinates to polar coordinates. Then you will be able to solve this system explicitly.

Nonlinear Systems near Equilibria—Linearization
Lecture 16—Transcript

Welcome back. Now for the next phase of nonlinear systems. Remember in the last lecture we talked about nullclines and we saw that nullclines usually gave us a good, broad overview of what was happening to solutions, except in certain cases, usually near equilibrium points. Remember we had several examples where at least numerically it looked like the equilibrium point was a saddle point, but the nullclines would only tell us that some solutions could come in there. Maybe infinitely many could come into the equilibrium point, rather than just one. Also, when we looked at the Predator-Prey system, we saw there was an equilibrium point that for some values of the parameter, solutions just went right into it like it was a sink, and for other values of the parameter, solutions spiraled into it. So the question is, what's happening near these equilibrium points? The idea of linearization will tell us what's happening.

So before moving to systems of differential equations, let me go to first order differential equations. Let me look at why prime is $f(y)$. Here, f will be our nonlinear function. Let's say we have an equilibrium point at 0. The question is what happens for our nonlinear first order differential equation near that equilibrium point? Here's the way to handle that. Remember from Calculus, we can expand our function $f(y)$, the right hand side of the differential equation, in a Taylor series about $y = 0$. That is, we can write $f(y)$ as the internet sum first; $f(0)$—remember that 0 because we get an equilibrium point there—$f(0) + f'(0)y + f''(0)/2!y^2 + f'''(0)/3!y^3$, etcetera, etcetera, etcetera.

So, here we have $f(0) = 0$. So the first term in this table series is 0. Then we have $f'(0)y$. And then we have a y^2 and a y^3 term, etcetera. Those terms are very small if y is close to 0. For example, if y is 0.1, then the y^2 term is 0.01, much smaller. The y^3 term is even smaller. So the tail of this Taylor series is actually very small. The first term, the term $y' = f'(0)y$ is what dominates this Taylor series. What we can do is approximate the right hand side of our differential equation, $f(y)$ by the new term, $f'(0)y$. So close to the equilibrium point, solutions to the nonlinear differential equation should be close to the solutions of the linear differential equation, y' is $f'(0)y$, and we know how to solve that.

If our equilibrium point happens to be at some other point that's nonzero, say at y equals maybe y_0, then a simple change of variable says that near this new equilibrium point y_0, solutions of the nonlinear differential equation resemble those of $y' = f'(y_0)y$. And $f'(y_0)$ is a number, so this is y' at some constant times y, it's a linear differential equation. Now we know what happens. If $f'(y_0)$ is negative, then we know the linear system has a sink. We know that the nonlinear system at least near the origin has a sink.

Similarly, if $f'(y_0)$ is positive, the linear equation has a source. We now know that the nonlinear equation must resemble it at least close to the equilibrium point, it too must have a source. But if $f'(y_0) = 0$, then we get no information about our nonlinear system. If $f'(y_0) = 0$, that means that our linear equation, which is $y' = 0$—everything is a constant. Obviously, that's not going to happen for the nonlinear differential equation.

For a nonlinear system, say $x' = F(x,y)$ and $y' = G(x,y)$, we have a similar linearization, but this time it involves the partial derivatives of the function F and G. First recall what the partial derivatives are. If we've got a function $F(x,y)$ depends on two variables, then we can take its partial derivative with respect to either of those variables. For example, the partial of F with respect to x is just the derivative of F with respect to x, where we think of the variable y as a constant. Similarly, the partial derivative of F with respect to y is differentiate all the y terms where we think of x as just a constant. For example, if you function $F(x,y) = 2x + 3y$, then the partial derivative of F with respect to x is differentiate the $2x$ get 2, but now differentiate the $3y$, that's a constant with respect to x, that's 0. So a partial of F with respect to x is 2. Similarly, the partial of F with respect to y is 3.

If, as another example, $F(x,y) = x^2 + 3xy + y^3$, then the partial of F with respect to x is the derivative of x^2 is $2x$, the partial of $3xy$ with respect to x is just $3y$. And then the derivative of y^3 is 0. And similarly, the partial derivative of F with respect to y would now be $3x$, plus now the derivative of y^3 is $3y^2$ with respect to y.

Here is the linearization for nonlinear systems. Suppose we had an equilibrium solution that x_0, y_0. Then we'll take a similar Taylor series for our functions F and G. Only these Taylor series now will include all kinds of different partial derivatives, maybe the second partial derivative of F with respect to x. Differentiate F twice with respect to x. Or maybe the second partial of F with respect to y, differentiate F twice with respect to y. Or the mixed partial derivative, the partial of F with respect to x and y, these will all be the higher order terms in our Taylor approximation. So the Taylor approximation will, just as in the one dimensional case, reduce to just the linear terms. What we get is the linearized differential equation, which looks like x' is the first derivatives, so that would be the partial of F with respect to x, evaluated at your equilibrium point times x and add to that the partial derivative of F with respect to y at your equilibrium point times y. Then throw away all the rest of the higher order terms in the Taylor series. Similarly, the linearized version of the y equation would be y' is first the partial of G with respect to x, evaluated at your equilibrium point, times x and then the partial of G with respect to y, again at the equilibrium point times y. So there is the linear system of differential equations whose solutions often dictate what happens to the corresponding solutions for the nonlinear system.

We're going to write this in matrix form just as we did with linear systems, as vector y' is equal to the matrix J times y, where J is what we call the Jacobian matrix. J is the matrix whose entries are in order, the partial of F with respect to x at the equilibrium point and the partial of F with respect to y at the equilibrium point, and then the second row is the partials of G with respect to x and then G with respect to y, again, at the equilibrium point. That's the Jacobian Matrix. It looks like there are a lot of letters there, but remember, we're evaluating each of these derivatives at the equilibrium point, so each of these entries actually are constants.

So, let's go with an example. Let's take the nonlinear system x' as $x - xy^2$; $y' = -y + xy$. It's nonlinear. We've got an xy^2 term in the first equation and an xy term in the second equation. We generally know how to solve systems like this explicitly, but we certainly do know we have an equilibrium point at the origin. So we can determine nearby behavior by realizing this differential equation at the equilibrium point 0,0. So let's do it. Let's compute the Jacobian matrix. The Jacobian matrix is, first take that derivative with respect to x in the first equation. You get $1 - y^2$. Then the partial of F with respect to y is $-2xy$. There is our first row of the Jacobian matrix. Then the second row is take the partial of y' with respect to x. You get just y and then the partial with respect to y is $-1 + x$. There is the full Jacobian matrix. We want the Jacobian matrix at the origin, but at the origin when x and y are 0, this matrix turns out to be just 1,0,0,−1. So we know that the eigenvalues are displayed along the diagonal are 1 and −1, so we know that this system has a saddle point at the origin.

Let's look and see what's happening to solutions of this system. Here is the direction field. We know we have an equilibrium point at the origin that's a saddle, and let's see what happens nearby. Yes, it's beginning to look like a saddle, but over here, things look a little different. This nonlinear system is much more complicated in the large than just the corresponding linear system, but near the origin, we certainly see what looks to be a saddle.

Let me see that by zooming in on the origin. Instead of letting x and y go from + to −2, let me go from say −0.1 to 0.1. In the y direction from −0.1 to 0.1. Now I've zoomed in on what's happening near this equilibrium point. What you see is yes, indeed, it does look like a saddle. Very close to the equilibrium point you often get a solution for a phase plane that looks exactly like the corresponding phase plane for the linear system.

That's not always the case. There are some troubling cases for linearized systems. For example, if you linearize your system at an equilibrium point and find that you end up in the trace determinant plane on one of those sort of bifurcation lines, like the place where we had repeated eigenvalues or the place where we had zero eigenvalues, then the nonlinear terms in your differential equation can have an effect. They can enter the fray and change slightly, ever so slightly what's happening to the linearized system, and then all things are different.

Let me give you a couple of examples of where linearization does not work. Here is one. Take the vector field $x' = x^2$ and $y' = y^2$. That's a very simple system to solve. That's a decoupled system. You only have x in the first equation and y in the second, but nonetheless, we certainly know that we have an equilibrium point at the origin. Take its Jacobian matrix. The partial of x^2 with respect to x is $2x$ and with respect to y is 0. Similarly, the partial of y^2 with respect to x is 0 and with respect to y is $2y$. So at the origin, our Jacobian matrix is all 0s. It's the linear system y' is the 0 matrix times y. We know what the solutions are there. They're all constant solutions. Obviously, that's not going to happen with our nonlinear system.

Let's go and look at what happens to solutions to this nonlinear system. Here is the nonlinear system. Notice that if x and y are negative, it looks like that equilibrium point is a sink. But if x and y are positive, it looks like this equilibrium point is a source, and if one is positive, the other is negative, it looks like this equilibrium point is a saddle. So this is a kind of crazy equilibrium point because our eigenvalues were on the trace axis, they were both 0. The nonlinear system does not at all behave like the corresponding linear system.

Here is another example. Let's take the nonlinear system that's more complicated now, $x' = y - x(x^2 + y^2)$; $y' = -x - y(x^2 + y^2)$, a very complicated nonlinear system. But we know we have an equilibrium point at the origin, both x' and y' vanish when x and y are 0. So compute the Jacobian matrix. We've got to take the partials of the first equation with respect to x and y. But the partial derivative of the first equation with respect to x is the first equation can be rewritten as $y - x^3 - xy^2$, so the partial with respect to x is $-3x^2 - y^2$. The partial with respect to y is $1 - 2xy$. Similarly, the second equation can be written as $y' = -x - x^2y - y^3$. So the partial derivatives with respect to x are $-1 - 2xy$. And with respect to y are $-x^2 - 3y^2$. So there is our Jacobian matrix. Evaluate that at the origin. All the xy terms disappear and you're left with just the matrix $0, 1, -1, 0$.

What are the eigenvalues for that matrix? We've actually seen that before. The trace is 0 and the determinant is 1. The characteristic equation is $\lambda^2 + 1 = 0$. So our eigenvalues are $\pm i$. That means we're on the determinant axis in the trace determinant plane near where we could have bifurcations. Go just to the left and you'll have a spiral sink. Go just to the right, you'd have a spiral source. Since we have a nonlinear system, these tiny nonlinear terms could force us off that line and give us either a spiral sink or a spiral source.

So here is the phase plane. What we see is, even though we have a center of a linearized system, all solutions, in fact, spiral in. All solutions spiral in to the origin. In fact, this equilibrium point is a spiral sink, despite the fact that the linearized system gave us a center. Why does that happen? Why do we have a spiral sink? Well, let's break this vector field into two pieces, the linearized piece and the nonlinear piece. So let me write $x' = y - x(x^2 + y^2)$ in two pieces. First, I'll write $x' = y$. Secondly, I'll write $x' = -x$, factor of $x^2 + y^2$. Similarly, with the y equation, break it into its linear and nonlinear parts. So first I'll put $y' = -x$. Then the second one will be $y' = -y$, factor of $x^2 + y^2$.

So, we've got two vector fields. What is the configuration of solutions for those vector fields? Well, the first one, $x' = y$ and $y' = -x$, that's just the linear system we just saw. That has a center. All solutions run around the origin in circles. The vector field is everywhere tangent to those circular solutions. But what about the other portion? What about the nonlinear portion? That is $x' = -x(x^2 + y^2)$; $y' = -y(x^2 + y^2)$. We're multiplying both of those components by $x^2 + y^2$, by that positive number. We've really got the vector field $x' = -x$ and $y' = -y$ but times that negative number of $x^2 + y^2$.

What happens to the vector field $x' = -x$ and $y' = -y$? Well, that's a vector field that's just pointing directly toward the origin at the point xy. Then the factor $x^2 + y^2$ just expands that. So, we have a vector field in the nonlinear term that's always pointing directly to the origin. We're adding to that a vector field that's running around on circles tangent to the circular solutions. If you add those two vectors, then you always get a vector that's pointing inside that circle. The vector field is always pointing inside the circle of radius R in the plane. That's why for our nonlinear systems, all solutions actually spiral into the equilibrium point. We have a spiral sink equilibrium point. If I change that minus to a plus in the nonlinear terms, we'd have a vector field that points out, points away from the origin. So now solutions would spiral away. We'd have a spiral source.

Now let's return to the examples we used in the last lecture, first the Predator-Prey system. That differential equation was $x' = x(1 - x) - xy$. That was our $F(xy)$ term. The second equation y' was $-y + axy$. That was $G(xy)$. We saw early on that we had an equilibrium point at the point $x = 1/a$ and $y = 1 - 1/a$, when a was greater than 1. When a was less than 1, this was negative, so it didn't count in our Predator-Prey equation. So we had that equilibrium point. Early on, we saw that sometimes solutions spiraled in. Sometimes solutions just went straight in. How do we determine what's happening there? Well, take the Jacobian matrix. Take the partial of F with respect to x. You'll get $1 - 2x - y$. Take the partial of F with respect to y and you'll get $-x$. Then take the partials of the second equation with respect to x and y, the partial of G with respect to y is that constant ay. The partial of G with respect to y is $-1 + ax$. Now we've got to plug in our equilibrium point, which is at the point $1/a$, $1 - 1/a$. When you do that, when you replace x and y by those terms, you get the Jacobian matrix $-1/a$, $-1/a$ and then $a - 1$ and 0.

What are the eigenvalues of this Jacobian matrix? What does the linear phase plane look like? This linear system depends on a. Our trace of this linear system is $-1/a$. The determinant is $a - 1/a$, which is just $1 - 1/a$. So our equilibrium point is going to depend on a, or at least the character equilibrium point is going to depend on a. So why don't we look at what this looks like in the trace determinant plane as a function of a? Remember the trace determinant plane? We had our repeated eigenvalues parabola $T^2 - 4D = 0$. We also had the trace axis. That's where we had determinant equals 0. That's where we had eigenvalues equal to 0. Those would serve sort of the funny lines. And then there was also the positive determinant axis where we had centers. These were places where a small nonlinear perturbation could throw us into a different regime.

Let's look and see where for the Predator-Prey system our eigenvalues put us. Again, the trace was $-1/a$. The determinant was $1 - 1/a$. When $a = 1$, the trace is then -1 and the determinant is 0. We're sitting on the trace axis. Now we are interested in what happens when a is positive, because that's when we have that new equilibrium point in our Predator-Prey system. What happens as a increases? As a increases, $-1/a$ goes to 0, as a goes off to infinity. Similarly, as a goes off to infinity, $1 - 1/a$, our determinant, goes to 1. In fact, since the determinant is $1 - 1/a$ or 1 plus the trace, we know that in the trace determinant plane each different value of a puts us on a straight line which connects the point $a = -1$ on the trace axis, up to a point one on the determinant axis.

We take a tour, a straight line tour, in the trace determinant plane going from the trace to the determinant axis. Now notice what that looks like. It looks like for awhile for certain values of a were in the region where have a real sink, but then you cross the repeated eigenvalue curve and enter the region where you have a spiral sink. So initially, our equilibrium point for low values of a is just a real sink. All solutions just head straight to the equilibrium point in the limit. Then beyond that, we have a spiral sink, all solutions spiral in. Just as we saw last time in the phase plane, sometimes solutions spiraled in, other times they did not. There is one particular a value where that change happens, where you hit the repeated eigenvalue curve. It turns out, a little computation says that that is $a = 2 + 2\sqrt{2}$.

There were a couple more questions that came up in the previous lecture. For example, remember the differential equation $x' = y - x^2$ and $y' = 1 - y$? We saw that we understood the totality of what solutions could do. Certain solutions could go into the equilibrium point at 1,1. Other solutions could go into the equilibrium point at $-1,1$. Other solutions could go off to infinity. We knew all possible behaviors.

The one question we had though was, at the equilibrium point $-1,1$, would it be possible for infinitely many solutions to actually come into it? When we looked on the computer, it kind of looked like a saddle, but using the techniques from nullclines we can't eliminate the possibility of a bunch of solutions actually coming in and others veering off. Remember that earlier differential equation $x - x^2$ and $y' = y^2$, we saw exactly that, infinitely many solutions coming in, infinitely many solutions going out, and nearby infinitely many solutions looking like saddle behavior. So our nullclines don't tell us what is happening in this particular case. So let's use linearization to see exactly what happens.

Our differential equation is $x' = y - x^2$. So the partial derivative with respect to x there is $-2x$. The partial with respect to y is 1. Our second equation $y' = 1 - y$. The derivative with respect to x is 0. The derivative with respect to y is -1. So our Jacobian matrix is $-2x,1,0,-1$. At the equilibrium point, $x = -1$; $y = 1$. This Jacobian matrix becomes the matrix 2,1,0,-1, an upper triangular matrix. Eigenvalues are 2 and -1. Our linear system has a saddle, so near the equilibrium point in our nonlinear system, we should also have a saddle. So the computer pictures were accurate. We did have an equilibrium point that was a saddle. Only one solution comes in, not infinitely many that could possibly happen for certain nonlinear systems.

We had another question about the differential equation $x' = y - x^2$ and $y' = x - y^2$. We had a problem near the equilibrium point at the origin. Was that a saddle? Handle it in exactly the same way. First linearize and determine the type, and hope that you're not on one of the bad points in the trace determinant plane. But, linearization, first equation is $y - x^2$. The partial with respect to x is $-2x$, with respect to y is 1. Second equation is $x - y^2$, the partial with respect to x is 1, with respect to y is $-2y$. So at the origin, when x and y are 0, our matrix is just 0,1,1,0. What's the type of that linear system?

Well, that trace here is 0, but the determinant is $0(0) - 1(1) = -1$. So our determinant is negative. Now go back to the trace determinant plane. Remember in the trace determinant plane, everything below the trace axis, that little half plane, consisted of phase planes that were saddles. So since the determinant here is negative, we know that the phase plane for the nonlinear system, also has a saddle, a positive, and a negative eigenvalue.

So again, we can rectify the problems that come up with nullclines often by this technique of linearization. That's two of the major techniques we can use to give us a qualitative idea for what's happening to nonlinear systems. Nullclines gives the big picture, which fails very often locally near equilibrium points. Linearization comes in sometimes and helps us understand what's going on there.

What we'll do next time is turn to another application, this time the Competing Species Model. It'll depend on various parameters, and we'll use both nullclines and linearization to understand the big picture again, the behavior of solutions, the bifurcations, the picture of what's going on in the parameter plane. Stay tuned for a summary of both of these techniques.

Bifurcations in a Competing Species Model
Lecture 17

Let's combine the previous topics to investigate what happens when we have 2 species that compete with each other for the same resources. Let x and y denote the population of these 2 species. We first assume that if the other species is absent, the population of the given species obeys the limited-growth population model. So when $y = 0$, we will assume that the x-population is given by $x' = x(1 - x/M)$. When $x = 0$, we have $y' = y(1 - y/N)$. So our phase plane thus far is the below.

Figure 17.1

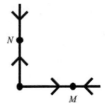

Our second assumption is that if the y population increases, it should have a negative effect on the x-population, and vice versa. So if y increases, x' should decrease. One way to model this is by adding the term $-(a/M)xy$ to the differential equation for x and $-(b/N)xy$ to the y-equation. So our competing species system is the below.

$$\frac{dx}{dt} = x(1 - x/M) - (a/M)xy$$

$$\frac{dy}{dt} = y(1 - y/N) - (b/N)xy$$

We have 4 parameters for this system: a, b, M, and N. The quantities a and b measure how competitive the 2 species are. For simplicity, let us assume that $M = N = 400$; the equations are below. (Note: There is nothing special about the number 400; it is simply the default value of M and N in the software we use.)

$$\frac{dx}{dt} = x(1 - x/400 - ay/400)$$

$$\frac{dy}{dt} = y(1 - y/400 - bx/400)$$

Now let's view the phase plane and nullclines. First assume that $a, b > 1$ (so our species are very competitive). The x-nullclines lie on the y-axis (where we know what happens) and along the straight line $1 - (x/400) - ay/400 = 0$, or $y = 400/a - x/a$. Similarly, the y-nullclines are given by the x-axis and the line $y = 400 - bx$. Note that the 2 nullclines that do not lie on the axes meet at the point $x = 400(1 - a)/(1 - ab)$, $y = 400(1 - b(1 - a)/(1 - ab))$. This is another equilibrium point for our system since $a, b > 0$. We call this the coexistence equilibrium point.

At the point (400, 400), we compute that $x' = -400a$ and $y' = -400b$, so the vector field points down and to the left. Similarly, at (1, 1) we have $x' > 0$ and $y' > 0$, so the vector field points up and to the right. We similarly compute the value of the vector field in the other regions bounded by the nullclines to obtain the following figure.

Figure 17.2

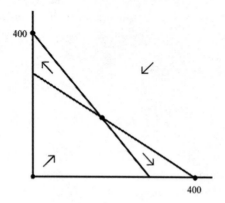

It appears that the coexistence equilibrium point is a saddle. Of course, we do not know this for sure from the picture. However, linearization at this point should give us the answer. For simplicity, assume $a = b = 2$; so our equations are

$$x' = x - x^2/400 - xy/200$$
$$y' = y - y^2/400 - xy/200.$$

The coexistence equilibrium point is found by solving the 2 equations

$$1 - x/400 - y/200 = 0$$
$$1 - y/400 - x/200 = 0$$

simultaneously. A little algebra yields $x = y = 400/3$. Then the Jacobian matrix is

$$\begin{pmatrix} 1-x/200-y/200 & -x/200 \\ -y/200 & 1-x/200-y/200 \end{pmatrix},$$

and at $(400/3, 400/3)$ this matrix is

$$\begin{pmatrix} -1/3 & -2/3 \\ -2/3 & -1/3 \end{pmatrix}.$$

The determinant of this matrix is $-1/3$. Since the determinant is negative, we know that the point in the trace-determinant plane that corresponds to this matrix lies below the trace axis. So we are indeed in the region where we have a saddle. The phase plane for $a, b > 1$ is shown in Figure 17.3.

Figure 17.3

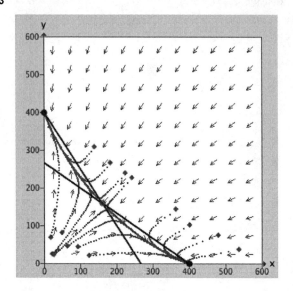

Any solution that starts at a point (x_0, y_0) with x_0 and y_0 positive (except for the special solutions that tend to the saddle) tends to either (400, 0) or (0, 400). That is, since both species are very competitive, one will almost surely prevail and the other will go extinct. Which species prevails depends on the initial populations of the 2 species.

Now suppose that species y is much less competitive, so the constant a decreases. Recall that one of the x-nullclines is given by $y = 400/a - x/a$. So the y-intercept of this line is $400/a$. Note that this y-intercept agrees with the y-intercept of the y-nullcline given by $y = 400 - bx$, which is 400 when a is equal to 1. So we see a bifurcation at $a = 1$ since the x- and y-nullclines that do not lie on the axes now meet at the point (0, 400). That is, the coexistence equilibrium point merges with the equilibrium point at (0, 400).

Figure 17.4

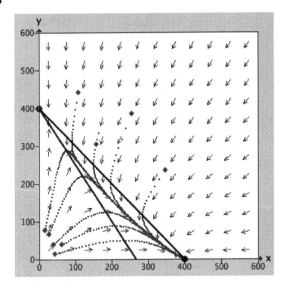

One can check that the Jacobian matrix at the equilibrium point (0, 400) is given by

$$\begin{pmatrix} 0 & 0 \\ -1 & -1 \end{pmatrix}.$$

This matrix has eigenvalues -1 and 0, so linearization does not give us an accurate picture of the phase plane. However, the configuration of the nullclines shows that all solutions that do not lie on the y-axis now tend to the equilibrium point on the x-axis at (400, 0). That is, the species y now goes extinct no matter what the initial point (x_0, y_0) is (as long as x_0 is positive).

When $a < 1$, the nullclines again indicate that all solutions tend to (400, 0)—the weaker species y remains extinct. Similarly, if b becomes less than 1 while a stays larger than 1, we see that the x-population goes extinct. Below is the phase plane for $a < 1, b > 1$.

Figure 17.5

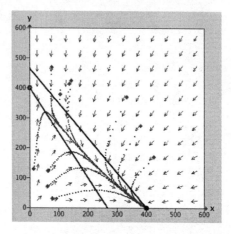

Next, let's look at what happens if both a and b are small ($a, b < 1$). Both of our 2 species are not that competitive. We see another bifurcation as b moves below 1 while a stays at that level. A new coexistence equilibrium point is born. The nullclines, however, indicate that this equilibrium is now a real sink. Indeed, the lack of strong competition leads to coexistence no matter where the 2 populations start out. The below is the phase plane when $a, b < 1$.

Figure 17.6

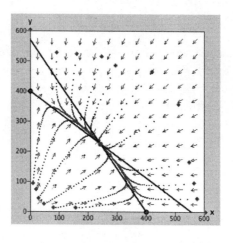

Finally, we can record all of this information in the analogue of the trace-determinant plane or the bifurcation diagram by plotting the different regions in the *a-b* plane where we have different behaviors. This also allows us to see where bifurcations occur.

Figure 17.7

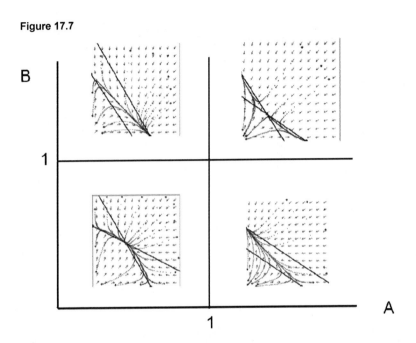

Suggested Reading

Blanchard, Devaney, and Hall, *Differential Equations*, chaps. 5.1–5.2.

Edelstein-Keshet, *Mathematical Models in Biology*, chap. 6.3.

Hirsch, Smale, and Devaney, *Differential Equations*, chap. 11.3.

Strogatz, *Nonlinear Dynamics and Chaos*, chap. 6.4.

Relevant Software

Blanchard, Devaney, and Hall, *DE Tools*, Competing Species, HPG System Solver.

Problems

1. **a.** Consider a simpler version of the competing species model where we do not include overcrowding:

    $$x' = x - xy$$
    $$y' = y - xy.$$

    Here both x and y are non-negative. First, find all equilibrium points of this system.

    b. Compute the Jacobian matrix at each equilibrium point.

    c. Determine the types of these equilibria.

    d. Determine the nullclines and sketch the regions where the vector field points in different directions.

    e. Sketch the phase plane for this system with $x, y \geq 0$.

2. **a.** Verify that the coexistence equilibrium point in the competing species model is indeed a real sink when $a, b = 1/2$.

    b. Use linearization to check the types of the equilibrium points that lie on the x- and y-axes for all a- and b-values.

3. a. A model due to Beddington and May for the competitive behavior of whales and krill (a small type of shrimp) is

$$x' = x(1 - x) - xy$$
$$y' = y(1 - y/x).$$

Here y is the population of whales. The carrying capacity for the whale population is not constant; rather, it is given by x and therefore depends on the krill population. Find the equilibrium points for this system.

b. Determine the types of the equilibria for the system above.

c. Use nullclines and the computer to paint the picture of the phase plane for this system.

Exploration

Here is a modified competing species model where we now allow harvesting or immigration (the parameter h).

$$x' = x(1 - ax - y)$$
$$y' = y(b - x - y) + h$$

First assume that $h = 0$ (no harvesting). Give a complete synopsis of the behavior of this system by plotting the regions in the a-b plane where different behaviors occur. Then repeat this for the parameters $h > 0$ and $h < 0$. Try to envision the complete picture by viewing the regions in the a-b-h space where different outcomes occur.

Bifurcations in a Competing Species Model
Lecture 17—Transcript

Hello, again. Time for another foray into non-linear systems of differential equations. In the last two lectures we've seen two major techniques to help us qualitatively understand what's going on for these solutions; the first was nullclines, the second was linearization. Now let's combine both of these techniques to analyze what happens when we have two species that compete for the same resources, two species that are after the same, say food supply. So here's another model.

Let me begin with some assumptions. Assumption number one is if one of the species is absent, there are no people in that species, then the current species obeys the limited growth population model. So for example, when species $y = 0$ the differential equation for species x would just be our old familiar limited growth or logistic population model, x' is say, $x(1 - x/M)$. M here is the carrying capacity for the x species. Similarly when $x = 0$ there's no x species. The y species obeys the logistic population model. Now y' is $y(1 - y/N)$. Now N is the carrying capacity for species y. That's our first assumption.

The second assumption involves the competition. What we'll assume is that if say the species y increases, this negatively impacts species x, so x population goes down and vice versa. There are a number of ways to do that. One way is to simply add a term to the differential equation for x that's of the form say, some negative constant times xy. If you've got a negative constant times xy, if the species y goes up, then minus that constant times xy means that the species x goes down. So that constant would roughly speaking measure how competitive the species y is. Traditionally the way we write this second term is not to just be a negative constant times xy but rather $-a(a/M)xy$. So that's our next term in the x differential equation. Similarly in the y differential equation we'll add a term $-(b/N)xy$. Basically b measures how competitive the species x is.

Altogether our differential equation for the competing species model is x' is $x(1 - x/M - a/M)xy$. Our differential equation for y is y' is $y(1 - y/N$ and then the new term $-b/N)xy$. Now we've got lots of parameters there; a and b, N

and M. For simplicity let me fix N and M to be 400 simply because that's the default in the software I always use. So our differential equation now reads x' as $x(1 - x/400 - ay/400)$, and similarly y' is $y(1 - y/400 - bx/400)$.

Let's first view the nullclines for this system. We've got the parameters a and b here. Let me assume that a and b are both larger than 1. Our species are very competitive. Remember a and b measures how competitive these two species are. So a and b we'll assume are larger than 1. What are the nullclines? Remember the x nullcline is where x' vanishes. So we need to set $x(1 - x/400$ then $- ay/400) = 0$. So we get $x = 0$ or that quantity $1 - x/400 - ay/400 = 0$. A, the first nullcline is $x = 0$. That's the y-axis. That's where we only have species y. That's where we have the logistic growth model. Then you've got that other nullcline.

What does that second nullcline look like? That nullcline $1 - x/400 - ay/400$? We could rewrite that as y is $400/a - x/a$. That is it's a straight line. It's a straight line in the plane with slope $-1/a$. Remember a is larger than 1, so the slope here is now between 0 and -1. It's y intercept when $x = 0$ is $400/a$ and it's x intercept when $y = 0$ is at 400. So our second x nullcline is a straight line with slope $-1/a$, and that's where $x' = 0$, so that's where I would get the field points vertically. That's the x nullcline.

It's similar for the y nullcline. y nullcline occurs when $y' = 0$. So that means either $y = 0$ or $1 - y/400 - bx/400 = 0$. Again, when $y = 0$ we're just looking at the x-axis, there we know we have the logistic population growth model for x. Then the other y nullcline simplifies to $y = 400 - bx$. So the y nullcline has slope $-b$, which is less than -1. Remember b is greater than 1. So slope less than -1. The x intercept is now $400/a$ and the y intercept is 400. So if we take that straight line with slope $-b$, a very vertical straight line, there we know that the slope field is horizontal. So there are our nullclines.

Where are the equilibrium points? They're at the places where the nullclines meet. So the x and y nullclines meet at the origin, at 0,0 of course no species, no differential equation. They also meet at 400,0 where the x species survives and y is extinct, and at 0,400 where the opposite happens. And there's also one equilibrium point in the middle that we'll find later. That's the coexistence equilibrium point. If our populations sit right there, then they coexist even though they're competing with each other.

To find out what's happening globally, let's get the direction of the vector field in the various regions between the nullclines. First, that far off region where x and y are large. To find the direction of the vector field, we need to just pick one point and see what the vector field is. So for example, take the point x and y are both = 400. Then what's the vector field at 400,400? Well, at 400/400 x' is $-400a$, that's negative. Similarly y' is $-400b$. That's also negative. So in that upper right region, the vector field always points down and to the left. You've got very large populations of x and y. They're both going to decrease.

What about if x and y are very close to 0 but positive? Well, if x and y are very close to 0 your differential equation is $x' = x$ times something that's very small, very close to actually something very close to 1. So x' is positive, and similarly y' is positive. So at the point very close to the origin, our vector field is pointing up and to the right. So all points within that region abutting the origin our vector field points northeast and so forth. If you go into the other two regions, the upper triangular region, the lower triangular region you'll see that the vector field points northwest in the upper region and southeast in the lower triangular region. So we just take one point in each of the regions between the nullclines, find out the vector field there, and we know what happens to globally all solutions.

Let's turn to the computer now and see what happens to this competing species model. So here I plotted our x and y nullclines. You see the vector field is horizontal along this nullcline there. You see the vector field is vertical along this nullcline here. As we said, all solutions that start out very large come in, x and y decline, and what you see is, well sometimes the population x goes extinct and sometimes the population y goes extinct, like right there. If we start out very low—it looks like we actually hit that equilibrium point there. That's a rarity.

We start out over here, it looks like the population of x goes extinct, same thing at that point, and now the population of y goes extinct. So it looks like we have syncs along the x and y-axis at our nonzero equilibrium point. It looks like we have a source down here at the origin, and it looks like this equilibrium point right in the middle, the coexistence equilibrium point is a saddle. So what happens is apparently for this system when a and b are big generally one of the two species is going to go to extinction, while the other one will survive.

So just to check on this, let's look at that coexistence equilibrium point and be sure that it's actually a saddle point. To do that, we use the method of linearization. Again for simplicity let me assume that both of our parameters a and b are some number bigger than 1, let's say 2. So our differential equation now is $x' = x - x^2/400 - xy/200$. And $y' = y - y^2/400 - xy/200$.

First find that equilibrium point that sits right in the middle, the coexistence equilibrium point. The right-hand side of our x equation can be rewritten as $x(1 - x/400 - y/200)$. The right-hand side of y' can be $y(1 - y/400 - x/200)$. Forget about the x and y terms there. Our coexistence equilibrium point can be found by solving the equation $1 - x/400 - y/200$ is equal to the corresponding term in the y equation that's $1 - y/400 - x/200$.

To solve that equation simply multiple by 400. Those equations reduce to $-x - 2y = -y - 2x$. And so therefore $x = y$. Then plugging that back into the first equation says that now $1 - x/400 - x/200 = 0$. So that's simplifying $1 = 3x/400$, or x and y equal to 400/3 is our coexistence equilibrium point.

So there's the equilibrium point, now we've got to linearize. Now we've got to compute the Jacobian matrix. That means taking the partials of both f and g, the first equation and the second equation with respect to x and y. Our first equation can be written as $x' = x - x^2/400 - xy/200$. So the partial derivative with respect to x of that equation is $1 - 2x/400$, that's $-x/200$ and $-y/200$. There's the partial with respect to x. Similarly the partial with respect to y, there's only one y term mainly $-xy/200$, so its partial is $-x/200$. Similarly for the partials of a second equation with respect to x and y, the partial with respect to x you can check is $-y/200$ and the partial with respect to y is $1 - x/200 - y/200$. There's our Jacobian matrix.

Now we've got to evaluate that Jacobian matrix at the value $x = y = 400/3$. That's our equilibrium point. A lot of numbers to throw into that matrix, but an easy computation says if you plug $x = y = 400/3$ into the Jacobian matrix, you'll get in the first term $1 - 2/3 - 3/3$, which is $-1/3$. You plug it into the second equation, you see $- 2/3$. Then for the second row of the matrix, the first entry is $- 2/3$. Finally when you plug x and $y = 400/3$ into the lower right entry, you get $1 - 2/3$ again $- 2/3$ which is $- 1/3$. There's our Jacobian matrix. It's the matrix $- 1/3 - 2/3 - 2/3 - 1/3$.

What's the type of this equilibrium point? Well, take the determinate. The determinant of this matrix is $1/9 - 4/9$, that's $- 1/3$. So this puts us in the trace determinate plane somewhere below the trace axis. The determinant of our matrix is negative. We're below the trace axis. We're in the region where indeed we have a saddle. So again here are the nullclines. There's our coexistence equilibrium point. We've just seen that that's a saddle; and yes, solutions that come in nearby veer off either to the left or to the right. There's exactly one solution that comes in exactly to that point. It's almost impossible to find that. Oh no, I didn't. The last time I did find that one solution.

What do we know about solutions in this case? We know that all solutions, except those very special ones that tend to the saddle point either tend to one of the equilibrium points on the x or the y-axis. That means one of our species dies out and one does not. Of course which species dies out, which species wins depends on which initial conditions we choose, whether x population is much larger or y population is much larger.

Now we have some parameters in this equation, so let's change these parameters and see what happens. Now suppose that the species y is much less competitive, so the parameter a decreases. Then remember our nullcline for y was y is $400/a - x/a$ and it had y intercept $400/a$. Remember, we had an equilibrium point up there on the y-axis too. So the question is: When is this y nullcline going to agree with the y intercept of the x nullcline? Well it agrees when $a = 1$. When $a = 1$, both of those nullclines hit a axis at exactly the point 400. What happens is we get a bifurcation. Let's watch what happens as we vary this x nullcline by letting a get smaller and smaller.

We started out with a and b relatively large, a bit larger than 1. Now if I let a get smaller, you see that the nullcline changes, and right there when $a = 1$ it looks like the two nullclines merge at the point 400 along the y-axis. When we're above, we have an equilibrium point, our coexistence equilibrium point. But when a goes down to 1, that coexistence equilibrium point merges with the equilibrium point along the y-axis.

Now what happens for this differential equation? Let's first look at the Jacobian matrix when $a = 1$. When $a = 1$ the Jacobian matrix turns out to be 0,0 − 1 − 1. Now will we get an upper triangular matrix? Actually this is a lower triangular matrix. So our eigenvalues are displayed on the diagonal. We get 0, −1 as our eigenvalues. We've got an eigenvalue 0. That means linearization gives us no information about what's happening near these equilibrium points. A small change in the non-linear terms can get us off the 0 eigenvalue axes in the trace determinant plane and cause major changes in what's happening to our linear system. But if we look at the nullclines and what happens in those nullclines, we can actually read off what's happening.

Here when $a = 1$, we know that solutions that start out with large x and y tend to down. When they hit this nullcline they become vertical, x' is 0. Then in this region the vector field points down into the right. In simile our nullcline analysis from before says in this triangular region solutions should come up, but when they hit this nullcline, the y nullcline, then it becomes horizontal, enters this region, and solutions should go down. So in fact, we don't need linearization for that equilibrium point. By our nullcline analysis, all solutions tend to the equilibrium point along the x-axis. Our x species always survives. Our y species always dies out, unless of course there were no x species to begin with.

What happens as a continues to decrease? As a continues to decrease, now these nullclines no longer intersect at a point. We still have an equilibrium point at 400 on the x and y axes and on the origin. But now the same nullcline analysis tells us everything that's happening here. Solutions that start out with x and y large, hit 1 nullcline and tend to the equilibrium point on the x-axis. Solutions that start out in between the nullclines go down and to the right. Solutions that start out near x and $y = 0$ increase until they hit the y nullcline and then tend to the equilibrium point along the x-axis. So what we see is when a gets less than or equal to 1, all solutions just tend to the equilibrium point $x = 400$, $y = 0$. That's what happens when a, a competition parameter for y, gets small. When that happens, the population of x survives and y dies out.

What happens if now b decreases while a stays positive? Basically the same thing happens. So here we're starting with a and b larger than 1. Now let b decrease, and you see now the other nullcline moves, and then when b = 1 the two equilibrium points merge, our coexistence equilibrium point now merges with the equilibrium point on the x-axis and then they separate. Essentially the same thing happens except now that lower triangular region has disappeared, whereas the upper triangular region has emerged. What happens for all solutions, after b passes through 1, decreases through 1 is, now the other species survives. Now species y survives; x species disappears.

Again, we could have used linearization at the case where $b = 1$, at the place where the bifurcation occurs, but again we'd see a 0 eigenvalue. Fortunately at that point we know what happens. Our nullcline analysis wins out and we can predict the behavior of all solutions in this case.

What happens if both a and b are less than 1? This means species x and species y are not very competitive. What we'll see is another bifurcation occurs as b moves below 1 when a had already been less than 1. Let's see this bifurcation. Here a is less than 1, b is larger than 1. Here we see all solutions tending to the equilibrium point on the x-axis, x survives.

What happens as we change b? One of these nullclines again begins to change. We see a bifurcation right there. We see the reemergence of that coexistence equilibrium point. Watch what happens. Now when a and b are both less than 1, all solutions seem to be coming into the coexistence equilibrium point. All solutions that start out with x and y positive seem to tend to that equilibrium point. So a combination of our nullcline analysis and equilibrium point analysis, linearization, should tell us that.

I won't go through computing the Jacobian matrix. We can see pretty clearly from the behavior of solutions in the phase plane that that point really is a sink. We've got lots of different situations depending on a and b. What I'm going to do is summarize all of this by drawing the analog of what we call for first order differential equations the bifurcation diagram, the analog of what we saw for linear systems, the trace determinant plane, this is the parameter plane.

I'm going to draw and single out different regions in the parameter plane where we get different behaviors. So my parameter plane will be the a,b plane. What we saw was that if a and b were both larger than 1, so we had fairly large competition, then what we saw was generally speaking one of those two species survives, not both. Except in that isolated instance where you come into the saddle. That's what happens in the region in the a,b plane where both a and b are larger than 1. In the case where a and b were less than 1, we saw we still had a coexistence equilibrium point, but now both x and y species survive. If we go into the region where a is less than 1 and b is bigger than 1, then what we saw was species x survived, whereas if we went into the region where a was greater than 1 and b was less than 1, then species y survives.

So we divide up our a,b plane into various regions where we have different behavior. That's our parameter plane. That's our visual summary of what's happening. We can also summarize that in terms of mathematics; in the region where a and b are both greater than 1 we saw we had a saddle coexistence equilibrium point. In the region where a and b were less than 1, we saw we had a sink for our coexistence equilibrium point. On the other hand, when a was small but b was large, we saw that one of the two species survived and so we had no coexistence equilibrium point. It had merged with one of the saddles. Similarly in the region where a is larger than 1 and b is less than 1, again a coexistence equilibrium point had disappeared. We don't have that anymore.

This parameter plane is also a nice way to view bifurcations. Remember in a trace determinant plane, we'd often take treks through the trace determinant plane, and when we crossed certain axis we found bifurcations. That's exactly what happens in this parameter plane. For example if I fix a to be less than 1 and let b increase, we saw that a sink and a saddle merged, meanwhile the species y went extinct. Let's look at that bifurcation one more time.

Here a and b were both less than 1. We see our sink in the middle, our coexistence equilibrium point. As we raise b, that equilibrium point descends.

It merges with the equilibrium point on the x-axis and then disappears. We see a bifurcation. The bifurcation in terms of the population goes from having coexistence to having the species y die out and x survive. That's the importance of the parameter plane. It allows us to see where we're going to get major changes.

If we were to go to the right, that is fixed b small and let a increase, a similar bifurcation occurs. The coexistence equilibrium point disappears and all of a sudden the species x dies out.

On the other hand, if we were to start out with both a and b large and go either horizontally or vertically, again we see the merging of two equilibrium points. A sink and a saddle would emerge, and then either x or y would go extinct, a record of the different behaviors as we travel around the parameter plane.

There's only one slight trek that we didn't take. What happens if we pass through that point right in the middle where a and b are both 1? What happens there? When a and b are both 1, whoa, look what happens. Both of those nullclines now merge and equal each other. Remember the intersection points of the x and y nullclines are our equilibrium points. So there are a bunch of equilibrium points here, very different behavior. All of a sudden all of our solutions are tending to some equilibrium point, but they're very different depending on your initial condition.

Why does that happen? Well if a and b are equal to 1, then our x nullcline in the middle is given by $1 - x/m - y/m$. I should say that m and n are the same here. Our y nullcline are given by $1 - y/m - x/m$. They're the exact same curves. So we get a whole straight line of equilibrium points for this system of differential equations. As we saw many solutions come into this. Now back to the parameter plane, any small change will throw us off into one of those 4 very different regions. So any small change in our parameters would destroy this.

That's a summary of what's happening for our competing species model. Next time we'll come back and look at a different kind of physical model, this one from chemistry, and we'll deal with it using the same two techniques.

Limit Cycles and Oscillations in Chemistry
Lecture 18

For most of the systems of differential equations we have seen thus far, the most important solutions have usually been the equilibrium solutions. But there is another type of solution that often plays an important role in applications: **limit cycles**. These are solutions $Y(t)$ that satisfy $Y(t + A) = Y(t)$ for all values of t and have the additional property that they are isolated. By isolated, we mean that there are no nearby limit cycles; all nearby solutions either spiral in toward the limit cycle or spiral away from it.

As an example of a limit cycle, consider the system

$$x' = -y + x(1 - (x^2 + y^2))$$
$$y' = x + y(1 - (x^2 + y)).$$

Note that this vector field can be broken into a sum of 2 pieces: the linear system $x' = -y$, $y' = x$; and the nonlinear system given by $x' = x(1 - (x^2 + y^2))$, $y' = y(1 - (x^2 + y^2))$. The linear system is a center with solutions traveling around circles centered at the origin. The nonlinear system is a vector field that points directly away from the origin inside the unit circle, points directly toward the origin outside the unit circle, and vanishes on the unit circle. So when we add these 2 vector fields, we find that there is one solution that travels periodically around the unit circle, while all other solutions (except for the equilibrium point at the origin) spiral toward this periodic solution.

So the solution lying on the unit circle is a stable limit cycle, since nearby solutions move toward this solution. If we change the $+x$ and $+y$ to $-x$ and $-y$ in the second vector field, the solutions now spiral away, so we have an unstable limit cycle.

Figure 18.1

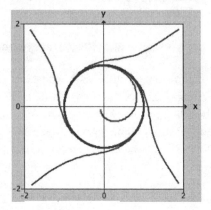

One of the principal tools for finding limit cycles is the Poincare-Bendixson theorem, which says the following: Suppose you have a ring-shaped region A in the plane where the vector field points into the interior of A along each of the 2 boundaries. Then provided there are no equilibrium points in A (and the vector field is sufficiently differentiable), there must be at least one limit cycle in A. In the figure below, we see a ring-shaped region A with limit cycle γ contained inside.

Figure 18.2

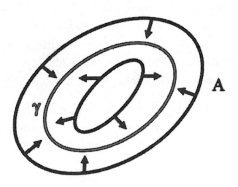

One of the most interesting discoveries in chemistry occurred in the 1950s, when the Russian biochemist Boris Belousov discovered that certain chemical reactions could oscillate back and forth between different states for long periods. Back then, most chemists thought that all chemical reactions settled directly to their equilibrium states. (For the interesting history, see the Winfree article in the Suggested Reading.)

Since that time, many chemical reactions have been found to oscillate. We'll look at a simpler model involving chlorine dioxide and iodide. The nonlinear system is

$$x' = -x + A - \frac{4xy}{1+x^2}$$

$$y' = B\left(x - \frac{xy}{1+x^2}\right),$$

where x and y are the concentrations of iodide and chlorine dioxide, respectively. There are 2 parameters here, A and B. We will simplify things by setting $A = 10$. A little algebra shows that there is a single equilibrium point at $(2, 5)$. Linearizing at this equilibrium point gives the Jacobian matrix

$$\begin{pmatrix} 7/5 & -8/5 \\ 8B/5 & -2B/5 \end{pmatrix}.$$

The trace of the Jacobian matrix is $T = \frac{7}{5} - \frac{2}{5}B$, and the determinant is $D = 2B$. Note that $T = 0$ and $D = 7$ when $B = 3.5$. So in the trace-determinant plane, we see that the linearized equation has a center when $B = 3.5$. When B is a little larger than 3.5, we have $T < 0$, so the equilibrium point is a spiral sink; when B is slightly smaller than 3.5, the equilibrium point is a spiral source. So we have some sort of bifurcation going on when $B = 3.5$. What happens both mathematically and chemically?

For the whole picture, we plot the nullclines. The x-nullclines are given by $x = 0$ and $y = 1 + x^2$, and the y-nullcline is given by

$$y = \frac{(10-x)(1+x^2)}{4x}.$$

Note that the y-nullcline has a vertical asymptote along the y-axis and crosses the x-axis only at $x = 10$. As usual, we compute the vector field in the regions between the nullclines to find what the phase plane looks like.

It is clear in this graph that solutions spiral around the equilibrium point. But how do they do this? When $B > 3.5$, we have a spiral sink at $(2, 5)$, and the computer shows that all solutions spiral into the equilibrium point.

Figure 18.3

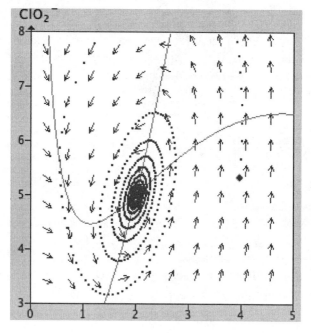

When the parameter B goes below 3.5, the equilibrium point changes to a spiral source—but something else happens. If we draw the rectangle bounded by the axes and the lines $x = 10$ and $y = 101$, we see that the vector field points into the rectangle R along all of the boundary points. Meanwhile we have the spiral source at $(2, 5)$, so we can find a small disk D around this point so that the vector field points outside of D at all points on the boundary of D. So we have a ring-shaped region $R - D$ on whose boundary the vector field points into the region $R - D$. By the Poincare-Bendixson theorem, there must be a limit cycle inside $R - D$.

In fact, what has happened is that this system has under gone a **Hopf bifurcation** at $B = 3.5$. Mathematically, when $B > 3.5$, all solutions tend to the spiral sink. But as B passes through 3.5, not only does the spiral sink become a spiral source, but we also see the birth of a limit cycle. In fact, all solutions (except the equilibrium point) now tend to this limit cycle, so this is a stable limit cycle.

Figure 18.4

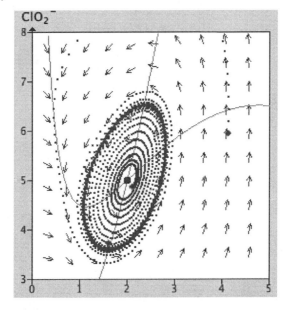

333

Chemically, when $B > 3.5$, the reaction oscillates down to its equilibrium state. But when $B > 3.5$, the reaction oscillates back and forth forever.

Important Terms

Hopf bifurcation: A kind of bifurcation for which an equilibrium changes from a sink to a source (or vice versa) and, meanwhile, a periodic solution is born.

limit cycle: A periodic solution of a nonlinear system of differential equations for which no nearby solutions are also periodic. Compared with linear systems, where a system that has one periodic solution will always have infinitely many additional periodic solutions, the limit cycle of a nonlinear system is isolated.

Suggested Reading

Edelstein-Keshet, *Mathematical Models in Biology*, chap. 8.8.

Hirsch, Smale, and Devaney, *Differential Equations*, chap. 10.7.

Strogatz, *Nonlinear Dynamics and Chaos*, chap. 8.3.

Winfree, "The Prehistory of the Belousov-Zhabotinsky Reaction."

Relevant Software

Blanchard, Devaney, and Hall, *DE Tools*, Chemical Oscillator, HPG System Solver.

Problems

1. **a.** Does the system
    $$x' = x^2 + y^2$$
    $$y' = x^2 + y^2$$
    have any limit cycles? Explain why or why not.

    b. What is the behavior of solutions of this system?

2. Show that the system
    $$x' = x - y - x(x^2 + y^2)$$
    $$y' = x + y - y(x^2 + y^2)$$
    has a periodic solution on the unit circle.

3. Does the system
    $$x' = x - x(x^2 + y^2)$$
    $$y' = x - y(x^2 + y^2)$$
    have a periodic solution on the unit circle?

4. Check that the equilibrium point for the oscillating chemical reaction in the lecture summary is indeed $(2, 5)$.

5. Describe the behavior of solutions of each of the following systems given in polar coordinates.

   a. $r' = r - r^2$, $\theta' = 1$

   b. $r' = r^3 - 3r^2 + 2r$, $\theta' = 1$

   c. $r' = \sin(r)$, $\theta' = -1$

6. Describe the bifurcations that occur in the system given by $r' = ar - r^2 + r^3$, $\theta' = 1$, where we assume that $a > 0$.

Exploration

A chemical reaction due to Schnakenberg is determined by the system

$$x' = x^2 y - x - 1/10$$
$$y' = -x^2 y + a,$$

where a is a parameter. Find the equilibrium points for this system, and compute the linearization. Do you observe any bifurcations? Can you show the existence of limit cycles for certain parameters?

Limit Cycles and Oscillations in Chemistry
Lecture 18—Transcript

Now it's time for some more nonlinear systems. For almost all of the differential equations with the first orders of systems that we've seen in this course, the most important solutions have often been the equilibrium solutions; but for nonlinear systems another type of solution that's important are so-called limit cycles. Today we'll look at limit cycles in nonlinear systems.

What are limit cycles? Basically they're periodic solutions of a differential equation like we saw in the undamped harmonic oscillator. However these periodic solutions are isolated. There are no nearby periodic solutions. A limit cycle is an isolated periodic solution. Remember saying it's periodic says that the solution $y(t)$ satisfies $y(t + A) = y(t)$. It's periodic.

Let me begin with an example. Let's take the nonlinear system $x' = -y + x(1 - (x^2 + y^2))$. Then x' is $x + y(1 - (x^2 + y^2))$. That's a nonlinear system. Pretty clearly we have an equilibrium point at the origin, but let's look elsewhere. Let me break this system, as I've done before, into two pieces; the linear piece and the nonlinear piece. So the linear piece will be $x' = -y$, $y' = x$. We know what happens to that linear system. We've seen over and over that all solutions except the origin are periodic. All solutions simply run periodically around the origin.

What about the nonlinear piece? The nonlinear piece is $x' = x(1 - x^2 + y^2)$, quantity $x^2 + y^2$, and y' is similarly. $y' = y(1 - x^2 + y^2)$. The linear system as I said is a center with solutions circling the origin, but what about that nonlinear system? That nonlinear system, it vanishes when $x^2 + y^2 = 0$. It's $x(1 - x^2 + y^2)$. So if $x^2 + y^2 = 1$, both of these terms vanish. Secondly this system is basically, well it's $xy(1 - r^2)$ where r^2 is $x^2 + y^2$. So we are taking the vector field xy, multiplying both terms by the quantity $1 - r^2$ and what happens to this? The nonlinear differential equation points directly away from 0 inside the unit circle and directly towards 0 outside the unit circle. On the unit circle where $x^2 + y^2 = 1$, that nonlinear vector field is just 0.

What do we have? Our linear system is always circling the origin. Our nonlinear system is pointing directly toward the unit circle, pointing out from inside the unit circle, pointing in from outside. So we would expect all of our solutions to spiral in to that one solution of the linear system that's the circle, the unit circles surrounding the origin. Let's look and see what happens with this nonlinear system.

There's our phase plane. If we start very close to 0, then our solutions quickly move away and tend to that periodic solution running around the unit circle. If we start far away, now our solutions tend in very quickly to that periodic solution. This is an example of a limit cycle because there's a periodic solution but, there are no nearby solutions. All of the nearby solutions spiral into this limit cycle. This is what's known as a stable limit cycle. In the previous example, if I changed my linear term from a plus to a minus, that would mean that the vector field was pointing away from the unit circle. Then solutions would spiral away from the limit cycle. That would be an un-stable limit cycle.

Actually finding the formulas for these limit cycles is generally impossible. There is one nice geometric theorem in mathematics that allows you to see that you do have a limit cycle for your system of differential equations. It's the Poincare-Bendixson Theorem. What this theorem says is, suppose you have some sort of ring-shaped region A in the plane. Think an annular region, but it doesn't have to be circular in the boundaries. It could be squares on the outside and maybe circles on the inside. You have some ring-shaped region in the plane. Suppose you know that the vector field points inside this ring-shaped region from both boundaries. On the outside it's pointing in. On the inside boundary it's pointing out. Then, first of all assuming there are no equilibrium points in this ring-shaped region and secondly that the vector field is sufficiently differentiable, then the Poincare-Bendixson Theorem says there has to be a limit cycle in the region, the ring-shaped region. So that's a very nice tool for actually proving the existence of limit cycles in nonlinear systems.

Now let me turn to another nonlinear system. This comes from chemistry. It revolves from one of the most interesting discoveries, I think, in chemistry when in the 1950s a Russian biochemist named Boris Belousov discovered that certain chemical reactions could oscillate back and forth between different states for long periods. Before that, all chemists thought that every chemical reaction would just tend down to equilibrium. It might take a while to get there but it would go to equilibrium.

Belousov found this example where it didn't, where it cycled, oscillated back and forth for a long time. There's a very interesting history regarding this. Belousov was simply dismissed by the rest of the then Soviet Chemical Society. He submitted his publication and the editor or somebody wrote back about his supposedly discovered discovery. In any event everyone just said this isn't true. He worked for the next 6 or 7 years trying to polish up his manuscript, trying to get more data that proved what he did, only to have his publication laughed at the second time. By then he was in his 60s. He decided to give up chemistry and just left, and there was this brilliant discovery gone. Luckily he left some of his papers around.

Meanwhile some other Soviet chemists had heard a little bit about what was going on with Belousov and they started investigating it. A young graduate student by the name of Zhabotinsky actually recreated the Belousov situation. It was right then in 1960s when one of the first joint U.S. Soviet chemistry meetings took place. Zhabotinsky presented his results and the chemical community went crazy. Wow, this can actually happen. Luckily as I said, Belousov kept his manuscripts around. People sort of knew he had done that. All of a sudden his name became very famous. In fact he and Zhabotinsky were awarded what was called the Lenin Prize back in 1980 for this discovery. Unfortunately by then Belousov had passed away. Nonetheless one of the biggest results in chemistry.

So here's the Belousov-Zhabotinsky reaction. Actually I'm going to simplify it a little bit so that the differential equation is more manageable, but basically this is the idea. Take two chemicals, let's say iodide and chlorine dioxide, and pour them into a vat at certain rates. They may change. Simultaneously instantaneously mix them together and meanwhile let the mixture drain out so that you have the same quantity of chemicals at all time.

Let me let x be the concentration of iodide and let y be the concentration of chlorine dioxide. Here's the differential equation for the rates of change of these two chemicals. Let's say x' is given by $-x + A$. Here A is a parameter. Then $-4xy/(1 + x^2)$. That's the differential equation for x. And $y' = B$—another parameter, basically the rate at which you're putting this chemical into the vat—$B(x - [xy/(1 + x^2)])$, a really complicated, nonlinear system of differential equations.

We have two parameters A and B. Let me just assume A is 10. The first job is find the equilibrium points. Look at the second equation. The second equation says don't worry about B. The second equation says $xy/(1 + x^2) = x$. But look at the first equation. We've got an $xy/(1 + x^2)$. So you have $-4xy/(1 + x^2)$. So that we can replace by just $-4x$. So in that case the first equation says that $-x + 10 - 4x = 0$ or $x = 2$. Then go to equation two, when $x = 2$ you see that equation two vanishes when $2 = 2y/5$ or $y = 5$. So for all of these parameters, for all the parameters B, we have our single equilibrium point at the point 2,5.

Now let's linearize at that equilibrium point. Remember our differential equation $x' = -x + 10 - 4xy/(1 + x^2)$ and $y' = B(x - [xy/(1 + x^2)])$. We've got to take the partial derivatives to get the Jacobian matrix; it's a lot of work. First, what's the partial derivative of the first equation with respect to x? Well, differentiate our first equation. The derivative of $-x$ is -1. The derivative of 10 is 0. Now we've got − times the derivative with respect to x of the quotient $4xy/(1 + x^2)$. So how are we going to differentiate that quotient? Remember the quotient rule from calculus. If you've got U/V then the quotient rule says the derivative of that quotient is $u'(v - v')u$ all over the denominator V^2.

So differentiating the term $4xy/1 + x^2$ gives $-4y(1 + x^2)$ the denominator. Then $-4xy$, the numerator, times the derivative denominator that's $2x$. All of that is over $1 + x^2$ quantity squared. So there's the partial of F with respect to x. We've got to plug in the values $y = 5$ and $x = 2$. When you do that you get the quantity -1, then $-20(5 - 40)4$ all divided by 25, which if you do the math adds up to 7/5. It's a kind of complicated, partial derivative, but elementary techniques from calculus shows you how to do it.

Similarly we've got to compute the partial of the first equation with respect to y, partial of the second equation with respect to x and y. When you do that you end up with the Jacobian matrix whose entries are 7/5 that we just computed, then $-8/5$, then $8B/5$ and $-2B/5$. The second role of your matrix depends on that parameter B.

What are the eigenvalues for our equilibrium point at the point 2,5 when B changes? We could compute all of the eigenvalues, but it's probably easier to then go back to the trace determinant plane. Our trace here is $7/5 - 2/5B$. The determinant is $7/5(-2B/5)$. That's $-14B/25$, and then $+ 64B/25$. So the determinant all together is just $50B/25$, or $2B$. So our determinant is $2B$. Therefore the trace is $7/5 - 2B/5 - D/5$, or if you simplify that, you see the determinant is given by $7 - 5T$. The determinant is given by $7 - 5T$. Go to the trace determinant plane. That's a straight line. That's a straight line with slope -5 and D intercept 7. When $B = 7/2$, when our parameter B is 3.5, then we have trace 0 and determinant 7. That's exactly where this straight line in the trace determinant plane crosses the D-axis. That's where our linearized system is a center. That's where linearization probably won't help us out.

To see what else happens let's go back to the other values in the trace determinant plane. In the trace determinant plane, these parameters give us a straight line given by the determinant is $7 - 5T$. So when B is a little larger than 7/2, then our trace is negative so we're off that center line in the trace determinant plane. We're in the region where we have a spiral sink. Similarly when B is a little smaller than 7/2, we're on the other side of that center line. We're in the region where we have a spiral source. We see some sort of bifurcation in this system as the value of B goes through 3.5. Mathematically we would go from a spiral sink to a spiral source.

In order to understand this more, let's plot the nullclines. Again, it's kind of complicated algebra. Let's first plot the y nullcline where $y' = 0$. That says that x must be equal to $xy/(1 + x^2)$. That says that our y nullclines are at $x = 0$, the y-axis, and at $y = 1 + x^2$. That's just a parabola opening upward, easy y nullclines. What about the x nullclines? We've got a set $0 = -x + 10 - (4xy/(1 + x^2))$. Bring the $-x + 10$ over to the other side and combine the fractions. You'll see that the x nullcline is given by y is $(10 - x)(1 + x^2)$ all divided by $4x$.

What do these nullclines look like? The y nullcline is easy. They're either $x = 0$, the x-axis, or the parabola $y = 1 + x^2$. I'm only interested in where the concentrations of iodide and chloride dioxide are positive. So we'll just look where x and y is positive. We see the vertical y-axis and the parabola opening upward. That's where our vector field is horizontal. Those are the y nullclines.

What about the x nullclines? The x nullclines we saw were given by $y = (0 - x)(1 + x^2)/4x$. What does that curve look like? Notice that the numerator only vanishes when $x = 10$. So this curve only crosses the x-axis at $x = 10$. Furthermore as x goes to 0 from the positive side, the numerator is always positive but the denominator goes to 0. That means that our x nullcline goes off to infinity as x approaches 0, as x comes closer to the y-axis. We also know that there's a single equilibrium point at the point 2,5. So our x nullcline must look like some curve that crosses the x-axis at 10, goes off to infinity near 0, and only meets that y nullcline at one point, namely the point 2,5. On that x nullcline our vector field is vertical.

Now let's find out what happens in between these regions and see what the direction of the vector field is. In the lower region I know the point 2,0 is there. Let me plug 2,0 into the differential equation. What you get is $x' = -2 + 10 + 0$. So $x' = 8$ and $y' = B(x)$. That's 2 again, $- 0$. That's $2B$ which is positive. So $x' = 8$ and $y' = 2B$. Both are positives in that lower region, the vector field must point up and to the right. Continue in all those other 4 regions. For example, $x = 2$, $y = 10$ is sort of in that upper region, the upper central region. At that point at $x = 2$, $y = 10$, -8; and $y' = B(2)$, and then if you compute it it's $-2B$. It's $-2B$. So that means that x' is negative and y' is also negative, so our vector field in that region points down and to the left.

Let's go on to the other regions. For example, a left-most region, pick the point say 1,3. Again, a simple computation says that $x' = 3$ and $y' = -B/2$. That's negative. So x' is positive, y' is negative. The vector field points down and to the right. If you go to the other region, let's say $x = 10$, $y = 1$, then again straight forward computations say that $x' = -40/101$. That's negative; and $y' = B(10 - 10/101)$, that's positive. So x' is negative and y' is positive. The vector field points to the left and up.

What we see from this nullcline analysis is that the solutions are just spiraling around that equilibrium point at 2,5, but how are they doing that? We've seen by linearization that when B is greater than 3.5 we actually have a spiral sink. On the other hand, when $B = 3.5$ the linearized system had a center. We'll have to see what happens there. Then when B was less than 3.5, we know we have a spiral source. Let's turn to the computer and see what happens to the entire system of differential equations.

So here is the system of equations governing the Belousov-Zhabotinsky reaction. Our plot is usual, the phase plane here, and down here I'll plot the concentrations of iodide and chlorine dioxide as a function of time. Then meanwhile I'll show you visually when we insert chlorine dioxide and iodide into this vat, instantaneously mix them, and let them flow out the bottom. Chlorine dioxide is painted purple and iodide is painted red in this picture. Now let's start at some place where our parameter B is greater than 7/2, where B is greater than 3.5. There we know we have a spiral sink. What happens to solutions? All solutions spiral into that equilibrium point at the origin. You might have seen the reaction turn a little bit yellow at the outset, but very quickly it settles down to equilibrium. As the Soviet scientists thought back in the 1950s and 60s, all chemical reactions should settle to equilibrium. No matter where I start this solution, this solution always tends to the equilibrium point at 2,5, always tends to our spiral sink.

What happens when we are at that point $B = 3.5$? What happens when $B = 3.5$? Remember that's where our spiral sink has become a center. That's where we don't know what's going to happen mathematically near our equilibrium point. Nonetheless what happens is solutions still spiral in. It takes a little longer to get there, but these solutions are still spiraling in to that equilibrium point at 2,5. It looks like it stopped, but if I keep going you see this solution is slowly making its way toward that equilibrium point at 2,5. It would take many, many time periods to actually get there. Nevertheless our solution would eventually tend to the equilibrium point, our chemical system would tend to equilibrium.

Now let me look at what happens when B is less than 3.5. Now we have a bifurcation. Now that equilibrium point at 2,5 suddenly becomes a spiral source. What happens? What happens is this: Start out nearby, you see the solution curve moves away, you see over in our vat the chemicals are oscillating between purple and yellow, purple and yellow, purple and yellow, and look mathematically at what's happening here? Solutions are actually tending to a limit cycle. Solutions from the outside are tending in to a periodic solution, and solutions from the inside are tending out also to that periodic solution.

We've got a solution to this differential equation that's a limit cycle. We've got a solution that corresponds in our Belousov-Zhabotinsky reaction to a solution that just varies periodically. The motion of these concentrations of these chemicals is a periodic function. It never settles down to rest. This, as I said earlier, was the discovery of Belousov way back when, which he was demeaned during his lifetime. But finally it's made a big difference in chemistry. I'm not a chemist but now I understand that not only can many of these reactions behave periodically but they can also behave chaotically as well.

One little term that comes up, we saw that the bifurcation was we went from a spiral sink when B was large to a spiral source, but something else happened there. When that equilibrium point changed from a spiral sink to a spiral source, suddenly we also saw the birth of a limit cycle. So two things happened, we get a change in the type of the equilibrium points and we also get the birth of a limit cycle. That's what's known in mathematics as a Hopf bifurcation.

Finally, how do we actually see that this limit cycle exists mathematically? That's a concept of the theorem we talked about before. Let's construct an annular region, a ring-shaped region where the vector field is always pointing inside. So if B is less than 3.5, let me first draw the rectangle. It's bounded by the x and y axes, and $x = 10$ and go all the way up that nullcline $y = 1 + x^2$. You'll reach there at $x = 10$, $y = 1 + 100$. $y = 101$ and then go left. $y = 101$. That's our annular region.

Using our nullcline analysis we see that the vector field points inside this rectangle at all points along its boundary. On the axes on the left it's pointing to the right, below it's pointing up and to the right. Along $x = 10$ we saw it pointed to the left and up, and above we saw it was pointing down. There's one slight problem here. At the point 10,101 there the vector field was right on the nullcline. There the vector field is pointing horizontally. But there's no problem because as soon as it moves horizontally, then the vector field is pointing down so the solution still immediately enters the rectangle bounded by those lines.

When B is less than 3.5 we know we have a spiral source at the equilibrium point. We know that nearby solutions are running away. We can find a small disc bounded by some closed curve around that equilibrium point where the vector field is always pointing outward. If we take the region, the rectangle minus that disc, we've got a ring-shaped region and the vector field is always pointing inside that region. We know there's no equilibrium points in there, so by the Poincare-Bendixson Theorem we must have a limit cycle in that ring-shaped region. Technically we could have more than one, but we saw that we have just one limit cycle inside that region.

That's what's going on with the Belousov-Zhabotinsky reaction. We'll just summarize it as usual. Mathematically when B was greater than 3.5 we saw that all solutions tended to a spiral sink, but when B goes below 3.5, that sink suddenly becomes a spiral source and we get the birth of that limit cycle, a stable limit cycle. Then chemically when B is greater than 3.5, our reaction just dies out. The reaction oscillates down to its equilibrium state. But when B goes below 3.5, the reaction oscillates forever. Again, it's a kind of interesting thing that occurs in the Belousov-Zhabotinsky reaction.

Next time we're going to move onto a couple of more specialized techniques for understanding nonlinear equations. Most of the equations arise in physics, mechanics, and things like that. Namely we'll look at Hamiltonian and Lyapunov functions. They are very different techniques but also very powerful nonlinear techniques.

All Sorts of Nonlinear Pendulums
Lecture 19

In this lecture, we introduce several new types of methods used to understand nonlinear systems of differential equations. These include Hamiltonian and Lyapunov functions. In each case, we will concentrate on various nonlinear pendulum equations.

The pendulum equations arise as follows. Suppose we have a light rod of length L with a ball (the bob) of mass m at one end. The other end of the rod is attached to a wall and is free to swing around in a circular motion. Our goal is to describe the motion of the ball. Suppose the position of the center of mass of the ball at time t is given by $\theta(t)$. Let's say that $\theta = 0$ is the downward-pointing direction and that θ increases in the counterclockwise direction.

We will assume that there are 2 forces acting on the pendulum. The first is the constant force of gravity given by mg (where $g \approx 9.8$ m/s^2). Only the force tangent to the circle of motion affects the motion, so this force is $-mg\sin(\theta)$. We determine this with trigonometry.

Figure 19.1

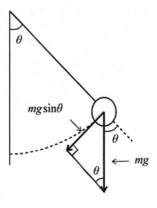

The position of the bob at time t is $(L\sin(\theta(t)), -L\cos(\theta(t)))$. The speed of the bob is then the length of the velocity vector, which is given by $(L\cos(\theta(t))\theta'(t), L\sin(\theta(t))\theta'(t))$. So the length of this vector is $L\theta'$. Similarly, the component of the acceleration vector in the direction of the motion is $L\theta''$. Our second force is then the force due to friction, which we assume as in the case of the mass-spring system is proportional to velocity. So this force is given by $-bL\theta'(t)$, where b is again called the damping constant.

Newton's second law, $F = ma$, then gives the second-order equation for the pendulum:

$$\theta'' + \frac{b}{m}\theta' + \frac{g}{L}\sin(\theta) = 0 \,.$$

As a system, we get

$$\theta' = v$$
$$v' = -\frac{b}{m}v - \frac{g}{L}\sin(\theta).$$

Note that because of the sine term in the second equation, this is a nonlinear system of differential equations.

Let's first consider the case where there is no friction. This is called the ideal pendulum. We assume as usual that $L = m = 1$, so the system of equations becomes

$$\theta' = v$$
$$v' = -g\sin(\theta).$$

As always, we first find the equilibrium points. Clearly, they are given by $v = 0$ and $\sin(\theta) = 0$. So our equilibria lie at $v = 0$ and $\theta = n\pi$ (where n is any integer). When n is even, this equilibrium point corresponds to the pendulum hanging at rest in the downward position. When n is odd, the pendulum is at rest in its perfectly balanced upward position. There is a conserved quantity for this system of equations. Consider what physicists call the energy function:

$$H(\theta, v) = \tfrac{1}{2} v^2 - g \cos(\theta).$$

In our solutions, both θ and v are functions of t, so we can consider the energy function also to be a function of t. And then we can compute its derivative with respect to t. We find that

$$\frac{dH}{dt} = v v' - g \sin(\theta)\, \theta' = v g \sin(\theta) - v g \sin(\theta) = 0.$$

Therefore H must be constant along any of the solution curves of the system. So we just plot the level curves of the function H (i.e., the curves given by $H = $ constant). Then the solutions must lie along these level curves.

Figure 19.2

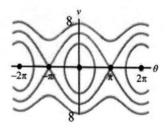

We know that our vector field is everywhere tangent to these level curves, so our solutions must therefore look like the below.

Figure 19.3

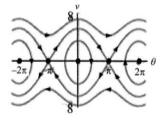

Besides our equilibrium solutions where the pendulum is either perfectly balanced in the upright position or hanging motionless in the downward position, we see 3 kinds of other solution curves labeled *A*, *B*, and *C*.

Figure 19.4

Close to the origin (or to any equilibrium point of the form $(2n\pi, 0)$), we see circular motions labeled *A*. These solutions correspond to the pendulum swinging back and forth periodically without ever crossing the upward position $\theta = \pi$. The second type of solution (labeled *B*) is where the pendulum moves continuously in the counterclockwise direction (or in the clockwise direction if $v < 0$).

The solutions labeled C tend to the upright equilibrium points as t tends to both $\pm\infty$. These upright equilibria are given by $(\pm\pi, 0)$. Using linearization, we can easily check that all of these points are saddle points. So this solution is what is called a **separatrix** (or saddle connection): It tends from one saddle point to another. These are the solutions that you would never see in practice. As time goes on, the pendulum tends to the upright equilibrium without ever passing through $\theta = \pi$.

The ideal pendulum is an example of a Hamiltonian system, named for the Irish mathematician William Rowan Hamilton (1805–1865). A Hamiltonian system in the plane is a system of differential equations determined by a function $H(x, y)$, which is called the Hamiltonian. The system is then given by

$$x' = \frac{\partial H}{\partial y}(x, y)$$
$$y' = -\frac{\partial H}{\partial x}(x, y).$$

H is a conserved quantity since, as before, if we differentiate H with respect to t, we find that

$$\frac{dH}{dt} = \frac{\partial H}{\partial x} \cdot x'(t) + \frac{\partial H}{\partial y} \cdot y'(t) = \frac{\partial H}{\partial x} \cdot \frac{\partial H}{\partial y} + \frac{\partial H}{\partial y} \cdot \left(-\frac{\partial H}{\partial x}\right) = 0.$$

We have seen another Hamiltonian system earlier, namely, the undamped mass-spring system:

$$y' = v$$
$$v' = -ay.$$

Here the Hamiltonian is given by

$$H(y,v) = \tfrac{1}{2}v^2 + \tfrac{a}{2}y^2$$

since

$$y' = v = \frac{\partial H}{\partial v}$$

and

$$v' = -ay = -\frac{\partial H}{\partial y}.$$

All level curves of H here are just ellipses or circles surrounding the origin (i.e., we have a center equilibrium point at the origin).

Now let's return to the original pendulum equation, but this time we will allow friction. Our second-order equation was

$$\theta'' + b\theta' + g\sin(\theta) = 0,$$

assuming $L = m = 1$ or, as a system,

$$\theta' = v$$
$$v' = -bv - g\sin(\theta).$$

This system is no longer Hamiltonian. But look at our former energy function from the ideal case

$$H(\theta, v) = \tfrac{1}{2}v^2 - g\cos(\theta).$$

We compute

$$\frac{dH}{dt} = v \cdot v' - g\sin(\theta)\theta' = v(-bv - g\sin(\theta)) + g\sin(\theta)v = -bv^2 \leq 0.$$

This says that $H(\theta(t), v(t))$ is now a nonincreasing function along solution curves. So our solutions must now descend through the level curves of the function H. So we again know what happens to our solutions. Recall that the level curves of H looked like the following.

Figure 19.5

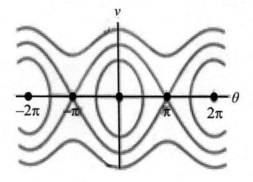

Now our solutions must look something like the below.

Figure 19.6

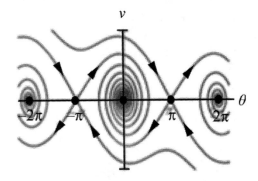

We see that now all solutions tend to the equilibrium point at which the pendulum hangs straight downward. A function like H is called a **Lyapunov function**, which is named for the Russian mathematician Aleksandr Lyapunov (1857–1918).

You may ask, when is a system of the form

$X' = F(x, y)$

$Y' = G(x, y)$

Hamiltonian? We would need $F(x, y) = \partial H/\partial y$ and $G(x, y) = -\partial H/\partial x$. If such an H exists and is twice continuously differentiable, we would need to see the following.

$$\frac{-\partial G}{\partial y} = \frac{\partial^2 H}{\partial y \partial x} = \frac{\partial^2 H}{\partial x \partial y} = \frac{\partial F}{\partial x}$$

Important Terms

Lyapunov function: A function that is non-increasing along all solutions of a system of differential equations. Therefore, the corresponding solution must move downward through the level sets of the function (i.e., the sets where the function is constant). Such a function can be used to derive the stability properties at an equilibrium without solving the underlying equation.

separatrix: The kind of solution that begins at a saddle point of a planar system of differential equations and tends to another such point as time goes on.

Suggested Reading

Blanchard, Devaney, and Hall, *Differential Equations*, chap. 5.3.

Guckenheimer and Holmes, *Nonlinear Oscillations*, chap. 2.2.

Hirsch, Smale, and Devaney, *Differential Equations*, chap. 9.4.

Roberts, *Ordinary Differential Equations*, chap. 10.7.

Strogatz, *Nonlinear Dynamics and Chaos*, chaps. 4.3 and 6.7.

Relevant Software

Blanchard, Devaney, and Hall, *DE Tools*, Duffing, Pendulums.

Problems

1. For a typical solution of the ideal pendulum differential equation, at which position is the bob moving fastest?

2. For a typical solution of the ideal pendulum differential equation, at which position is the bob moving slowest?

3. For the ideal pendulum, use linearization to determine the type of the upward equilibrium point.

4. For the ideal pendulum, use linearization to determine the type of the downward equilibrium point.

5. Use linearization to determine the type of the downward equilibrium point for the damped pendulum.

6. Use linearization to determine the type of the upward equilibrium point for the damped pendulum.

7. Is $x' = x^2 + y^2$, $y' = -2xy$ a Hamiltonian system?

8. Consider the linear system $Y' = AY$, where

$$A = \begin{pmatrix} a & b \\ c & d \end{pmatrix}.$$

Determine conditions on a, b, c, and d that guarantee that this system is Hamiltonian. What types of equilibrium points can occur for such a linear Hamiltonian system?

9. Find the Hamiltonian function for the linear system in problem 8.

Exploration

Duffing's equation is a model for a mass-spring system with a cubic stiffness term. It is also a model for the motion of a magnetoelastic beam mounted between 2 magnets. The second-order equation is given by $y'' + by - y + y^3$. First, assuming the damping constant $b = 0$, show that this is a Hamiltonian system. Second, what happens when b is nonzero? Third, use a computer to investigate the behavior of the Duffing equation when a periodic forcing term, say $F\cos(wt)$, is introduced.

All Sorts of Nonlinear Pendulums
Lecture 19—Transcript

In the last few lectures we viewed several different techniques to understand nonlinear systems of differential equations, namely nullclines and linearization. In this lecture we'll investigate several other ways to come to grips with nonlinear systems, namely Hamiltonian and Lyapunov functions. In those cases here we'll concentrate on one particular example, namely the nonlinear pendulum.

The nonlinear pendulum equation arises as follows. We have a light rod of length L and we attach a ball of mass m at one end. That ball is often called the "bob". I don't know why they named it after me but they did. In any event, then we let the bob swing as a typical pendulum does. We'll measure the position of the bob by $\theta(t)$ where the angle $\theta = 0$ is the downward position and where increases in the counter clockwise direction. So that means that the position of the bob at time T measured in the plane is given by the vector $L\sin(\theta(t)) - L\cos(\theta(t))$. And when $\theta = 0$ the position is $0 - L$ directly downward. As θ increases the number $L\sin\theta$ turns positive, so indeed the motion is counter clockwise.

For differential equations' sake we need the velocity. The velocity vector here though is the derivative, namely $L\cos(\theta(t))\theta'(t)$ and then $L\sin(\theta(t))\theta'(t)$. Notice the magnitude of that velocity vector is just $L\theta'$. Similarly the acceleration is given by $L\theta''(t)$. For this nonlinear pendulum system there'll be two forces. The first force will be the force of gravity which is given by mg, that constant—the gravitational constant g, so g is a constant throughout. Only the force that's tangential to the circle of motion affects the motion. So this force isn't mg, rather it's $-mg\sin(\theta)$.

Where does that come from? That actually comes from a little trigonometry. What we'll do is draw the straight line down from position θ at time t of length mg. Our pendulum weren't attached to the wall or anything. The force would just be that downward force mg, but it isn't. Our pendulum is required to swing around the circle. What we've got to do is take the component of this vector tangent to the circle. That's the force on the bob.

So we've got to compute that. How do we do that? As I said, at position θ we take the downward force factor of length mg. Then we draw the vector tangent to the circle of motion and say pointing downward. That's going to be the force acting on the bob. So we connect the tips of that tangential vector and the downward vector so that the resulting triangle is a right triangle. Then that side of the triangle that's tangent to the circle of motion is our actual force vector. To compute its length what we do first is to extend the rod through the pendulum a little bit. When we do we see that the angle between this extended rod and the downward force vector mg is exactly the same as our position angle, namely it's θ.

One side of the triangle turns out to be parallel to this extended rod. So the lower angle in that triangle is also given by the same number, namely θ. From a little bit of trigonometry look at that triangle. We know that the sign of that lower angle θ is equal to the length of the opposite side divided by the length of the hypotenuse. So the $\sin(\theta) = f/mg$. That says then that the length of the force vector is indeed as I said earlier $mg \sin(\theta)$ since θ is increasing in the counter clockwise direction we want the force acting in the downward direction, in the opposite direction. So we take the force to be $-mg \sin(\theta)$.

recall that the acceleration of the bob is given by $L\theta''$. When there's no other force present we therefore have the differential equation for the nonlinear pendulum. Newton's second law, force = mass(acceleration) says that the second order equation for the nonlinear pendulum is $ML' =$ force, $-mg \sin()$. Or written out $\theta'' + g/L(\sin(\theta)) = 0$. This is the differential equation for what's called the ideal pendulum. There's no friction present.

Our second force though is then the force that's due to friction, which we'll assume as usual as in the case of a mass spring system for example, to be proportional to the velocity. this second force is given by $-b$, a damping constant b, times the velocity, $L\theta'$. Again, b is called the damping constant.

So we have it. Newton's second law force = mass(acceleration) then gives the second order differential equation for the pendulum. It's $\theta'' + b/m\theta' + g/L \sin(\theta) = 0$. Or as a system of differential equations, it's $\theta' = v$, as usual of velocity. Then $v' = \theta''$. Bring all of the other stuff over to the right.

$V' = -b/m(v) - g/L(\sin(\theta))$. Because we have a $\sin(\theta)$ term on the right-hand side this is a nonlinear system of differential equations.

Let's first consider the case where there's no friction. So it's the damping constant $b = 0$. This is what we call the ideal pendulum. I'm a mathematician so I'll assume as all mathematicians do that most of those constants are 1. I'll assume the length L and the mass m is 1. So our system of differential equations becomes $\theta' = v$ and $v' = -g\sin(\theta)$.

First where are the equilibrium points? Pretty clearly the equilibrium points are when $v = 0$ and $\sin(\theta) = 0$. So θ must be $n\pi$ where n is any integer. If n is even, that means that the pendulum is pointing in the down position, right, for when say $n = 0$. When n is odd, say $\theta = \pi$, then the pendulum is at rest at its perfectly balanced upward position, an equilibrium solution that you'd never see in practice, or very rarely see.

How we're going to solve this equation is that from physics there is a conserved quantity for this system of equations. This is what the physicists call the energy function, the total energy function. It's given by the function of θ and v. I'll call it, $H(\theta,v) = \frac{1}{2}v^2 - g\cos()$. Now let's think of H as a function of t. We can do that because for our solutions $\theta(t)$ would be one solution and $v(t)$ would be the other. So let's compute the derivative of H with respect to t. H consists of $\frac{1}{2}v^2 - g\cos(\theta)$. The derivative by the chain rule is the derivative $\frac{1}{2}v^2 = vv'(t) +$ the derivative of $-g\cos(\theta) = g\sin(\theta)$ now times the derivative of θ, $\theta'(t)$. Or all together plugging in the differential equation you see $-vg\sin(\theta) + vg\sin(\theta)$ and that adds up to 0. So the derivative of H with respect to t is equal to 0. That says that H must be a constant along any of the solution curves of this system.

What do we do to understand these solutions? We just plot the level curves of our function H. That is the curves given by $H(\theta,v)$ equal to a constant. Then our solutions must lie along these level curves.

What do the level curves look like? Let's do a couple of examples. What if $H = $ constant $-g$? So $v^2/2 - g\cos(\theta) = -g$. The only places where that happens is $\theta = 0$ and $v = 0$. Or also $\theta = \pm 2n\pi$. If θ is not equal to $\pm 2n\pi$ then $\cos(\theta)$ is less than 1 so H is larger than $-g$. What about the level curve say

H = the constant g? Well there we need $v^2/2$ to be equal to $g\cos(\theta) + g$. This holds at two points given any θ value, except when $\theta = \pm\pi$. When $\theta = \pm\pi$ we then have $v^2/2 = 0$, so we need v to be $= 0$. That says that the level curve corresponding to $H = g$ consists of two points, one plus and one minus in the v direction except over $\pm\pi$ and its two π multiples.

Now what if we take H to be between $-g$ and g? Let's say for simplicity $H = 0$. Then we need $v^2/2$ to be equal to $g\cos(\theta)$. Again, we've got two such values for v, namely $v = \pm\cos(\theta)$ to the ½, unless $\theta = \pm\pi/2$, and here again we get $v = 0$. If θ is on the other side of $\pm\pi/2$ then $\cos(\theta)$ is negative. We have no such values. Then finally, what are the level curves for H greater g look like? Here we need $v^2/2 - g\cos() =$ to some constant greater than g. Then that happens for, you can find two such v values for any given θ. So this tells us what the level curves of the function H look like.

Here we've plotted the level curves for this function H. You see when H is greater than g, our curve extends like so. When $H = g$ our curve has two v values; one above, one below. Okay. But only one v value at these points $\pm\pi$. And then when H is between $\pm g$ our level curves are just these ellipses. Now we know that our solutions are constant along all of these level curves. Let's see what the solutions look like.

If H is greater than g, that's the case where our pendulum just continually rotates maybe in the counter clockwise or the clockwise direction. When H is smaller than g, that's where our pendulum just swings back and forth without rotating. If we're at $-g$, that's the equilibrium point where the pendulum is pointing downward. Then where $H = g$, well there's an equilibrium point right there at $\theta = -\pi$ or $+ \pi$ and also there's a solution that sort of connects it. Well, technically you can never find that. There is a solution that goes from the upward equilibrium to the upward equilibrium swinging through two π units as it does. So there's the picture of all of our solutions. There's the picture of the motion of the pendulum. What we see is that function H determines exactly what's happening.

This situation comes up a lot mostly in physics and in mechanics. This is an example of a Hamiltonian system, named for the Irish mathematician William Rowan Hamilton. What is a Hamiltonian system? Well, given a function of two variables, let's now say x and y, say H is a function of x and y, a Hamiltonian system is a system of differential equations of the form $x' = \partial H$ with respect to y, and $y' = -\partial H$ with respect to x. If you have such a system that function H is a conserved quantity, because compute the derivative of H with respect to t. The derivative of H with respect to t is first take the partial with respect to x and then multiple by $x'(t)$. Then take ∂H with respect to y and multiple by $y'(t)$ by the chain rule. Remember though $x'(t)$ is $dH\, dy$ and $y'(t) = -\partial H$ with respect to x.

All together you get the $dH\, dt$ is the partial derivative of H with respect to x times the partial with respect to y and then minus the partial with respect to y times the partial with respect to x or 0. In our case for the ideal pendulum our Hamiltonian function H was just the function we just saw; $\frac{1}{2}v^2 - g\cos(\theta)$. θ' indeed was v. That's ∂H with respect to v. The partial of $\frac{1}{2}v^2$ with respect to v, and v' was $-g\sin(\theta)$. And that's minus the derivative of H with respect to θ.

We've actually seen another Hamiltonian system in this series of lectures before. For example, the undamped mass-spring system given by $y' = v$ and $v' = -ky$. Here the Hamiltonian function is $H(y,v) = \frac{1}{2}v^2 + k/2y^2$. The reason for that is $y' = v$. Yes, that's the ∂H with respect to v, and $v' = -ky$ and yes, that's $-\partial H$ with respect to y. So let's just look at these level curves. Here's our undamped mass spring system. If you plot the level curves of H, you see that they're all ellipses. Yes, $\frac{1}{2}v^2 + k/2y^2$ is in fact equal to a constant and gives these level curves. This too is a Hamiltonian system. As we've seen, all of these solutions run around on ellipses. All of these solutions run around on the level curves of the Hamiltonian function.

Let's go back to the nonlinear pendulum. Now let's include damping. Remember our differential equation was $\theta'' + b\theta' + g\sin(\theta) = 0$ and here the damping constant b is positive. Again I'm assuming that the mass and the length of the rod is 1. So as a system of differential equations this system now reads $\theta' = v$ and $v' = -bv - g\sin(\theta)$. We've got a new term, the term $-bv$.

This system is no longer Hamiltonian. But look at our old Hamiltonian function, $H(\theta,v)$ given by $\frac{1}{2}v^2 - g\cos(\theta)$. If we differentiate that with respect to t, we find the following: dH/dt is of course the derivative of $\frac{1}{2}v^2$.

That's $v(v') + g\sin(\theta)\theta'$, just as before. But now putting in v' and θ' we get $v(-bv - g\sin(\theta))$ and then $+ g\sin(\theta)v$. So the $g\sin(\theta)$ terms cancel out and what we're left with is $-bv^2$. The derivative of H with respect to t is $-bv^2$. Remember b is positive. That number, that derivative is less than or equal to 0. This says that our Hamiltonian function is now a non-increasing function along the solution curves of this system. The derivative with respect to t is always less than or equal to 0. So our solutions must now descend through those level curves of the function H that we plotted earlier.

The only place where H' with respect to t, the derivative of H with respect to t can be 0 is when $v = 0$. Along the v-axis the derivative of H with respect to t is 0. Along that axis we've got our equilibrium points; $\theta = 0$, $\theta = \pm\pi$, etcetera, H is constant at those equilibrium points. But at all other points v is not 0 or π or whatever, so v' is not $= 0$, $v' = -g\sin()$ so it's not equal to 0. So solutions actually pass through $v = 0$ and keep going. This says that all non-equilibrium solutions for the nonlinear damped pendulum equation must decrease through those level curves and then just keep going to equilibrium.

This function H is now called a Lyapunov function, named for the Russian mathematician Aleksandr Lyapunov. Let me show you what this looks like. Remember what our level sets look like. For the case where $v = 0$ we had a level set up top and down below. We had circular level sets in the middle. We had level sets connecting the equilibrium $\pm\pi$ either above or below.

When we included damping, our solutions must be descending through all of these level sets, and they've got to tend either to one of the equilibrium points say at π or at 0. So let's see that by including some damping here. Let me let B be nonzero. What we see is, just as in the real world, our solution winds down to equilibrium. Mathematically it's descending through those level curves of the function H. Physically the pendulum is coming to rest at the lower equilibrium point. What we see is again, plotting the level curves of this now Lyapunov function tells us essentially everything about what's happening with this nonlinear system.

Now let's go back to the mass-spring system. Remember when there was no damping we had a Hamiltonian system. What happens if we include damping? The Hamiltonian function in the undamped case was H of y,v.

We're in the v, v situation now. H of y, v is $\frac{1}{2}v^2 + k/2y^2$. We saw the level sets were all ellipses. Now if we go to the undamped case and compute dH, dt we see that the derivative of H with respect to t is $v(v' + y(y'))$. Now that gives us v times not only $-ky$ but also $-bv$ and then $+ k(y,v)$. So again, the $K(y,v)$ terms disappear and we're left with again, $-bv^2$. That's the derivative of H with respect t. Again we see dH, dt is negative. So again, solutions decrease through those level curves for our old Hamiltonian function, again we have a Lyapunov function.

Let's see this for the mass-spring system. So here I've plotted those old level curves for our Hamiltonian function in the case where the damping was 0. Now when damping is nonzero we know that our solutions must decrease through these level curves and tends to an equilibrium point, in fact the equilibrium point at the origin. If we add a little bit of damping exactly as we've seen before solutions now decay right to the origin. If the damping is not that large, if it is very close to 0 it takes a long time or a longer time to decay to 0. The fact is again, mathematically we know what's happening. In the case where there's no damping we've got our level curves. As soon as there's damping solutions must always go down, down, down through those level curves.

Finally how do we know if a system is Hamiltonian? If I give you $x' = f(x,y)$ and $y' = g(x,y)$ how do you know you can find a Hamiltonian function? Well you've got to find a function H such that $x' = \partial H$ with respect to y and $y' = -\partial H$ with respect to x. If such a function exists, you have such a function H, and this function is sufficiently differentiable, twice continuously differential, then you've got to have that the mix partial derivatives of H with respect to y and x and x and y are equal. That is the second partial derivative of H first with respect to x, then with respect to y must be equal to the second partial derivative of H, first with respect to y and then with respect to x. Think about it though. The second partial of H with respect to y with respect to x would then be just minus the partial derivative of g with respect to y. Similarly the second partial of H with respect to x and then with respect to y would be the partial of F with respect to x.

If we are going to have a Hamiltonian function we must have that the partial derivative of F, our first equation, with respect to x is equal to the negative of the partial derivative of g, our second differential equation, with respect to y. For example is the differential equation say $x' = x + y^2$ and $y' = y^2 - x$ is that a Hamiltonian system? The partial of the first with respect to x is just 1, whereas the partial of the second equation with respect to y minus that partial derivative is $-2y$. One is not equal to $-2y$. That's not a Hamiltonian system.

How about the differential equation $x' = y$ and $y' = x - x^2$? Now the partial derivative of F with respect to $x = 0$ and the partial derivative of g with respect to y, technically minus that is also equal to 0. So this system is a Hamiltonian system. We have a Hamiltonian function. What is that Hamiltonian function? We must have that y, that's the first equation, must be equal to the partial of H with respect to y. So let's integrate that equation with respect to y. We get the interval of $y = y^2/2$ with now plus a constant. But that constant could involve x because we're integrating with respect to y. So the partial of H with respect to y tells us that Hamiltonian function must be of the form $y^2/2$ + some function of x. What's that function of x? We also know that $x - x^3$ is minus the partial of H with respect to x. But if we differentiate the H function with respect to x, we just found, we get $-Fe'$, the derivative of Fe our missing function, Fe' of x. So $x - x^3$ must be equal to minus the derivative Fe' of x.

Or integrating that with respect to now x we see that Fe of x must be given by $x^4/4 - x^2/2$. So a little bit of integration gives us our Hamiltonian function. Our Hamiltonian function is $x^4/4 - x^2/2 + y^2/2$ + a constant. If you now differentiate that function with respect to t, you'll see that the derivative of H with respect to t is equal to 0. The level sets of that function determine the behavior of this differential equation.

Just as an application this differential equation is what's known as the Duffing oscillator. What you do is you take a very slender steel beam clamped to the roof between two magnets equally distant apart and no friction in between and pull the little narrow steel beam out of it and let it move, and then the magnets force it to bounce back and forth. This is then our system of differential equations $x' = y$, $y' = x - x^3$. Let's just look at the motion of this Duffing oscillator.

Here it is. Here's our differential equation. Here on the right is the motion of our slender steel beam with magnets to the left and to the right. Here are the level curves of that Hamiltonian function. Solutions may run around the origin, that is the steel beam can oscillate back and forth between the two magnets. Solutions could run around on one side of the origin. Our magnet forces the steel beam to stay to one side. Solutions can look again, are running around the origin.

So again, those are the pictures of the level curves of the Hamiltonian function which we could plot almost by hand. That would tell us exactly that this system of differential equations behaves in that way. There's an equilibrium point at the origin which we'll never see, a saddle point. But again, the level curves tell us everything that's happening. Finally if I include some damping into this equation, we get a Lyapunov function. Now all of our solutions eventually run down to one of those equilibrium points. Maybe the equilibrium point on the left or the equilibrium point on the right, but again all solutions just go down through these level sets and tend to equilibrium.

So Hamiltonian and Lyapunov functions give us another grasp on the types of systems that often arise in physics, specifically in mechanics. Next time we're going to sort of change things a little bit. Next time we'll take the pendulum and add a little bit of forcing. What you'll see is out comes the phenomenon called chaos.

Periodic Forcing and How Chaos Occurs
Lecture 20

In this lecture, we turn to one of the most interesting developments in the modern theory of differential equations: the realization that solutions of systems of differential equations may behave chaotically. Our first example is the periodically forced pendulum. We can force the pendulum in one of 2 ways: moving the pendulum up and down periodically or forcing the bob alternately in the clockwise and counterclockwise directions.

In the first case, the system of differential equations is given by

$$\theta' = v$$
$$v' = -bv - g\sin(\theta) - F\cos(\omega t)\sin(\theta).$$

Here F measures how far we move the pendulum up and down, and $2\pi/\omega$ is the period of the forcing.

In the undamped case, we see that the solution curves meander around the phase plane in a very complicated fashion. Sometimes the pendulum moves around in the clockwise direction; at other times, it moves around in the opposite direction. The question is can you predict 5 minutes from now which way the pendulum will be swinging? The answer is very definitely no; this unpredictability is one of the hallmarks of chaotic behavior.

Figure 20.1

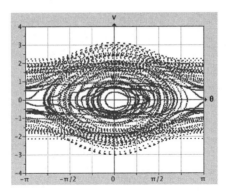

Another ingredient of chaos is what is known as sensitive dependence on initial conditions. Basically, this means the following: If we start the pendulum out at 2 different but very nearby initial points (θ_0, v_0) and (θ_1, v_1), then by continuity, the corresponding solutions start out in very similar fashion. But before long, the 2 solutions diverge from one another, and the corresponding motions of the pendulum are vastly different from one another.

In order to see how we understand this chaos, let's consider the Lorenz system of differential equations. This was the first system of differential equations that was shown to behave in a chaotic fashion. Edward Lorenz (1917–2008) began his career as a mathematician but turned his attention to meteorology while in the army during World War II. In an attempt to understand why meteorologists had a hard time predicting the weather, Lorenz suggested a simplified model for weather. Basically, Lorenz suggested trying to predict the weather on a planet that was surrounded by a single fluid particle. This particle is heated from below and cooled from above, and like the entire atmosphere on Earth, the particle moves around in convection rolls.

The Lorenz system of differential equations that describes this motion is given by

$$x' = a(y - x)$$
$$y' = Rx - y - xz$$
$$z' = -bz + xy.$$

Here a, b, and r are parameters. For simplicity, let us set $a = 10$ and $b = 8/3$. The remaining parameter $R > 0$ is called the Rayleigh number. We can easily check that there are 3 equilibrium points for this system. One is at the origin, and the other 2 are given by $(x_*, x_*, R-1)$ and $(-x_*, -x_*, R-1)$, where $x_* = \sqrt{8/3(R-1)}$. When R is relatively small, most solutions tend to one of the 2 nonzero equilibrium points that correspond to the fluid particle moving periodically in either the clockwise or counterclockwise direction.

But as R increases, suddenly the solutions no longer tend to the equilibria. Rather, they tend to circle around these 2 points in a kind of haphazard fashion. If we choose 2 nearby initial conditions, just as in the forced pendulum case, the corresponding solutions very quickly diverge from one another, and we again have sensitive dependence on initial conditions. The figures below are 2 views of the same orbit: one projected into the x-z plane (left), the other into the y-z plane (right). This type of sensitive dependence has been called the butterfly effect.

Figure 20.2

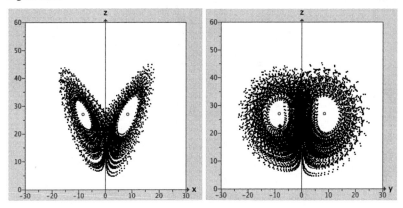

Since our solutions lie in 3-dimensional space, it is difficult to see what is actually happening. It appears that the solutions are tending toward a 2-dimensional object called the Lorenz template (or the Lorenz mask). Note that 2 lobes of the template are joined along a straight line, one lobe in front and the other in the rear.

Figure 20.3

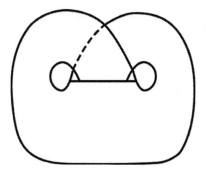

Then it appears that solutions wind around this template, looping around the 2 holes and crossing the central line over and over again.

Figure 20.4

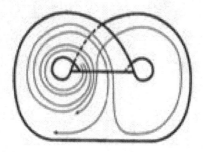

Technically, this cannot happen because we would then have different solutions that merge into the same solution, contradicting the existence and uniqueness theorem (which we discussed earlier). But the fact is that solutions do come very close to an object that is similar in shape to this template—only there are infinitely many leaves that are bundled closely together. This object is known as the Lorenz attractor.

Most solutions on this template return over and over again to the central line as depicted above. So instead of looking for the entire intricate solution of the Lorenz system, what we can do is look at how this solution returns over and over again to this line denoted by L. So we can think of these solutions as being determined by the first return function defined on L. This function assigns to each point p on L the next point along the solution through p that lies on L. So this gives us the first return function $F: L \to L$. The approximate shape of the graph of F is below.

Figure 20.5

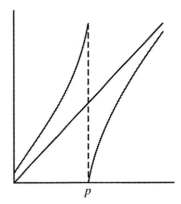

Now the solution starting at p can be tracked by iterating this function. That is, given p, we first compute $F(p)$ to find the next return to L. Then we compute $F(F(p))$ to get the second return, and so forth. We denote this second iterate of F by $F^2(p)$. Then $F^3(p)$ gives the third return, and so forth. The collection of points $F^n(p)$ is called the **orbit** of p under F, and determining how this orbit behaves gives us a good idea of what is happening to the orbits on the Lorenz attractor.

Amazingly, many simple functions have chaotic behavior when iterated. For example, look at what happens to the orbits of 0 and 0.0001 for the quadratic function $x^2 - 2$ or the orbits of .5 and .5001 for another quadratic, $4x(1-x)$.

Important Term

orbit: In the setting of a difference equation, an orbit is the sequence of points $x_0, x_1, x_2, \ldots$ that arise by iteration of a function F starting at the seed value x_0.

Suggested Reading

Blanchard, Devaney, and Hall, *Differential Equations*, chaps. 2.5 and 8.5.

Guckenheimer and Holmes, *Nonlinear Oscillations*, chap. 2.3.

Hirsch, Smale, and Devaney, *Differential Equations*, chap. 14.

Strogatz, *Nonlinear Dynamics and Chaos*, chap. 9.

Relevant Software

Blanchard, Devaney, and Hall, *DE Tools*, Butterfly Effect, Lorenz Equations, Pendulum Sensitive Dependence, Pendulums.

Problems

1. a. Consider the Duffing oscillator equation with no forcing

$$y' = v$$
$$v' = y - y^3.$$

What are the equilibrium points for this system?

b. Show that this system is Hamiltonian with Hamiltonian function

$$H(y, v) = v^2/2 - y^2/2 + y^4/4.$$

c. Sketch the level curves of this Hamiltonian function.

d. Describe the motion of the slender steel beam in the different cases that arise.

2. Show that for the Lorenz system, there is a straight line solution converging to the origin along the z-axis.

3. Show that the function
$$L(x, y, z) = x^2 + 10y^2 + 10z^2$$
is a Lyapunov function for the Lorenz system when $R < 1$.

4. Given the results of the problem above, what can you say about solutions when $R < 1$?

5. Let $V(x, y, z) = Rx^2 + 10y^2 + 10(z - 2R)^2$. Note that $V(x, y, z) = c$ defines an ellipsoid centered at $(0, 0, 2R)$ in 3-dimensional space. Show that we may choose c large enough so that V is decreasing along any solution curve that starts outside the ellipsoid given by $V(x, y, z) = c$.

6. What does this say about solutions of the Lorenz system that start far from the origin?

Exploration

Consider the Rossler system given by

$x' = -y - z$
$y' = x + y/4$
$z' = 1 + z(x - c)$

where $0 < c < 7$. First find all equilibrium points for this system. Describe the bifurcation that occurs at $c = 1$. Then use a computer to investigate what else happens for this system as c increases. Are there any similarities to the Lorenz system?

Periodic Forcing and How Chaos Occurs
Lecture 20—Transcript

Welcome back. In this lecture we're going to delve into one of the biggest developments in the field of differential equations in the last 40, 50 years, namely the fact that differential equations can behave chaotically. What I'll do in this lecture is first show you a number of different examples of chaotic differential equations. I'll begin with some non-autonomous, two dimensional differential equations like our pendulum example from the last lecture and then some autonomous, non-time dependent three dimensional differential equations. Before that though a little bit of history.

Actually the origin of chaos goes back to the work of Henri Poincare, one of the greatest mathematicians in the 19th century, the French mathematician Poincare. He worked in all sorts of areas of mathematics. You may have heard of the recent resolution of the so-called Poincare conjecture that generated a lot of interest in mathematics. In any event, he stimulated so many people to work in mathematics. One thing that happened was among his other areas of interest were differential equations, specifically the differential equations that arise in celestial mechanics, the so-called n body problem.

See what happens when you have n planets running around crazily with mutual gravitational attractions. In any event, people since the time of Kepler had known how to solve the two body problem, two planets, two masses rotating around with mutual gravitational attraction, but nobody could solve the three body problem. What happens when you have three masses moving? It's a major open problem still today.

In any event, in the year 1887 King Oscar of Sweden established a prize for someone to try to solve the three body problem. Well of course Poincare was going to win it. He was the greatest mathematician of the day, and he did win and wrote a paper, but some people from Sweden started looking over this paper and said, "Wait a minute, Henri. You made an assumption there that may not be correct." What he did was assume that what we call the stable and unstable manifolds join up together smoothly. They just come together not at an angle. Let's not worry about the stable and unstable manifolds, what they are, but that's what he assumed. The reason he assumed it was in those days all we had were analytic techniques to solve differential equations. When you solve the differential equation that had stable and unstable manifolds, they always came together perfectly smoothly. They never met at an angle.

Poincare after hearing this went back and said wait a minute, what's happening here? Can they meet at an angle? When he thought about it, when he looked into it, he realized that if that happened the solutions would go crazy. If that happened there'd be chaos in that solution. What he decided to do was he immediately used his prize money to buy back all of the journal articles where his erroneous paper had been published and then published a new paper, which basically founded dynamical systems theories. He also said at the time, wait a minute. To understand this crazy behavior we're going to have to go on and develop much more geometry and topology in these qualitative methods in dynamical systems. That essentially gave birth to all of these areas of mathematics which began to flourish in the 20th century and then allowed us finally, more recently, to understand this chaos.

In any event, what I'd like to do is to begin by showing you some chaotic differential equations. I'll give you three examples of pendulums in chaotic situations. Here's the first one. Here's a gift that was given to me by one of my former students. Over here is a little pendulum system. What I have on the outside is a pendulum that's going to rotate around. On the inside is a pair of smaller pendulums, and now there's a magnetic force in the major pendulum that's going to affect the motion of the smaller pendulums.

Let me show you what happens when I let this go. The outside pendulum is swinging just as we expected before. There's some magnetic force in there, but when the speed is large that magnet doesn't work, but when the speed is slower the magnetic force kicks in, and look what happens to that small pendulum inside. It's moving around very erratically. What you're seeing is basically the beginnings of chaotic behavior. Can you predict what the motion of that internal pendulum will be? just when you think it's settling down to rest then all of a sudden, boom, it goes crazy again. This is a kind of pendulum differential equation.

It's true that that outer pendulum is settling down to rest, so technically speaking this is not chaotic. On the other hand, if we were to eliminate the damping I wouldn't be able to breathe, but this system would be chaotic. Let me show you a couple of more pendulum examples. Let me take that previous nonlinear pendulum equation. Now instead of just holding the pendulum attached to the wall, let me move it up and down periodically. So this differential equation is $\theta' = v$. θ position of our pendulum, v = velocity. $v' = -\sin(\theta)$ and now forcing it periodically up and down I add in a $-F\cos(\omega t)\sin(\theta)$. So that's a two dimensional, non-autonomous differential equation, non-autonomous because there's a t term there.

Let's look and see what happens to solutions. Here in the phase plane below the graph of θ with respect to t and on the right our forced undamped pendulum. You see I'm moving the pendulum up and down which causes the pendulum to run around, it looks periodically for a while, slowing down. Sometimes it moves a little bit more wildly and sometimes more smoothly, but what's going on here? Look at the picture in the phase plane, very complicated. Oh, and now things are happening. Now it's starting to swing around 360 degrees, very complicated behavior going on here. The picture in the phase plane looks crazy. The motion of the pendulum, can you predict what this pendulum will be doing 5 minutes from now? No way. This is chaotic behavior. This is one of the big things that's happening in this area of mathematics these days.

So there's one example of a pendulum. Now let me turn to another example of a forced pendulum. In this case instead of taking the pendulum and moving it up and down periodically, this time let me grab that little knob in the middle and force it periodically. I'll move that around circularly. Again a periodic change in the pendulum equations. So these equations are $\theta' = v$, as before and $v' = -\sin(\theta)$. I've eliminated the g term. But now just $+F\cos(\omega t)$. So I'm forcing this with period $2\pi/\omega$.

So let's look and see what this differential equation looks like. It's a little bit different here though. Instead of just taking one initial condition what I'm going to do is take two initial conditions very close to each other. One of the θ value will be off in the third decimal place. So here's the picture for this circularly forced pendulum. Our pendulums are here on the right. Remember that we're going to start out at essentially the same initial condition. They'll be off in the third decimal place, and here's our picture in the phase plane.

Here is what the solutions look like. At first they do essentially the same thing, but then the solutions start to separate, and now notice they go their own separate route. We started out with two pendulums very close together. We have a nice continuous differential equation, but look what happens very soon thereafter. The pendulums each take their own identity and go crazy. This is one of the basic ingredients of chaotic behavior. This is sensitive dependence on initial conditions. If I take any two initial conditions for these pendulum equations no matter how close together if the θ value is often the 10th decimal place, then the two corresponding solutions of this differential equation will behave vastly differently. This means that this kind of behavior is totally unpredictable.

This chaotic phenomena of sensitive dependence occurs in all sorts of differential equations. Let's go back to the Duffing oscillator that we talked about last time. Remember we had a slender steel beam. It was clamped in between two magnets with no friction going around. Now let's add a forcing term. Let's take this contraption and shake it back and forth horizontally so both the beam and magnets are moving horizontally. Again we get a second order differential equation that's non-autonomous. The differential equation is $x' = y$, $y' = x - x^3$ as before with a damping term maybe $-by$. And then our forcing term is $F\cos(\omega t)$. Again I'm periodically forcing with period $2\pi / \omega$. Let's look and see what happens to the Duffing oscillator.

Here again is our Duffing equation. Now I'm going to add some forcing by shaking everything back and forth to the right. When I have a relatively large amount of damping this is what happens to solutions that start out to the right. It looks like they're kind of wandering around in a periodic solution, but no, just moved way out of that. Actually it's not moving in a periodic solution. What we have here is what's known as a strange attractor. This solution is wandering around in a certain confined region, but it's certainly not a periodic solution. It's not a limit cycle. It's going a little bit crazy in that region.

If we were to start off over here, the same thing. The beam would be attracted more toward the left magnet. Because we're shaking it back and forth the behavior over there would be kind of chaotic. It's moving around in this annular region without any definite path. Now if I change the damping a little bit, then the regions in which these move around change. They get much larger. But still, this particular solution is moving around toward the area defined by the right-hand magnet. Again, it's behaving chaotically. How do we understand this motion? Go over to the left and again, the motion is kind of crazy, behaving chaotically, but still confined to that region over there. That's another example of what's known as a strange attractor.

Now if I let the damping be even smaller, now the solutions go crazy all over the plane. Sometimes your beam is being attracted to the left and sometimes it's being attracted to the right. Sometimes it just doesn't know where to go. Again we're seeing chaotic behavior in this simple two-dimensional non-autonomous differential equation. If we get rid of all of the damping then it just goes crazy. The solution goes crazy; sometimes it winds around in the phase plane, sometimes it does not.

So no matter what we put in for a damping constant here, as long as it's relatively small we see chaotic behavior, kind of different chaotic behavior. Sometimes it's confided just to the right or to the left, and sometimes it runs around everywhere. So the question is: How do we understand what's happening for this system of differential equations?

Now let me turn to perhaps the most important chaotic differential equation, the Lorenz equation from meteorology. This turns out to be a three-dimensional differential equation that's autonomous. There's no t terms there. Here's the differential equation. It's $x' = 10(y - x)$ and $y' =$ some constant r known as the Rayleigh number, $r(x - y - xz)$. Then our third equation is $z' = -8/3z + xy$. Where does this equation come from?

Well Ed Lorenz back in the 1940s was a mathematics graduate student under George Birkhoff who worked in the area of dynamical systems, differential equations. He was drafted into the army during World War II, and the army said, oh you're good in mathematics; we're putting you in our meteorology department. So he changed his career path and went on into meteorology. He was basically a mathematician, and after World War II he was out of the army at MIT wondering why meteorologists could not predict the weather. Why could they not do that?

In order to predict the weather you've got to know the motion of every single molecule of air around earth, billions and gazillions of little particles moving around, fluid particles as they call them. Obviously that differential equation would be impossible to solve. So what Lorenz did, being a mathematician he said, why don't we simplify this process. That's what he came up with. That's the Lorenz equation. The Lorenz equation basically describes the motion of a single fluid particle, a single molecule of air in a planet; one molecule of air in a planet. Can you predict what will happen?

What happens to this molecule of air is it's heated from below and cooled from above. So the molecule of air tends to run around in convection roles. Lorenz thought that well maybe we can understand what's happening to the convection roles in this case. So let's see if we can understand this differential equation.

The equation is $x' = 10y - x$, $y' = rx - y - xz$, and $z' = -8/3z + xy$. What first are the equilibrium points? The first equation: $xy - x = 0$ says that y must be equal to x. Then equation two says, well $y = x$, so equation two reads $0 = r(x - x - xz)$. So that factors into x factor of $r - 1 - z$. That says that either x must be 0 or z must be $r - 1$. Then finally the third equation says that xy must be equal to x^2 which—well it is equal to x^2 because $y = x$, which must then be equal to $8/3z$. So we've got it. Either $x = 0$ in which case y must be 0, and by the third equation z must be 0. So we have an equilibrium point at the origin in three-dimensional space. The other possibility is we have an equilibrium point at the point say $kkr - 1$ where k is either $\pm$ the square root of $8/3(r - 1)$. So we've got three equilibrium points, at the origin and at these other two points, but because we have a square root of $r - 1$ we need r to be greater than 1.

So we've at least found the equilibrium points. Can we find anything else? Let's turn to the computer and see what happens to this three-dimensional system of differential equations. Here's the Lorenz system. I'm drawing here the phase space. That should be three dimensional, but of course I can only draw it in two dimensions. So I'm taking a slice. I'm taking the yz plane. Over here we'll mimic the motion of that fluid particle that's heated from below and cooled from above that should be running around in convection roles, and down below we'll plot the graphs of some of these solutions.

We have a parameter, the Rayleigh number. Let me let the Rayleigh number at first be small, let's say $r = 20$. Then what happens is, well you see this solution is spiraling into an equilibrium point. That's one of those two nonzero equilibrium points that we just found. What that corresponds to in our differential equation is actually a circular motion of that fluid particle. The fluid particle is heated from below so it rises, cooled from above so it goes back down. In the equilibrium state it's just moving around periodically. If I go over to the left side of the phase plane, basically the same thing happens. Our solutions are tending to the equilibrium point. Again this corresponds to a periodic motion of our fluid particle. It looks pretty tame. It looks like this simple meteorological model will be predictable.

But now let's change the Rayleigh number. Let the Rayleigh number be 28. Now start off your solution some place. Now it no longer spirals into that equilibrium point. It runs away and goes over to see the other equilibrium point, and then back to see the right-hand equilibrium point. It spirals away, over to see the left-hand equilibrium point back to the right. This is running around the phase space kind of crazily. What you're seeing again is chaotic behavior. This solution is just going crazy.

If you look though over in that phase space, you're kind of seeing a pattern. You're kind of seeing something that looks like a mask. You're kind of seeing too that solutions keep returning to that region between the two equilibrium points. That will become important in a minute, but before that let me look at another example of what's happening in the Lorenz equations.

Let me take the same differential equation, again Rayleigh constant 28, but now as we did before let's start out with two different initial conditions very close to each other. Start out with two very close initial conditions say right there. If you look at it my, x coordinate was 0 in the first and 0.001 in the second. Now very, very quickly these solutions move apart. I'm drawing a straight line between them to show you the separation. We again have sensitive dependence on initial conditions.

This is what Lorenz later called the butterfly effect. He said that well, what does this mean about the weather? Any small change in initial conditions will create a dramatic change in the behavior of solutions. In terms of meteorology, if a butterfly flaps its wings somewhere in South America, that could cause a hurricane in North America. Sensitive dependence means all bets are off in terms of your initial conditions.

Now there are a number of wonderful stories about Ed Lorenz who sadly passed away a few years ago. How he actually discovered this was kind of interesting. It was back in the 1960s in the first days of computers. What Ed Lorenz did was he figured out this differential equation, put it into his computer, let the computer solve it, it took a long time in those days. In those days the computer didn't print out the graphs of solutions like we're seeing here. Rather in those days the computer would print out that long list of numbers generated by the numerical algorithm. With that, Lorenz painstakingly drew the solution and said, "Wow. I see some crazy behavior here."

He then went out of his office and went to I don't know where, but he met a colleague and said, "Boy, you should see what I'm seeing." He brought his colleague back into the office and showed him and said, "Here I'm going to put this differential equation into the computer." He put it in again and printed out the results and said, "Oh my God. These are completely different. What's going on?" He had mistyped the initial conditions in the fourth or fifth decimal place. What he found was that the solutions behaved completely differently. He saw sensitive dependence in initial conditions. That was really the birth of modern chaos theory.

One of the stories I like to tell my students when I'm teaching my differential equations course—I'm up on the fifth floor in Boston overlooking the Charles River. Across the river is Cambridge where Ed Lorenz used to live. When I would be teaching this I'd always say to the students, out there, there is Ed Lorenz's house. The students would all say, huh? There's a mathematician who's alive. Well, that's what all my students thought of mathematicians. In any event, sometimes I'd actually point out the window and say, there he is. There he is walking his dog. So it's interesting to students to see this eminent—well meteorologist right there, right across the river from them.

So let's go back to the computer here. Notice that we've got some sort of pattern here. Again, even though we have sensitive dependence on initial conditions, things are kind of going crazy. Again, as I showed earlier, it looks like these solutions are crossing through this line over and over again. See how these solutions keep crossing over and over through that line? Stop there, come back. I don't quite get it, but come back, come back, etcetera. These solutions cross over that central line over and over again.

That's not really a line because we're in three dimensions. It's really sort of a rectangular area, but it's very narrow. It's very short in one direction. So these solutions keep coming back to this rectangular area. What mathematicians now do is say, well these continuous curves in three-dimensional space are kind of hard to understand. Let's look at what we call the first return map to that rectangular area. Take that little rectangular area between the two equilibrium points that all our solutions are returning to and figure out. Start at this point, rotate around until you come back, record that. Rotate around until you come back again, record that, etcetera, etcetera, etcetera. So you would record a list of points given by essentially the first returns to that rectangular area.

As I said, that rectangular area is very small. So what we do to model the Lorenz system is in fact contract it right down to a line. What we basically have is a collection of differential equations that are sort of running around a piece of cardboard that goes down and then comes up. One piece goes in front of the other, the other piece goes behind. Solutions sometimes come behind. Solutions sometimes go in front. But they return to where that cardboard box or cardboard, whatever it is, template meets at some point. So solutions keep returning to a line. Let's see that again.

There we go. Here's that line right there. Each time a point returns we'll record that point. We'll record a sequence of points. That will give us a good idea of what's happening to this differential equation. So what we'll do is turn to, not differential equations but iterated functions. We'll get a function on that line that records what happens as the solutions come back over and over again. That'll be a simple one-dimensional function. Then what we'll do is iterate that function. What I mean by "iterate" is we'll start with some c, some number. Plug it into the function, out will come a number. Then we'll take that number, plug it into the function and do that over and over again. We'll produce a list of numbers. That list of numbers is what we call the orbit of the iterated function. Then assuming we can understand that, that will tell us something about the corresponding differential equation.

In higher dimensions what we'd do is, if we had a periodically forced differential equation we might just record the numbers say, in the plane at every time 2π units, every time 2π, record that point. We'd get an iterated function in the plane. That's the way we turned to finally understand chaotic behavior. Let me give you a couple of examples of iterated functions.

Let's just take the simplest possible case say, x^2 on the realign. What happens when you iterate x^2? Well if you start with 0, $0^2 = 0$, $0^2 = 0$, $0^2 = 0$, not much. Start with 1, $1^2 = 1$, $1^2 = 1$, $1^2 = 1$, not much. Start with 2, $2^2 = 4$, $4^2 = 16$, $16^2 =$ big, big$^2 =$ bigger, goes off to infinity. Not much is happening with iteration of x^2. How about $x^2 - 1$? Now let's 0. $0^2 - 1 = -1$, $-1^2 - 1 = 0$, $0^2 - 1 = -1$, back to 0. This is what we call a cycle of period 2. Not much is happening there. How about say, $x^2 - 2$? Well, start with 0, $0^2 - 2 = -2$, $-2^2 - 2 = 2$, $2^2 - 2 = 2$, $2^2 - 2 = 2$. We're stuck at 2, not much is happening there. Instead of starting with 0 let's start, like Lorenz did with 0.0001.

So let me turn to the computer to iterate $x^2 - 2$ starting with seed 0.0001. And as we saw with Euler's method a long time ago, the easiest way to do these iterative processes is by a spreadsheet. So here let c say b, let's go to -1, -1, start with the c at 0 and what we see is just as we saw before $0^2 - 1 = -1$, square that -1 you're at 0, iterate, iterate. You just see these numbers bouncing back and forth between 0 and -1. But now let's go to $c = -2$. What happens is just as we just saw, $0^2 - 2 = -2$, $-2^2 - 2 =$ up to 2. Then we're stuck at two forever, not much happens. Now let's put in 0.0001 and what you see is completely different behavior. You're seeing chaotic behavior. The simple function $x^2 - 2$ on the realign behaves chaotically.

This is where we're going now, to understand the chaos in differential equations we'll reduce to iterated functions and make use of all sorts of different qualitative techniques to try to comprehend that chaos.

Understanding Chaos with Iterated Functions
Lecture 21

As we discussed in the previous lecture, the way mathematicians have begun to understand the chaotic behavior that occurs in higher-dimensional systems of differential equations is often by iterating a first return map.

Here is another way iteration arises. Recall the limited population growth model (a.k.a. the logistic population model) that we saw earlier. This was the differential equation $y' = ky(1 - y/N)$ that we solved in multiple different ways. Here is a variation on this theme—the discrete logistic population model. Suppose x_n denotes the population of some species in year n (or, for example, generation n). The discrete logistic population model is given by

$$x_{n+1} = kx_n(1 - x_n/N).$$

That is, the population given in year $n + 1$ is just $kx_n(1 - x_n/N)$, where k and N are constants that depend on the particular species. Here, N is the maximal population rather than the ideal population. For simplicity, let's look at the equation

$$x_{n+1} = kx_n(1 - x_n).$$

Here x_n represents the fraction of the maximal population. This is an example of a **difference equation**. To find the population in ensuing years, we simply start with an initial population (or seed) x_0 and plug it into the function $f(x) = kx(1 - x)$. Out comes x_1. Then we do the same with x_1 to generate x_2. We do this over and over to produce $x_0, x_1, x_2, \ldots$, the so-called orbit of the seed x_0.

For difference equations, let's consider more generally $x_{n+1} = f(x_n)$ for some function f. We iterate f to produce the orbit of some seed x_0. For example, suppose $f(x) = x^2$. Then the orbit of 0 is 0, 0, 0, ..., so we say that 0 is a **fixed point** for f. We see that -1 is an eventually fixed point since $f(-1) = 1$, which is also fixed. If $0 < |x| < 1$, then the orbit of x tends to the fixed point at 0. But if $|x| > 1$, repeated squaring sends the orbit of x off to infinity. So we understand the fate of all orbits in this simple case.

As another example, if $f(x) = x^2 - 1$, the points 0 and −1 lie on a cycle of period 2 since $f(-1) = 0$ while $f(0) = -1$. We also say that 0 and −1 are periodic points with prime period 2. More generally, x_0 is a periodic point of prime period n if $x_n = x_0$ and n is the smallest positive integer for which this is true.

As with earlier parts of this course, we can visualize orbits in several different ways. One way to do this is to plot a time series for the orbit. For example, for our cycle of period 2 for $f(x) = x^2 - 1$, the time series representation of the orbit would be the following.

Figure 21.1

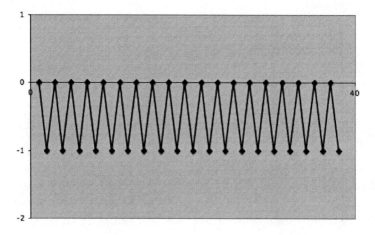

A second way to visualize orbits is to use graphical analysis. Given the function f, we plot its graph and then superimpose the line $y = x$ (the diagonal line) on this graph. Given a seed x_0, we first draw a vertical line from the point (x_0, x_0) on the diagonal to the point $(x_0, f(x_0)) = (x_0, x_1)$ on the graph. Then we draw a horizontal line back to the diagonal yielding (x_1, x_1). We continue to (x_1, x_2) on the graph and over to (x_2, x_2) on the diagonal. For example, below is the graphical analysis representation of the orbit of 0.9 under $f(x) = x^2$ (left) and the orbit of -0.7 under $f(x) = x^2 - 1$ (right). Note that the orbit of -0.7 in this second graph tends to the 2-cycle at 0 and -1.

Figure 21.2

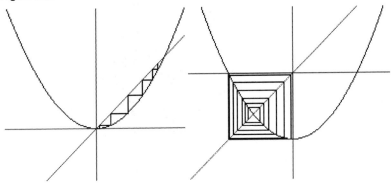

Unlike the logistic differential equation where all solutions essentially do the same thing, for the discrete logistic model, orbits can do many different things. We will use the orbit of the critical point for the function $f(x) = kx(1 - x)$ from now on. A critical point is a point where the derivative of the function vanishes, so when $f(x) = kx(1 - x)$, we have $f'(x) = k - 2kx$, so the only critical point is $x = 1/2$. When $0 < k < 1$, the critical point tends to 0, which is a fixed point.

Figure 21.3

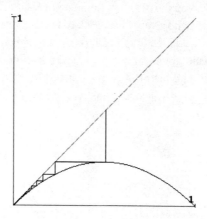

When $1 < k < 3$, graphical analysis shows that the critical orbit now tends to a nonzero fixed point. This point is the nonzero root of the equation $kx(1 - x) = x$, namely $(k - 1)/k$.

Figure 21.4

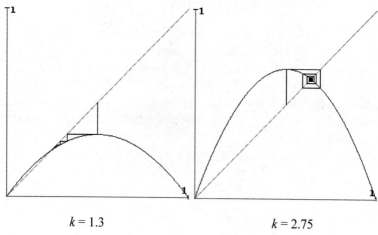

$k = 1.3$ $k = 2.75$

For $k = 3.3$, the critical orbit now tends to a 2-cycle, but when $k = 3.47$ the critical orbit tends to a 4-cycle.

Figure 21.5

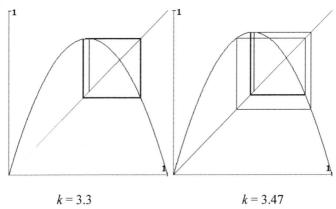

$k = 3.3$ $\qquad\qquad\qquad\qquad$ $k = 3.47$

When $k = 3.83$, the critical orbit tends to a 3-cycle. The time series for this critical orbit is below.

Figure 21.6

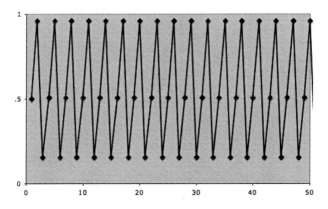

When $k = 4$, things go crazy. The orbit of 1/2 is quite simple: It is 1/2, 1, 0, 0, 0, …, so the orbit is eventually fixed. But any nearby orbit behaves vastly differently. The orbit of 0.5001 as a time series is below.

Figure 21.7

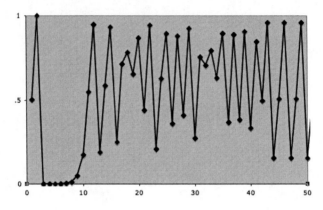

And here it is using graphical analysis.

Figure 21.8

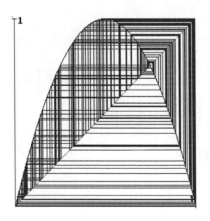

We find similar behaviors when we iterate the quadratic function $x^2 + c$. When $c = 0$, the orbit of 0 is fixed; $c = -1$, a 2-cycle; $c = -1.3$, a 4-cycle; $c = -1.38$, an 8-cycle; $c = -1.76$, a 3-cycle. And again we get chaotic behavior when $c = -2$.

Clearly, lots of things are happening when we change the parameter k in the logistic function or c in $x^2 + c$. To see all of this, we plot the **orbit diagram** for this function. In this graphic, we plot the parameter k horizontally with $0 \leq k \leq 4$. Along the vertical line above a given k-value, we plot the eventual orbit of the critical point $1/2$. By "eventual," we mean that we iterate, say, 200 times but only display the last 100 points on the orbit. That is, we throw away the transient behavior.

Figure 21.9

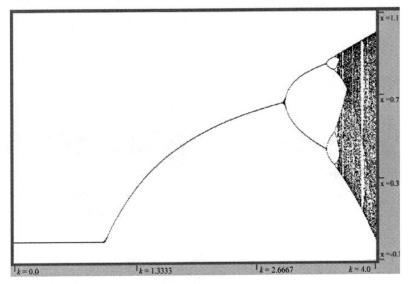

And here is a magnification of the orbit diagram with k in the interval $3 \leq k \leq 4$.

Figure 21.10

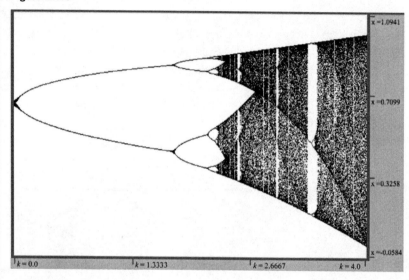

Important Terms

difference equation: An equation of the form $y_{n+1} = F(y_n)$. That is, given the value y_n, we determine the next value y_{n+1} by simply plugging y_n into the function F. Thus, the successive values y_n are determined by iterating the expression $F(y)$.

fixed point: A value of y_0 for which $F(y_0) = y_0$. Such points are attracting if nearby points have orbits that tend to y_0; repelling if the orbits tend away from y_0; and neutral or indifferent if y_0 is neither attracting or repelling.

orbit diagram: A picture of the fate of orbits of the critical point for each value of a given parameter.

Suggested Reading

Alligood, Sauer, and Yorke, *Chaos*, chap. 1.5.

Blanchard, Devaney, and Hall, *Differential Equations*, chap. 8.1.

Devaney, *A First Course in Chaotic Dynamical Systems*, chap. 3.

Hirsch, Smale, and Devaney, *Differential Equations*, chaps. 15.1–15.3.

Relevant Software

Nonlinear Web, http://math.bu.edu/DYSYS/applets/nonlinear-web.html

Orbit Diagram for $Cx(1-x)$, http://math.bu.edu/DYSYS/applets/bif-dgm/Logistic.html

Problems

1. a. Find all fixed points for the function $F(x) = x^2$.

   b. What is the fate of all other orbits for this function?

2. What is the fate of all orbits of $F(x) = x^2 + 1$?

3. a. What is the fate of all orbits of $F(x) = ax$ where $0 < a < 1$?

   b. What is the fate of orbits of this function for other values of a?

4. Find all fixed points for the logistic family $kx(1 - x)$ with $k > 0$.

5. Find the cycle of period 2 for the function $F(x) = -x^3$. Is there a 2-cycle for the function $F(x) = x^3$?

6. Let $D(x)$ denote the doubling function defined on the interval $0 \leq x \leq 1$ and given by

   $2x$ if $0 \leq x < 1/2$ and

   $2x - 1$ if $1/2 \leq x \leq 1$.

   Show that the seeds 1/3, 1/7, and 1/15 lie on cycles; and compute their periods.

7. Find all points that lie on cycles of prime period 2, 3, and 4 for the doubling function.

8. Draw the graph of $D(x)$, $D^2(x)$, and $D^3(x)$. How many times does the graph of $D^n(x)$ cross the diagonal? Find all points that lie on cycles of period n for the doubling function (not necessarily prime period n).

Exploration

Using the applets below, investigate the orbit diagrams for the functions $x^2 + c$, $cx(1 - x^2/3)$, and $c\sin(x)$. Do you observe any similarities in these diagrams as you zoom in?

Orbit Diagram for $x^2 + c$:

http://math.bu.edu/DYSYS/applets/bif-dgm/Quadratic.html

Bifurcations of the cubic map:

http://math.bu.edu/DYSYS/applets/bif-dgm/Cubic.html

Bifurcations of the sine map:

http://math.bu.edu/DYSYS/applets/bif-dgm/Sine.html

Understanding Chaos with Iterated Functions
Lecture 21—Transcript

As we discussed last time, the way we finally came to understand certain chaotic differential equations is to reduce to an iterated function. For example, remember in the Lorenz equation we watched our solution as it came through a certain area, came through a certain area, returned, returned, over and over again. What we would do is figure out a function that measured that return and then iterate it.

What does it mean to iterate a function? We're going to start with some seed. For simplicity we'll take the seed to be real numbers. We'll just work for now on the real number line. So we start with a seed, that's a number I'll call it x_0. And then we produce what's called the orbit of $x0$. That's a sequence of numbers x_0, x_1, x_2...where x_n is obtained by applying your function to the previous number $x_n - 1$.

For example let's do one that we did last time; x^2. What happens when we iterate x^2? If we start with 0 then $0^2 = 0$, $0^2 = 0$, $0^2 = 0$. Our orbit is just 0, 0, 0, 0. That's what we call a fixed point analog of an equilibrium point for a differential equation. How about 1? Well $1^2 = 1$, $1^2 = 1$, $1^2 = 1$. That's also a fixed point for this iterative process. How about -1? Well $-1^2 = 1$, $^2 = 1$, $^2 = 1$. Our orbit here is $-1, 1, 1, 1, 1$ repeated. We say that -1 is eventually fixed. What if we choose a number that's an absolute value less than 1 but not 0, say ½? $½^2 = ¼$, $¼^2 = 1/16^2$, 1/256, small, smaller, smaller, etcetera. Those orbits all go to 0. Those orbits tend to the fixed point at 0.

Similarly what happens if we take a number that's larger than 1? Then we know what happens there. Take 3, $3^2 = 9$, $9^2 = 81$, $81^2 =$ big, big$^2 =$ bigger. Those orbits just go off to infinity, very simple behavior. Again, as we saw last time, what about $x^2 - 1$? There the orbit of 0 is $0^2 - 1 = -1$; $-1^2 - 1$ is back to 0. So the orbit is $0, -1, -1, 0, 0, -1, 0, -1$, etcetera, etcetera. That's a cycle of what we call prime period 2. What do I mean by prime period? I'll say that the seed x_0 is prime period n. If the n point on the orbit $x_n = x0$ and n is the smallest integer for which this happens.

Here's another way of thinking about iterating functions. Let's start with a function $f(x)$. What do I mean by iterating f? What I'll first do is start with $f(x)$ and then I'll apply f to $f(x)$. I'll compute the composition f of $f(x)$. Then I'll iterate a third time f of f of $f(x)$, three applications of f to itself. Then I'll right $f(N)$ of x to mean the N fold composition of f with itself. $f(N)$ might be a little confusing. I don't mean raise $f(x)$ to the N power. I mean iterate f N times. If I wanted to raise $f(x)$ to the N power I would have written it quantity $f(x)$ inside parentheses raised to the N power.

For example, how about $x^2 + c$? Well $f(x)$ is $x^2 + c$. What's $x^2(x)$ the second iterate of x? We take $x^2 + c$ and apply the same function to it. So we take $x^2 + c$ and square it and add c. What's the third iterate of this function? I've got now $x^2 + c^2 + c$. I've got to plug that into my quadratic function. I've got to take $x^2 + c^2 + c$ and first square it and then add c and so forth. We keep composing this function with itself over and over again. That gives us the orbit of the point x.

Let me turn to this quadratic function and do a couple of more examples of what happens when we iterate $x^2 + c$. As I said earlier, it's usually most easy to use a spreadsheet. Here is my spreadsheet for $x^2 + c$. I started with a c value 0 and my seed was 0. If we go to $c = -1$ with seed 0 then what we see is in terms of the list of numbers in the orbit, of course $0 - 1$, $0 - 1$ over and over and over. Then here I've plotted the time series $0 - 1$, $0 - 1$, just as we did with Euler's method way back when.

If I go up to -1.3, so $c = -1.3$ and now we see a different behavior for the orbit of 0. The orbit of 0 cycles with period 4. If I'm right here and I iterate once, twice, three, four times, I come back where I started. We get a cycle of prime period 4. If I go up to -1.38, if you look at the dots here, you actually see a cycle of period 8, little hard to see but if you count them there's actually a cycle of period 8. If I go to say -1.76 you now see a cycle of period 3. So lots of different things can happen when we iterate functions, including this very simple quadratic function. As we saw last time, if I take c to be -2, the orbit of 0 goes to -2 then to 2 where it's fixed. So this is eventually fixed and we see that in the time series. But if I change my c just a little bit to 0.00001 then the orbit does something completely different. We get chaotic behavior. A small change in that initial condition creates a dramatic change in what's happening to the orbit. If I change this seed to say, 0.002 then again, you see the time series is very different, dramatically different behavior. That's chaos appearing in the realm of simple iterated quadratic functions.

Just as a reminder, it's easy to compute these orbits using a spreadsheet. Let me just remind you how to do that. Let me take the function $x^2 + c$ where $c = -1$ and compute the orbit of say 0. Well we know what that is but let's do it using a spreadsheet. How will we do it? Well we've got our function $x^2 - 1$. We've got our seed 0. We've got to put in a formula for computing successive iterates of $x^2 - 1$. Well that's easy to do with a spreadsheet. To enter the formula you type in equals and then now we're going to compute the previous number squared. That previous number lies right above here in cell B3. We've got to square that. We've got to square that, and then subtract off 1. Or what I could do is add in a constant reference to cell E1. Remember how to do that. That's \$E\$1. That gives me a constant reference to that c value. So when $c = -1$ yes, the next point on the orbit is -1, and then I can just fill this down as far as I want to go and we get the successive iterates of $x^2 - 1$.

If I were to change my seed to 0.2, so the seed is 0.2 you see the orbit initially changes a little bit but eventually we tend to that cycle of period 2. If I change c to say -1.3, then the orbit of 0.2 does something completely different. But in fact as we saw before, that tends to a cycle of period 4. Then finally we can just plot this time series by, let me highlight all the data—not all of the data but enough of the data—and then insert a chart. Let me choose a chart that's a scatter plot with the dots connected by straight lines. There we've got our chart. Let me expand it. There's the time series for the orbit of 0.2 under $x^2 - 1.3$.

So iterating functions is one way to understand differential equations, but there's another very different way that these iterations come up. Let me refer you to what's called the discreet logistic population model. Remember way back at the beginning of the course we dealt with the logistic population model or what we call the limited growth population model. That was the differential equation $y' = ky(1 - y/n)$ where n was the carrying capacity. Here's another model for a population. Let me let x_n be the population of some species, again living in isolation in year n. So x_n is the population of that species in year n. Then the discreet population model says the population in year $n + 1$, so that's $x_n + 1$, is given by some constant times $x_n(1 - x_n/m)$. So here k and m are constants.

M is a little different from the n we had in the logistic differential equation; n was the carrying capacity, the ideal population. Here m is the maximum population. Notice what we have here. We have a very different kind of equation. The logistic equation involved derivatives. It was a differential equation. Here we've got what we call a difference equation. The population next year, $x_n + 1$, is given just by k(population this year times $(1 -$ the population this year)$/m$.

For simplicity let me assume that that parameter m, the maximum population is 1. That doesn't mean that the maximum population is 1. That means what we're going to do is we'll measure the population as a percentage of the maximum population. Population 0.5 means that your species is 50% full.

Now here's a little word here. The way we got into the iterating the simple quadratic functions, these difference equations—notice by the way that the discreet logistic model is $x_n + 1 = kx_n(1 - x_n)$. That's just a quadratic expression. It's just like the one we did a moment ago with the spreadsheet. We're iterating a quadratic function. The way we got into this was back when we first starting seeing chaos, when the Lorenz system burst forth, and many other systems like that, they only appeared in higher dimensions. The Lorenz equation is a three-dimensional equation. Those pendulum equations were two dimensional but we also had the time in there, so they were three dimensional.

What happened was many mathematicians and physicists were looking at higher dimensional chaotic systems. Then Robert May a biologist from England, in fact he was the science advisor to the Queen of England for a while; he came to mathematicians and said, "Why are you working in such high dimensions?" He said, "Well because we have lots more tools up there." He said, "But wait a minute. Down in one dimension you can see the chaotic behavior just as well." He said, "Look at this logistic population model. That's just as chaotic as everything else you're seeing." That really caught people's attention, made people start focusing on just one-dimensional iterated functions. Amazingly we only understood this logistic population model, the iteration of a quadratic function on the realign in the middle of the 1990s. That's how complicated it was. Thanks to Robert May we did come down to lower dimensions.

So let me turn to some other basically qualitative methods for understanding iteration of function, in particular the logistic population model. Here is our logistic population model. Now I'm writing as $P_{n+1} = kP_n(1 - P_n)$. When I start with the value of $k = 0$ and my initial condition 0.5 what we see is, well 0.5 you plug that into the function, you plug that into $kP(1 - P)$. You immediately go to 0. If I change that k to say 0.7 then start with a population 0.5. You see that this population also goes extinct. It takes a little bit more time to go out but the population still dies if k is small.

On the other hand, if I go up to $k = 1.5$, again with population 0.5 to start with, now you see that this population tends to some limiting value, tends to what we'd think of as an equilibrium point or what we called earlier a fixed point. If I go up to $k = 2.5$, again when I iterate starting at 0.5 the orbit again tends to level out at one fixed point. But if I go up to $k = 3.2$, now all of a sudden we see a cycle of period 2. The population goes up and down, up and down every succeeding year. If I go to $k = 3.47$ now we see that the cycle cycles back and forth every 4 years. The population returns to where it was every fourth year, a different behavior. If we go to $k = 3.55$, again we see different behavior.

If you look at the dots they're not the same every 4 years. They're the same every 8 years. We've actually now cycled with period 8. So lots of different behavior for this logistic population model. If I go up to 3.83 we see that the population cycles every third year. And if I go up to $k = 4$, start with initial population 0.50002 we see—well look at that. It's behaving kind of crazily. I put in 0.5, look what happens. 0.5 after one year we go up to 1 when we put it into the logistic population model, but 1 is our maximum population. That means we're going extinct and yeah, we do. The population dies out the next year. But as I just showed, if I put in 0.50001 then something very different happens. We see again chaotic behavior. So chaos comes up in the simple quadratic functions.

Let me in fact show you this a little bit more continuously. Let me vary that parameter continuously from 3 up to 4. We're starting here at 3. Again I'm taking the initial seed 0.5 and watch what happens. As I vary this parameter, at first we see a cycle of period 2, but then all of the sudden we see a cycle of period 4, and then a cycle of period 8, and then all of a sudden we're into the chaotic regime. Well sometimes you see sort of periodic behavior. Like right there, it looks like we have a cycle of some period. Go up a little bit higher, now the cycle of period 3, and then again we're into the chaotic regime. Lots going on for this logistic population model; lots going on when you iterate the simple function $kx(1 - x)$. If you recall, the same sort of things were happening when we iterated $x^2 + c$.

Here's now a different way of understanding what happens when you iterate functions. It's what we call graphical analysis, a graphical iteration. Here I'm starting with the value of $k = 2$. My seed is 0.1. What I'm going to do to see this in terms of the graph is first I'll plot the graph of $2x(1 - x)$. There it is. Secondly I'll superimpose on that graph the line $y = x$. Whenever the graph hits that line $y = x$ we have a fixed point for our iteration. Now here's what I'll do to see the iterative process graphically. I'll start say at 0.1. When I iterate once I get 0.18. Graphically I do it this way. I start down here on the x-axis at 0.1, draw a vertical line up to the graph, there my y value is 0.18, and then I run horizontally back to the diagonal line $y = x$. There my y and x value is 0.18.

Now I do that again. Iterate by going vertically to the graph, horizontally to the diagonal. Then do it again and again and again. What you see is this particular staircase leads right up to that fixed point. Yes, my orbit seems to be converging 2.5. Now as I change k for a while you see that this orbit just tends to that fixed point. It's now cycling around that fixed point, but eventually it moves away and eventually we see a cycle of period 2. If I keep iterating, it gets kind of hard to see what's happening when I include all this transient behavior. So what I'll do is throw away the first few iterations of this function and then record graphically what's happening later.

Let me throw away again the first few iterations. Here my k value is approximately 2.7 and as before my seed is 0.5 and you can see I've got a long list of orbits here. You don't see it because in fact I've thrown away the transient behavior. Now let me increase the constant, the parameter k, continuously. Then all of a sudden you see a bifurcation. All of a sudden we see a cycle of period 2 appearing. Keep raising k, then all of a sudden a cycle of period 4 arises, because if we were to start at this point and iterate once, twice, three times, and four times, we'd come back exactly where we started. Keep iterating. Now you see if you count the little lines there, a cycle of period 8. Keep going and we're into the chaotic regime. So things are behaving pretty chaotically, mostly, though sometimes you actually see a cycle, chaotic and then a cycle of period 3. So as we saw by hand earlier, there's lots going on when we iterate this simple quadratic function $kx(1-x)$.

Let me put all of these pictures together and draw both the graphical analysis picture of orbits and the time series picture of orbits. Let me take a large set of iterations. There's my list of the orbit. Again when $k = 2$ you see from graphical iteration that the orbit tends to this point right here, a fixed point where the graph hits the diagonal $y = x$. You see from the time series that the time series levels off at that fixed point, 0.5. Now as I vary k, you see all of a sudden things are changing. In the time series you see a cycle of period 2. It takes a while to get there from the graphical iteration. Then we see another change. All of a sudden you see a cycle of period 4 and into the chaotic regime.

What we see is there are lots of things going on for this logistic population model, this iteration of a quadratic function. Remember in many of the situations in this course we've tried to get the big picture of what's happening. Let's do that now with what we call the orbit diagram. The orbit diagram is going to be a summary of all possible behaviors. When we iterate $kx(1-x)$. What I mean by that is we'll plot k horizontally, a bunch of different k values. Then we'll take the orbit of 0.5 and plot it vertically over each k. I won't plot the entire orbit of 0.5. I'll plot the eventual orbit of 0.5. Say I'll iterate it 500 times, but I'll throw away the first 100 iterations. That means I'm going to throw away the transient behavior.

A natural question is why am I using 0.5 here. Well if you differentiate $kx(1-x)$, if you differentiate $kx - kx^2$ you get $k(1-2x)$ and that equals 0 when $1 - 2x = 0$, when $x = 0.5$. So 0.5 is the critical point for this function. Similarly when I did $x^2 + c$ the derivative of $x^2 + c = 2x$. The critical point there was 0. I always use 0 as my starting point for seed. The question is why. We'll see as we go on that the critical point knows everything. The orbit of 0.5 for this logistic iteration tells us everything that's going on. That's why I use 0.5 to plot the orbit diagram. Let's see what this looks like.

Here is the orbit diagram for the logistic family. Remember I'm plotting k horizontally. Here $k = 0$ and I'm running all the way up to $k = 4$. Remember I'm iterating 0.5 a bunch of times and throwing away the first iterations so what I see is the fate of the orbit of 0.5. What you see is just as we saw with some of the examples we did earlier, initially when k is small, the orbit of 0.5 dies out. The population goes extinct. This is the value of the population equal to 0. Then suddenly it emerges. Suddenly the population goes to some fixed value. Above this k point we see just one point. My orbit of the critical point has tended to one fixed value. Then a little bit later we see another change. For these k values the orbit of 0.5 goes to two points, a cycle of period 2. If you peer in closely, a little later it goes to a cycle of period 4. Then lots of stuff is happening here.

I seem to see a little window here. Let me magnify that window. When I magnify that window, what you see is what we call the period 3 window. At this particular k value you see directly above it only three points. Our orbit is tending to a cycle of period 3. Now if I magnify this region right here, look what you see, essentially the same picture as we saw before. Magnify it again, it looks like a copy of the original logistic orbit diagram. One point goes to 2, goes to 4, goes to 8. Over here you see what's a period 3 window. Of course there are actually other pieces above it and below, so it's not really period 3, but it does look exactly like our previous diagram.

Let me go back to the original diagram. We start out with population going extinct, then leveling off, then going to a cycle of period 2. Let me magnify this region right here. Our initial point lies on a cycle of period 2, which branches into a cycle of period 4, cycle of period 8. Let me magnify. They're all down here. We see a window that looks like it has period 3, looks exactly like what we saw before. If I go and magnify this region starting with a cycle of period 4, we see again the exact same thing. Cycle of 4 goes to 8, goes to 16, into the chaotic regime. There's another window of period 3. The more you zoom, the more you see the same structure here. There's a lot going on when we're iterating this function, but there is some interesting geometric behavior behind it that's in fact eventually going to tell us everything that's happening here.

Here's one more example of that; here's the orbit diagram for $x^2 + c$. Now I'm using the initial seed 0. That's the critical point for $x^2 + c$. Here you see essentially the same thing as we saw before. Now however c is going down from around 0.25 to −2. You see initially one fixed point which turns into a cycle of period 2, cycle of period 4, and just as in the logistic case way down here a period 3 window. Again, if you start zooming in here you see the replication of the orbit diagram over and over again. Period 1 goes to the 2 cycle, goes to period 4 cycle, goes to 8 cycle, the so-called period doubling route to chaos.

So that's where we're going to turn now to investigate a little bit more of what's happening near all these cycles. How are they arranged? Where do they come from? What are the bifurcations? We'll see lots of amazing things going on.

Periods and Ordering of Iterated Functions
Lecture 22

As we saw when we studied first-order differential equations, there are 3 different types of equilibrium points—sinks, sources, and nodes. The same is true for iterated functions: There are 3 different types of fixed points—attracting, repelling, and neutral (or indifferent).

A fixed point is attracting if nearby seeds to the fixed point have orbits that tend to the fixed point. It is repelling if nearby orbits tend away from the fixed point. And it is neutral if neither of the above cases occurs. For example, $f(x) = x^2$ has an attracting fixed point at $x_0 = 0$ and a repelling fixed point at $x_0 = 1$. This is easily seen with graphical analysis.

Figure 22.1

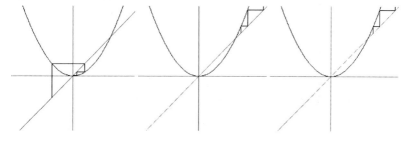

Also, $f(x) = x^2 + 1/4$ has a neutral fixed point at $x_0 = 1/2$ since nearby orbits to the left of the fixed point tends toward it, while orbits to the right of the fixed point tend away from it.

Figure 22.2

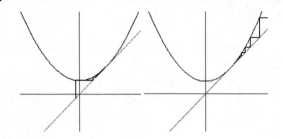

We can use calculus to determine the types of fixed points. Suppose we have a fixed point x_0, so $f(x_0) = x_0$. If $|f'(x_0)| < 1$, then graphical analysis shows that this fixed point is attracting. If $|f'(x_0)| > 1$, then x_0 is repelling. If $f'(x_0) = \pm 1$, then we get no information; x_0 could be attracting, repelling, or neutral.

If we have a periodic point x_0 of period n, then we know that $f^n(x_0) = x_0$. So this periodic point is attracting or repelling depending on whether $|(f^n)'(x_0)|$ is less than or greater than 1. For example, for the periodic points 0 and -1 of period 2 for $f(x) = x^2 - 1$, we have $f^2(x) = x^4 - 2x^2$, so $(f^2)'(0) = (f^2)'(-1) = 0$. Therefore these periodic points are attracting. Graphical analysis of both $f(x)$ and $f^2(x)$ shows this as well.

Figure 22.3

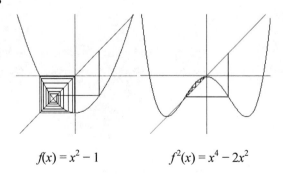

$f(x) = x^2 - 1$ $\qquad f^2(x) = x^4 - 2x^2$

Bifurcations also arise when we iterate functions. For example, consider $f(x) = x^2 + c$. There is a bifurcation when $c = 1/4$ as is seen in the graphs when we pass from $c > 1/4$ to $c < 1/4$. There are no fixed points when $c > 1/4$; a single (neutral) fixed point when $c = 1/4$; and 2 fixed points (1 attracting, 1 repelling) when c is slightly below 1/4. This type of bifurcation is called a saddle-node bifurcation, just as in the case of first-order differential equations.

Figure 22.4

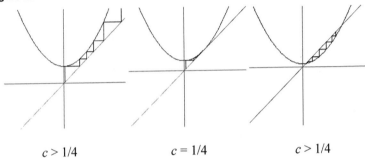

$c > 1/4$ $c = 1/4$ $c > 1/4$

There is a very different type of bifurcation that we saw many times in the orbit diagram. This is the period doubling bifurcation. What happens at this bifurcation is an attracting periodic cycle of period n suddenly becomes repelling, and at the same time, an attracting cycle of period $2n$ branches away. For example, consider again $f(x) = x^2 + c$. When $c < 1/4$, we always have 2 fixed points at

$$x_\pm = \frac{1}{2} \pm \frac{\sqrt{1-4c}}{2}.$$

The fixed point at x_+ is always repelling. At x_- we have

$$f'(x_-) = 1 - \sqrt{1-4c}.$$

So this fixed point is attracting if $-3/4 < c < 1/4$ and repelling if $c < -3/4$. We have $f'(x_-) = -1$ if $c = -3/4$. More importantly, graphical analysis shows that a new periodic point of period 2 is born when c goes below $-3/4$. Equivalently, one can check that the equation $f^2(x) = x$ has only 2 real roots when $c > -3/4$ (the 2 fixed points) whereas there are 4 roots when $c < -3/4$. This is the period doubling bifurcation.

Figure 22.5

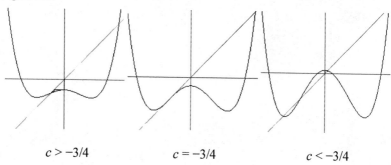

$c > -3/4$ $c = -3/4$ $c < -3/4$

One of the most amazing theorems dealing with iteration of functions on the real line is a result proved by Alexander Sharkovsky in the 1960s. This theorem says the following. Consider the following ordering of the natural numbers: First list all the odds (except 1) in increasing order, then list 2 times the odds (again except 1) in increasing order, then 4 times the odds, 8 times the odds, and so on. When you have done this, the only missing numbers are the powers of 2, which we then list in decreasing order. Below is the Sharkovsky ordering.

$3 \Rightarrow 5 \Rightarrow 7 \Rightarrow 9 \Rightarrow ...$

$2 \bullet 3 \Rightarrow 2 \bullet 5 \Rightarrow 2 \bullet 7 \Rightarrow 2 \bullet 9 \Rightarrow ...$

$4 \bullet 3 \Rightarrow 4 \bullet 5 \Rightarrow 4 \bullet 7 \Rightarrow ...$

8•3 ⇒ 8•5 ⇒ 8•7 ⇒ ...

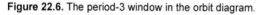

... 2^3 ⇒ 2^2 ⇒ 2 ⇒ 1

Sharkovsky's theorem says that suppose you have a function that is continuous on the real line. If this function has a periodic point of prime period n, then it must also have a periodic point of prime period k for any integer that follows n in the Sharkovsky ordering. In particular, if a continuous function has a periodic point of period 3, then it must also have a periodic point of all other periods!

Now go back to the orbit diagram for the logistic function. Toward the right, we see a little window that had, at least initially, what appears to be a cycle of period 3. By Sharkovsky, there must be infinitely many other periodic points in this window. Why do we not see them?

Figure 22.6. The period-3 window in the orbit diagram.

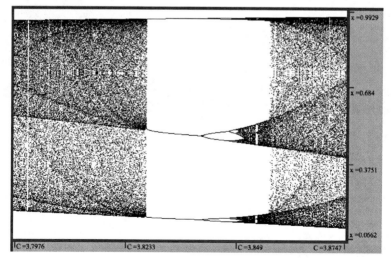

Well, the answer is that only the period-3 cycle is attracting; all the other cycles are repelling. The reason for this is another amazing fact about iteration of polynomials (and, in fact, what are called analytic functions). If you have an attracting or neutral periodic cycle, then the orbit of one of the critical points must be attracted to it. This means that for the logistic map (where there is only one critical point, $x = 1/2$), we can have at most one attracting cycle. Everything else must be repelling.

Suggested Reading

Alligood, Sauer, and Yorke, *Chaos*, chaps. 1.3–1.4.

Blanchard, Devaney, and Hall, *Differential Equations*, chaps. 8.2–8.3.

Devaney, *A First Course in Chaotic Dynamical Systems*, chap. 5.

Hirsch, Smale, and Devaney, *Differential Equations*, chap. 15.2.

Relevant Software

Nonlinear Web, http://math.bu.edu/DYSYS/applets/nonlinear-web.html

Orbit Diagram for $Cx(1-x)$, http://math.bu.edu/DYSYS/applets/bif-dgm/Logistic.html

Problems

1. Find the fixed points for $F(x) = x^3$ and determine their type.

2. **a.** Is the cycle of period 2 for $F(x) = x^2 - 1$ given by 0 and -1 attracting or repelling?

   b. Illustrate this using graphical analysis.

3. **a.** For the function $F(x) = ax$, what is the type of the fixed point at 0? (This type will depend upon a.)

   b. At which a-values does this family of functions undergo a bifurcation?

4. Determine the values of k for which the fixed points in the unit interval for the logistic family $kx(1 - x)$ with $k > 0$ are attracting, repelling, or neutral.

5. Determine the type of the cycle of period 2 for the function $F(x) = -x^3$.

6. Let $D(x)$ denote the doubling function given by $2x$ if $0 \leq x < 1/2$ or $2x - 1$ if $1/2 \leq x \leq 1$. Show that all of the cycles are necessarily repelling.

7. Determine the k-value for which the logistic function undergoes a period doubling bifurcation at the fixed point.

8. Describe the bifurcations that the function $F(x) = ax + x^3$ undergoes when $a = 1$ and $a = -1$.

Exploration

Using the applets below, investigate the orbit diagram for the functions $kx(1-x)$ and $x^2 + c$. In particular, magnify the successive regions that begin where period doubling bifurcations occur as displayed below. Do you see any similarities? Explain what you observe.

Orbit Diagram for $Cx(1-x)$:

http://math.bu.edu/DYSYS/applets/bif-dgm/Logistic.html

Orbit Diagram for $x^2 + c$:

http://math.bu.edu/DYSYS/applets/bif-dgm/Quadratic.html

Periods and Ordering of Iterated Functions
Lecture 22—Transcript

Hello, again. Last time we saw that in order to understand the chaotic behavior of differential equations we needed to move to a slightly different area, the area of difference equations. Remember what a difference equation was. Basically it was iteration of a function. For the most part that will be an iteration of a function of just one real variable, an iteration of a function of just say x. Recall what that meant. We started with an initial value. I often call it a seed. It's a little confusing $x^2 + c$ and seed. But anyway, we started with initial value, plugged it into the function, got a new number out, plugged that into the function, a second number out, etcetera, etcetera, etcetera. We generated the orbit of our seed, our initial value.

As we saw for differential equations, particularly first order differential equations there are actually three different types of equilibrium points. There were sinks, sources, and nodes. Now for iteration of functions it is an analog of equilibrium points. Then what we saw were fixed points. For example for the function x^2 we saw that 0 was a fixed point; $0^2 = 0$. 1 was a fixed point; $1^2 = 1$. The orbit of 1 is just 1, 1, 1, 1, 1. Now exactly as in the case with differential equations there are three different types of fixed points. They are in analogy to sink, source, and node called attracting, repelling, or neutral, or sometimes indifferent.

What's an attracting fixed point? A fixed point is attracting if nearby points tend to that fixed point. A fixed point is called repelling if nearby points have orbits that tend away from that fixed point, just like a source. Finally, a fixed point is neutral if it's neither attracting nor repelling. For example, think of x^2. x^2 has an attracting fixed point at 0 and a repelling fixed point at 1. One way to see that is by graphical analysis. Here's the graph of x^2. By the way, in this lecture and several succeeding lectures I'll be using tools that are java applets that are available on the Web at a Web site that's listed in the company book. Here anyway is the applet called Nonlinear Web.

I've drawn the graph of x^2. I've superimposed the diagonal $y = x$. We saw last time that we could visualize the orbit of any given initial value by just going vertically from the graph to the diagonal, then to the diagonal to the graph, etcetera, etcetera. For example, there's the orbit of that seed. Go to the graph then over, to the graph then over, to the graph then over, you see that this orbit goes to 0. Zero is an attracting fixed point. Start to the left, then again nearby orbits to the left. If you go to the graph and over, graph and over, etcetera you tend to that fixed point at 0. Zero again is an attracting fixed point.

On the other hand, if I go up near 1, then nearby orbits go far away; 1 is repelling. Nearby orbits exit a neighborhood of fixed point. Now if I change this function from x^2 to $x^2 + 0.25$, so $x^2 + 0.25$ again you can see a fixed point right here. Now that fixed point is attracting from the left and repelling from the right. So this is an example of a fixed point that's neutral.

How are we going to tell the types of these fixed points? Just as in the case of sinks and sources we'll use the derivative to determine the type of a fixed point. For example we'll see in a minute that if you have a fixed point and the derivative of your function at that fixed point is an absolute value of less than 1, then that fixed point is attracting. That's a little bit different from the notion of a sink for an equilibrium point for a differential equation. Remember there the derivative had to be less than 0. Here the derivative has to be between −1 and 1 for a fixed point, for an iterated function to be attracting. Similarly we'll see if the derivative is at larger than 1 in absolute value the fixed point is repelling. If the derivative is equal to 1 in absolute value that is either ±1 you get no information.

Let's see that again, using the computer. What I'll do this time is just look at graphs of a linear function. So here are the graphs of a linear function $y = ax + b$. Of course the derivative here is always a, so we see a fixed point for this particular graph where the graph meets the diagonal. Here a is less than 1, the derivative is less than 1, and that fixed point is attracting. Nearby points have orbits that tend to that fixed point. In fact if I change that a value a little bit, as long as it's less than 1 you see that that fixed point remains attractive. Here's the graph of $ax + b$. If I change b a little bit, again, since the derivative is less than 1 that fixed point is always attracting. If I let a go negative, as long as it's not beyond −1, again, when you go to the graph and over, to the graph and over, to the graph and over you tend to that fixed point. If the derivative is between ±1 you always have an attracting fixed point.

On the other hand, if the derivative becomes larger than 1, then what you see is that that fixed point is now repelling. If I let a be larger than 1, nearby points are repelled away from that fixed point. Nearby points move further and further away. The derivative is larger than 1; that fixed point is always repelling. Similarly if the derivative is larger than −1 again, the fixed points now move away. The orbits of points nearby rather, move away. This point iterates further and further and further away. So what we see is if the derivative is larger than 1 in absolute value; repelling. Derivative is less than 1 in absolute value, then you get attracting. If the derivative is equal to 1 in absolute value, you could have either as we'll see.

What about cycles? Remember a cycle for an iterated function is a point at 0, such that when you iterate $f(n)$ times at x_0 you come back to x_0. Remember f_n is what we call the n-fold iterate of f. For example $f^2(x)$ is just $f(f(x))$. Now if we're going to have a cycle of period n, that means f_n of that point at 0 is at 0. So that means that x_0 is a fixed point to the n iterate. So we'll say that the cycle is attracting or repelling or neutral depending on whether that fixed point is attracting, repelling, or neutral for the n iterate function. That is it depends on the derivative of f_n. For example let's take the function $f(x) = x^2 - 1$. We've seen that this function has a two cycle at 0 and −1; $f(0) = -1$ and then $f(-1)$ is back at 0. So the orbit of $0 = 0 - 1, 0 - 1$, etcetera, a 2 cycle.

How do we decide whether it's attracting or repelling? Let's take the second iterate of our function $x^2 - 1$. The second iterate is you take $x^2 - 1$, square it and subtract off 1. So $x^2 - 1^2 = x^4 - 2x^2 + 1$. Then we subtract off −1 and get $x^4 - 2x^2$. Now, 0 and −1 are clearly fixed points for that second iterate. To determine the type of cycle we just compute the derivative of that second iterate at both 0 and −1. But the derivative of $x^4 - 2x^2 = 4x^3 - 4x$. So at 0 and at −1 we see that the derivative is 0, so this cycle is an attracting cycle. Nearby orbits should be tending to that cycle, oscillating back and forth between 0 and −1.

There is the graph of $x^2 - 1$. You see by graphical analysis that there's a two cycle at 0 and −1. Points near this fixed point move away and tend to that 2 cycle. In fact all other nearby points do something, but then eventually start cycling around to that cycle of period 2. That's what happens when you have an attracting cycle. Let me see that by looking at the graph of the second iterate of this function. Here's the graph of the second iterate of $x^2 - 1$. There's the graph of $x^2 - 1^2 - 1$. We have two fixed points here that we saw for the graph of $x^2 - 1$ but there now are two additional fixed points at 0 and at -1. They're the points that lied on our cycle of period 2. What you see is under iteration of the second iterate all points tend near to that attracting fixed point. Nearby points to 0 tend to 0. Nearby points to -1 tend to -1 by graphical iteration.

Just to harken back to a lot of the things we did with differential equations the concept of bifurcations also arrives for iterated functions. Let's consider just say $x^2 + c$. And remember for $x^2 + c$ when $c = ¼$ we had a neutral fixed point. That's where bifurcations begin to happen. Let's look and see what happens to $x^2 + c$ for c near ¼.

Here's the graph of $x^2 + c$. When c is greater than ¼, our graph lies above the diagonal. That means that any seed or initial value when you iterate it, it gets larger. So in fact all orbits just go off to infinity. For $x^2 + c$ where c is larger than 0.25, no fixed points, all orbits go off to infinity. But when we go down to $c = ¼$, that's where this graph just touches the diagonal, and we saw that that fixed point was attracting from the left, repelling from the right. So as c lowers down to 0.25 we get a bifurcation. Suddenly orbits no longer go to infinity, rather some of them are trapped. Some of them tend to the neutral fixed point. Now when c goes below 0.25, suddenly we see the appearance of two fixed points. The one on the right is repelling, nearby orbits go far away. The one on the left is attracting nearby orbits come into it.

This is the kind of bifurcation that we saw actually for first order differential equations. This is what we call the saddle node bifurcation but now in the setting of an iterated function. We can actually analyze this analytically not just graphically. We can find the fixed points. For example we know that $x^2 + c$ will have two fixed points graphically when c is less than ¼. But we can see that analytically by solving the equation $x^2 + c = x$. Or bring the x to the other side; $x^2 - x + c = 0$. That's a quadratic equation and we can solve it. The two routes of that quadratic equation are $½ \pm \sqrt{1-4c}/2$. I'll call those two routes our fixed points $x+$ and $x-$. Clearly there's $\sqrt{1-4c}$ there. We need $1 - 4c$ to be positive. So $x+$ and $x-$ are our fixed points only if c is less than 0.25.

What are the types of these fixed points? Well analytically we compute the derivative, but the derivative of $x^2 + c = 2x$. So at the fixed point $x+$, well $x+$ we have ½ + that square root over 2. Both of those numbers are positive, $2x+ = 1+$, $\sqrt{1-4c}$. That's always greater than 1. $x+$ is repelling just as we saw in the graph. What about $x-$? Well the derivative at $x- = 1 - \sqrt{1-4c}$. So that number is always less than 1, so that's good. But we also need that derivative to be greater than -1. So we need $1 - \sqrt{1-4c}$ to be larger than -1 in order for that to be attracting. Or doing the math we need 2 to be greater than $\sqrt{1-4c}$. Or squaring both sides 4 must be greater than $\sqrt{1-4c}$. Or c must be greater than $-3/4$. So we have it. We know when we've got a bifurcation. It occurs when $c = ¼$. We see orbits that no longer run off to infinity but now tend to an attracting fixed point in some cases when c is between ¼ and $-3/4$, another example of a bifurcation.

We saw several other kinds of bifurcations for differential equations. They also occur here. Remember the pitchfork bifurcation. Pitchfork bifurcation was when we had an equilibrium point that suddenly split into three. In the bifurcation diagram it looked like we had a pitchfork. Let me show you a pitchfork bifurcation when we iterate functions. this time for the function a constant times sign x. A constant times sign x, clearly we have a fixed point here at 0. What's the type of that fixed point? Well, the derivative of $k \sin(x)$ when we let the constant be k, is $k \cos(x)$ and at 0 we get the derivative is just k. So that fixed point for $k \sin(x)$ is attracting if k is between + and − 1, and that fixed point is repelling if x is either greater than 1 or less than -1.

So let's see what happens when k passes through one of those values. Here's the graph of $k \sin(x)$. If k is small we see a fixed point at the origin, which is attracting just as we saw. If k here is between 0 and 1, that fixed point is attracting, but as k increases the derivative becomes 1, so the fixed point is neutral. Then all of a sudden we see the appearance of two new fixed points. Suddenly 0 has become repelling, nearby orbits move far away. But these two new fixed points are attracting, nearby orbits come in. So this is an analog of the pitchfork bifurcation that we saw earlier.

There's one other bifurcation that we saw very often in the orbit diagram we looked at last time. This is what's called a period doubling bifurcation. Remember we saw the orbit diagram. We went from fixed point to a 0.2

point, then to a 0.4 point, then to a 0.8 point. These were period doubling bifurcations. So let's look at a special case of that.

Again let's look at $x^2 + c$. We've seen that $x^2 + c$ has a pair of fixed points if c is less than 0.25. Those fixed points were routes of the quadratic equation $x^2 - x + c = 0$. To find cycles of period 2 we've got to solve the equation the second iterate of $f(x) = x$. We've got to find fixed points for $f(f(x))$. So how are we going to do that? Well $f(f(x))$ is take $x^2 + c$, square it, add c; that's $f(f(x))$ and now we've got to set that equal to x. What do we get? Well $(x^2 + c)^2$ gives us $x^4 + 2cx^2 + c^2$. We've also got $a + c$ and equal to x on the right. Let me bring that x over to the left. Our equation for the cycles of period 2 now reads $x^4 + 2cx^2 - x + c^2 + c = 0$.

That looks pretty complicated, a fourth degree polynomial equation, not so easy to solve. But we know something here. We know some routes of that fourth degree polynomial. The fixed points also satisfy $f(f(x)) = x$. So the fixed points are routes of that equation too. But the fixed points satisfy the equation $x^2 - x + c = 0$. So what we can do is from way back when. We can take that fourth degree polynomial and divide the quadratic polynomial $x^2 - x + c$ into it. What we find is that that fourth degree polynomial factors into the first quadratic $x^2 - x + c$ times another quadratic $x^2 + x +$ quantity $c + 1$. So for that second quadratic formula we find our cycles of period 2. They're given by the routes of that formula which is $-1 \pm$ the square root of now $1 - 4$ factor of $c + 1$ all divided by 2. Or $-1 \pm$ square root of $-3 - 4c$ divided by 2.

In order to have these two cycles what's under that square root sign must be positive. We need $-3 - 4c$ to be greater than 0 or -3 to be greater than $4c$. That is we need c to be less than $-3/4$ in order to get that two cycle. We get another bifurcation when c passes through $-3/4$. Suddenly we go from just having two fixed points to all of a sudden there's a period 2 point. So let's see what that bifurcation looks like graphically.

Here's the graph of $x^2 + c^2 + c$. Here's the second iterate of $x^2 + c$. Now as we let c become more negative, notice that we just have two fixed points for the second iterate of f. Those are our original fixed points for the first iterate. But as we go down, suddenly we see some new fixed points emerge.

Suddenly we go from having just two fixed points to having 4 fixed points. But we know that there are only two fixed points for the original function. So these fixed points must be that 2 cycle we just saw. In fact they arise exactly when $c = -3/4$. We go to have two new fixed points that in fact are initially attracting.

The period doubling bifurcation is a little different from what we saw in our differential equations part of the course. The period doubling bifurcation is this. We have a fixed point that may be attracting or repelling, but at some value it goes from attracting to repelling or repelling to attracting. Meanwhile we get the birth of a new cycle of period 2. That's what we saw was happening in the orbit diagram as suddenly a fixed point would break into 2 and then that 2 cycle would break into 4, etcetera. Those were all period doubling bifurcations.

As you can see, a lot of the stuff we did with first order differential equations carries over immediately to iterated functions or difference equations. But there's a lot more going on. We don't have time to tell you all these amazing things that happen with iteration of functions, but here's my favorite theorem of all time. It's a theorem due to Alexander Sharkovsky in the 1960s. Alexander Sharkovsky was a Ukrainian mathematician. He found this theorem in the 1960s and really impressed a lot of the then Soviet mathematicians. But he published it in Russian I think or maybe Ukrainian, whatever, so the west did not know about this theorem until a lot later. A funny story comes up there.

Before I tell you that though let me just tell you Sharkovsky's theorem. Suppose you have a function that's a continuous function on the realign. That's my only hypothesis. You just need a continuous function on the real line. Let me list all the natural numbers in a kind of strange order. First I'll list all the odds, excluding 1. So 3, followed by 5, followed by 7, followed by 9, etcetera. Then I'll list all 2 times the odds, again excluding 1. So next I put down 2×3, 2×5, 2×7, etcetera. Then 4 or 2^2 times all the odds, again excluding 1. So 4×3, 4×5, 4×7, and on and on. Then 2^3 times the odds. So 8×3, 8×5, 8×7. When I continue doing that when I'm done I've listed all the natural numbers except 1, 2, 4, etcetera; the powers of 2. So what I'll do is list them at the end but in reverse order; 2_n, $2_n - 1 \ldots 2^2$, 2, and 1.

So here's Sharkovsky's amazing theorem. If you have a continuous function on the realign and it has a cycle of prime period n, then it must also have a cycle of prime period k for any k that follows n in the Sharkovsky ordering. Think about that. You find a cycle of some period, then that number appears in the Sharkovsky ordering. Automatically you know that you have cycles of all other periods below that in the Sharkovsky ordering. For example if you find a cycle of period 12, that's 4×3, then you know you have a cycle of period 4×5, 4×7 ... all those other numbers below 12 in the Sharkovsky ordering. But not necessarily cycles of period 3 and 5 and 7, the numbers would come above.

Notice that you have a cycle that's not a period, a power of 2. It's not of the form 2_n. Then you're somewhere above that last line in the Sharkovsky Theorem. So you must have infinitely many cycles for your function. Now look, suppose your function has a cycle of period 3. Then you've got to have cycles of prime period everything. If you find one cycle of period 3, your function must have cycles on every single period, lots going on there.

Now a couple of comments. First this theorem only holds to functions on the realign. It doesn't hold for functions in other spaces. For example just think of a function on the circle. That's rotate the circle by a third of the turn. You rotate the circle by a third of the turn, do it again, and a third time. Then all points are periodic with period 3 and there are no other periodic points. So this is a theorem that's very special, but incredibly interesting on the realign, and it has also had some ramifications.

Back in 1975 two American mathematicians, Tien-Yien Li and a good friend Jim Yorke, proved that for functions on the realign, if you have a cycle of period 3 you have cycles of all other periods. That is they proved the first very special case of the Sharkovsky Theorem. That was really greeted with brouhaha in the U.S. And of course the Soviet mathematicians were saying, "Wait a minute. Sharkovsky did this whole thing and you're approving just one special case and saying this is great." Well that's true. We didn't know about Sharkovsky's Theorem. On the other hand, what really made this paper resonate was the title of the paper was *Period Three Implies Chaos*. *Period Three Implies Chaos*, the first use of the word "chaos" in the scientific literature by Li and Yorke in 1975. Once that happened these fields exploded.

Let me turn back to the orbit diagram. There's the orbit diagram for the logistic function, a constant times x times $(1-x)$. We saw last time a period 3 window. Here it is. There's the period 3 window. You see a cycle of period 3. Remember what this means. For these C values we're plotting the orbit of the critical point, 0.5 and it's seeing a cycle of period 3. But wait a minute. Sharkovsky says there has got to be infinitely many other cycles going on there. Why don't we see them? Where are they?

Well, here's another important fact. It'll become very important in the last lecture. The reason is if you've got a polynomial function, or more generally what we call an analytic function, then if you have an attracting or a neutral cycle, then some critical point must be attracted to it. So what's a critical point? Again that's a place where the derivative is 0. If you've got an attracting cycle, like in this case it looks like we have a cycle of period 3, that critical point must find it. Can you then have other attracting cycles? Well no. The only place where the derivative of the logistic function is 0 is at that one maximum. You have only one critical point. So you can have at most one attracting cycle.

So that's why we don't see any of the other cycles in the orbit diagram for these values. There's actually though a lot more going on in that region that we're not seeing. There's chaotic behavior. That's what we'll attempt to understand in the next lecture.

Chaotic Itineraries in a Space of All Sequences
Lecture 23

Thus far we have encountered chaos in a variety of different settings, including Euler's method, the periodically forced pendulum equation, the Lorenz system, and iteration of the logistic map. In this lecture, we describe how mathematicians begin to understand chaotic behavior.

To keep things simple, let's concentrate on the logistic map $F(x) = kx(1 - x)$, where $k > 4$. Here we choose $k > 4$ so that the maximum value of F is greater than 1. When we use the computer, it appears that all orbits go to infinity. Of course that is not the case, as there are clearly 2 fixed points. Also, the graph of F^n indicates that F^n has exactly 2^n fixed points. How are these cycles arranged, and what are their periods?

The graph of F shows that all orbits of points x, with $x < 0$ or $x > 1$, go to $-\infty$. Also, there is a small open interval A_1 surrounding 1/2 and containing points that are mapped to points to the right of 1. So these points have orbits that escape from the unit interval after 1 iteration and then go to ∞. Then, by graphical analysis, there is a pair of intervals that we call A_2 that contains points that map to A_1, so these points also escape after 2 iterations. Then there are 4 intervals A_3 whose points map to A_2 and hence also escape. Next, there is a set of 2^n open intervals that contain points that escape after n iterations.

Let X be the set of points that never escape from $I = [0, 1]$. So $X = I - \cup A_n$. We want to understand the behavior of the orbits of all points in X. We see that this set is similar to the famous Cantor middle thirds set. This set is obtained as follows. Start with I. Whenever you see a closed interval, remove its open middle third. So we first remove from I the open interval (1/3, 2/3). That leaves us with 2 closed intervals, so we remove the open middle third of each, meaning (1/9, 2/9) and (7/9, 8/9). Then take this process to the limit to obtain the Cantor set.

The Cantor set is remarkable in many ways. First, it is a fractal set because it is self-similar. Second, even though it looks like it contains very few points, it actually contains exactly as many as there are in the entire unit interval! And the remaining points are not just endpoints of the removed intervals. In fact, most points in the Cantor set are not such endpoints. This is a strange set indeed.

To understand the behavior of orbits in X, we move out of the realm of the real line and quadratic functions and into a seemingly much more complicated environment, the space of all possible sequences whose entries are either 0s or 1s together with the shift map. Toward this end, we break up the unit interval I into 2 subintervals: the left half given by I_0, lying to the left of A_1, and the right half given by I_1, on the other side of A_1. (This is similar to what we observed as part of the Lorenz attractor.) Then, given any point x_0 in the unit interval, we associate an infinite sequence $S(x) = (s_0, s_1, s_2, ...)$ of 0s and 1s to x_0 via the rule $s_n = j$ if $F^n(x_0)$ lies in the interval I_j. The sequence $S(x_0)$ is called the **itinerary** of x_0.

For example, the itinerary of 0 is $S(0) = (0, 0, 0, ...)$ since 0 is a fixed point that lies in I_0. Similarly, $S(1) = (1, 0, 0, 0, ...)$ since $F(1) = 0$. Note that if $S(x_0) = (s_0, s_1, s_2, ...)$, then $S(F(x_0)) = (s_1, s_2, s_3, ...)$. That is, to obtain the itinerary of $F(x_0)$, we just drop the first entry in the itinerary of x_0. This is what we call the **shift map** σ. So we have $\sigma(s_0, s_1, s_2, ...) = (s_1, s_2, s_3, ...)$.

Now let Σ denote the set of all possible sequences of 0s and 1s modulo the identifications mentioned above; Σ is called the sequence space on 2 symbols. We say that 2 itineraries are close if they agree on the first n digits. Two itineraries are even closer together if they agree on the first $n + d$ entries. For example, the itineraries $(0, 0, 0, 0, 0, ...)$ and $(0, 0, 0, 1, 1, 1, ...)$ are fairly close together, but the sequences $(0, 0, 0, 0, 0, ...)$ and $(0, 0, 0, 0, 0, 0, 0, 1, ...)$ are even closer together.

While the sequence space looks a little crazy, note that we know a lot about the fate of orbits of σ. For example, we can immediately write down all of the cycles of period n for σ. Just take any string of digits of length n and repeat this string over and over infinitely often. The resulting sequence in Σ then lies on a cycle of period n for σ.

The amazing fact is that the dynamics of F on the unit interval are the same as that of σ on Σ. That is, each point x in I has a unique associated itinerary and vice versa: Given any itinerary in Σ, there is a unique point in I with that given itinerary. The proof of this is not too difficult and is usually given in introductory courses on iterated functions and chaos.

Because of all this, we now have a very good idea about sequences whose orbits under the shift map have very different types of behavior. Consider the sequence that begins (0, 1, 0, 0, 0, 1, 1, 0, 1, 1, ...); in other words, we first list all possible strings of 0s and 1s of length 1. Then we list all 4 possible strings of length 2, then we continue with the 8 strings of length 3, the 16 strings of length 4, and so forth. This sequence lies in Σ and so corresponds to a unique point x_0 in I. But look at what happens to the orbit of $S(x_0)$ under the shift map. If you shift this orbit a large number of times, you can arrange that this orbit comes arbitrarily close to any point in Σ! As a consequence, the orbit of x_0 must come arbitrarily close to any other point in X. This is what we call a dense orbit. These are exactly the types of orbits that seemed to occupy so much of the orbit diagram, and these are the typical orbits we see when we choose a random seed in X.

Do you see sensitive dependence on initial conditions? Given any x in X, x has the associated sequence $S(x) = (s_0, s_1, s_2, ...)$. But any nearby sequence has an itinerary that must differ from that of x at some iteration, so the 2 orbits eventually reach different intervals, I_0 and I_1. Indeed, if we just change the tail of the itinerary of x at each stage after the nth, we find a nearby orbit whose eventual behavior is vastly different.

This process can be used to analyze chaos in lots of different settings. For example, consider the function on the unit circle that simply doubles the angle of a given point, that is, $\theta \to 2\theta$. We can also write this as the complex function $F(z) = z^2$. Note that $1/3 \to 2/3 \to 1/3$, so $1/3$ lies on a 2-cycle; $1/7$ lies on a 3-cycle; and $1/15$ lies on a 4-cycle. We can do symbolic dynamics as before by breaking the circle into 2 sets, $I_0 = [0, \pi)$ and $I_1 = [\pi, 0)$. Similar procedures to the above show that this function is chaotic on the unit circle. We could show that $\theta \to 3\theta$ is chaotic on the unit circle by breaking the circle into 3 arcs—$[0, 2\pi/3)$, $[2\pi/3, 4\pi/3)$, and $[4\pi/3, 2\pi)$—and then doing symbolic dynamics on the 3 symbols 0, 1, and 2.

Important Terms

itinerary: An infinite string consisting of digits 0 or 1 that tells how a particular orbit journeys through a pair of intervals I_0 and I_1 under iteration of a function.

shift map: The map on a sequence space that just deletes the first digit of a given sequence.

Suggested Reading

Alligood, Sauer, and Yorke, *Chaos*, chaps. 1.5–1.8.

Devaney, *A First Course in Chaotic Dynamical Systems*, chaps. 9–10.

Hirsch, Smale, and Devaney, *Differential Equations*, chap. 15.5.

Relevant Software

Iteration Applet, http://math.bu.edu/DYSYS/applets/Iteration.html

Nonlinear Web, http://math.bu.edu/DYSYS/applets/nonlinear-web.html

Problems

1. **a.** In the Cantor middle-thirds set, list the 4 intervals of length 1/27 that are removed at the third stage of the construction.

   b. What is the total length of all of the removed intervals in the first 3 stages of this construction?

   c. What is the length of all the intervals that are removed at stage 4?

   d. What is the formula for the length of all of the intervals removed at stage n?

   e. Using an infinite series, add up the lengths of all removed intervals to determine the length of the Cantor set.

2. Using symbolic dynamics, describe the itineraries of all points that eventually land on 0. How many such points are there?

3. Which points in I have orbits that eventually land on 0?

4. Give an example of a different point whose orbit fills the set X densely.

5. Give an example of a sequence whose orbit under the shift map comes arbitrarily close to the sequence $(0, 0, 0, \ldots)$ but never lands on this sequence.

6. Consider the shift map on the sequence space of 3 symbols—say, 0, 1, and 2. How many periodic points of (not necessarily prime) period n are there? Can you write down a sequence that has a dense orbit?

Exploration

Consider the doubling function on the interval [0, 1) defined by $D(x) = 2x$ if $0 \leq x < 1/2$ and $D(x) = 2x - 1$ if $1/2 \leq x < 1$. (Here we exclude the point 1 from the domain.) First use a computer, a calculator, or the Iteration Applet at http://math.bu.edu/DYSYS/applets/Iteration.html to compute various orbits for the doubling function. What do you see? Is this correct behavior? Can you explain what the computer is doing?

Actually, the doubling function behaves just as chaotically as the function $4x(1 - x)$ does. Use symbolic dynamics to see this by setting $I_0 = [0, 1/2)$ and $I_1 = [1/2, 1)$. Incidentally, there is another name for the itinerary associated to each x. What is this symbolic representation?

Chaotic Itineraries in a Space of All Sequences
Lecture 23—Transcript

In the last several lectures, we've seen a lot of chaotic situations. Earlier we saw that chaos could arise in Euler's method. We've seen it in the Lorenz equations, the periodically forced pendulum, when we iterated certain logistic functions. In this lecture, we're going to see how mathematicians can sometimes completely comprehend this chaotic behavior. They can understand all possible fates of solutions, may not understand what happens to this x or that y, but they know the totality of things that can happen.

Let's begin with a simple example of this. Let me look at iterating the function, I'll call it $f(x)$ is $kx(1-x)$, our logistic function where k is now greater than 4, and I'll look at just x values on the unit interval that'll call I, the interval from 0 to 1. Here's the graph of $kx(1-x)$ when k is greater than 4. You see that this graph pokes above the line $y = 1$.

First, by graphical iteration, if you take an x that's less than 0, you see the graph of the logistic function dives down toward minus infinity outside of the unit interval, so if x is less than 0, by graphical analysis, that orbit goes to infinity, no chaos on the negative real axis. Similarly, if x is greater than 1, that orbit under 1 iteration, is by graphical analysis less than 0, then it too goes off to infinity. There is no chaos to the right of 1. All these orbits just go to infinity. All of the chaotic behavior that we're going to see occurs in the interval between 0 and 1.

What's happening there? Well, we saw that the graph of this function poked through the line $y = 1$. If you draw the graph of the second iterate of this function, you'll see that that graph will poke through the interval $y = 1$ twice, and if you draw the third iterate, you'll see that that graph will poke through 2^2 times, and on and on and on. In particular, the graph of the third iterate crosses the diagonal exactly 8 times. There are 8 fix points for the third iterate. Two of them are the original fix points that you see in the graph of $kx(1-x)$, but the other 6 lie on cycles of period 3, recall Sharkovsky.

In any event, now let's look at where the chaotic behavior occurs. If you go to the unit interval, you see that since the graph pokes above $y = 1$, there's a tiny, little interval in there, let me call it A_1 that consists of points whose first iterate lies to the right of 1, so the second iterate lies to the left of 0 as we just saw, and then off to infinity. There's an interval, A_1, in which all points leave the unit interval after 1 iteration, no chaos there.

Go a little bit further down, and if you look closely, by graphical analysis, there are a pair of intervals, I'll call them both A_2, that have the property that any point in A_2 under one iteration goes into A_1, under the second iteration goes to the right of 1, and then off to minus infinity, so no chaos there.

One more step, if you check there are actually 4 intervals, I'll call them all A_3, two on each side of the intervals in A_2 and any point in A_3 has the property that under one iteration, it goes into one of the intervals, A_2, the second iteration, under A_1, third iteration to the right of 1, and off to infinity. There's no chaos there. All those points go to infinity.

In general, what you can check is there are 2^n disjoint intervals. I'll call them all A_n, in which all points in the intervals A_n leave the unit interval after exactly n iterations. We have a central one, A_1, flanked by a pair of intervals, A_2. They're flanked by a pair of intervals each, so 4 intervals, A_3, then 2^n intervals, A_n, as we go on. There's no chaos in all of the intervals that I've called A_n.

Where is the chaos? The chaos is in the set of points that don't leave the unit interval. Let me let x, be the set of all points whose orbits never leave I. x is all those points that lie in I but never leave, so all those points that lie in I, but take away the union of all those little intervals A_n. Points in x never leave the unit interval. They stay there forever. What's happening on that set, x. First off, x is what's called a Cantor set. That's one of the most important sets in mathematics. It's one of the most basic fractals. What a Cantor set is, is any set that's basically the same as the Cantor middle thirds set. Here is the Cantor middle thirds set.

What I'm going to do is start with the unit interval, the interval from 0 to 1, and then whenever I see an interval, I'm going to pull out the middle third of that interval, the open middle third of that interval. I see the unit interval; I immediately move the open middle third. That is, I take out all points between 1/3 and 2/3. That leaves me with two closed intervals, namely the intervals 0 to 1/3, and the interval 2/3 to 1. There are two closed intervals. Remove the open middle third, that is, throw away the interval (1/9, 2/9) in the left interval and the interval (7/9, 8/9), that's the middle third of the right interval.

That leaves us with 4 closed intervals, and we continue to remove from each of them the middle 27th. That leaves you with 8 closed intervals and on and on and on. In the limit, you get what's known as the Cantor middle thirds set. It looks pretty simple, but this set is pretty crazy.

As I said, it's the most basic fractal. If you've heard of fractals, those are sets that are self similar that when you zoom in on them, you see the same structure over and over again. Look at the Cantor middle thirds set. Basically, we have a left piece, to the left of the interval (1/3, 2/3) that we threw out, and a right piece. But if you put on a magnifying glass and zoom in on that left piece, it looks exactly like the whole Cantor middle thirds set, and then we broke that into two little pieces, much smaller pieces on each side. Zoom in on them, and they'll look exactly like the Cantor middle thirds set if you keep magnifying, magnify by 3, magnify by 9, etcetera. That's still similarity. That's sort of the hallmark of these sets known as fractals.

That's one thing. More importantly, this Cantor middle thirds set consists of what we call uncountably many points. What that means is we could put this set, which looks like a scatter of points, into 1-to-1 correspondence with the entire unit interval. Think about that. That set where we've thrown away millions, billions, infinitely many open intervals, in fact, has the same number of points as the set we started with, the unit interval.

It's kind of crazy, Cantor really went crazy when he discovered this object. A little bit more, what are the points in this Cantor middle thirds set? Zero is in there because we've never removed 0. We always took out the middle third interval, so 0 would be there, so would 1, and so would 1/3 and 2/3 because we never removed them. We just removed the open middle intervals, so all those endpoints are there, all the endpoints of the removed intervals are in the Cantor middle thirds set. That is nothing. Most points are not endpoints.

There's a lot more going on in the Cantor middle thirds set. Most points are what we call buried.

If you remember our construction of the set x, we went precisely the same way as the Cantor middle thirds set. We removed that open interval, A_1, that escaped after 1 iteration, those two open intervals, A_2, that escaped after two iterations, the 4 open intervals, A_3, escaped after 3 iterations, and what you can show pretty easily is that the set x, where we're looking for chaotic behavior is the same as the Cantor middle thirds set.

That's step one. How do we understand the chaos? What we do is we move into the realm of symbolic dynamics. Here's the way we'll do it. Take that interval, I, the unit interval, and we know there's a middle interval where there's no chaos. That's the interval A_1. Let me call the interval to its left, the interval containing 0, I_0, and the interval to its right I_1. Any point in x has orbit that stays for all time in the unit interval, so it stays for all time in either I_0 or I_1, so what we're going to do is associate an itinerary to each point in this set x where we want to understand the behavior of the orbits.

Take a point, x_0, in the set capital X. I'll say its itinerary is what I'll call $S(x_0)$, and what's that going to be? The itinerary is going to be an infinite string of 0s and 1s. It's going to be an infinite string of bits. I'll call it $(s_0, s_1, s_2, ...)$ where each of those numbers, $s_0, s_1, s_2, ... s_j$ is either 0 or 1. I'm going to associate to each point, I'm going to realign, this infinite string of bits.

How do we do that? I'll say the jth entry, s_j, is 0 if the jth iterate of x_0 lies in I_0. Remember the orbit of x_0 is bouncing around either between I_0 or I_1. The jth iterate of x_0 lies in I_0, then that j^{th} it, $s_j = 0$. On the other hand, if the jth iterate of x_0, the jth point on the orbit is in I_1, I'll say that $s_j = 1$.

Here are a couple of examples, back to the real line, we know that 0 is in the set X. Why is that? Zero is a fixed point. It never escapes to infinity, so it's in the set of points that stay for all time in the unit interval. What's its itinerary? Zero lies in I_0, $F(0) = 0$ lies in $\underline{I_0}$, and $F^2(0) = 0$ lies in I_0, that is, the entire orbit stays right in I_0. The itinerary of the point 0 is that string of bits that are all 0s, so 0, 0, 0, ... I'll write that that as $\overline{0}$, overline, which means I'm taking 0 and repeating it infinitely often. What about 1? One also lies in the

431

set X because by the graph, $F(1)$ is 0, and then 0 stays in x for all subsequent iterations. What's the itinerary of 1? One starts out in that right interval, what we called I_1, so the first digit in the itinerary of 1 is 1, and then the next digit, in fact, all the rest of the digits are 0 because 1 lands on 0 and then stays forever in I_0. The itinerary of 1 is 1 and then overbar 0.

By the graph, we see that there's a 2 cycle in the unit interval. You see that square, which represents, by graphical iteration, a 2 cycle. Let me call the left point x_0 and the right point x_1. So x_0 is sent to x_1 and x_1 is sent to x_0. The orbit doesn't escape. It's in our set X. What's the itinerary of x_0? Well, x_0 lies to the left, so its first digit is 0. It maps to x_1, lies to the right, next digit is 1, x_1 comes back to x_0, next digit of 0 then 1 and 0 and 1. The itinerary of x_0 is just 0 1 repeated. What about the itinerary of x_1? Exactly the same, x_1 starts in I_1, goes to I_0, goes back to I_1 and so forth. The itinerary of x_1 is (1 0 1 0 ...) exactly 1 0 repeated.

Let me let Σ, capital sigma, Greek for sequence space, be the set of all possible sequences of 0s and 1s. This is, as I said, the sequence space. An element of this set is an infinite string of bits, 0s and 1s, maybe 0, 1, 0, 0, 1, 0, 1, 1, 0, 1, 1, ... any possible sequence of 0s and 1 is a point in Σ.

What we're going to do to understand the chaos is move away from the real line, how ugly, how complicated, and move to this new set, our sequence space where we'll be able to understand the dynamics completely. It looks kind of crazy right now. Wait a minute. The real line? That's a bad set? This space of all possible sequences is a good space. Yes, wait until you see.

The question is: What does this space, Σ look like? I'm not going to go into too much detail here. We can actually put a distance function on it. We can actually write down a distance function meaning if you give me two sequences, maybe 0, 0, 0 repeated, and 1, 0, 1, 0, 1, 0, 1, 0 repeated. Those are two points in Σ, two infinite sequences, but that we think of as a point in Σ. You can actually write down a distance. You can tell exactly how far they are in this crazy sequence space.

Just to keep it easy, what I'll let you know is that two points in sigma, two of these infinite sequences, are close if they agree in the first n spots, the first n

digits are the same. They are closer if the first $n + k$ digits are the same, and they're even closer if the first $n + k + j$ digits are the same. The more of the initial digits that are the same, the closer those points are. For example, take the sequence 0 repeated. That's pretty close to the sequence 0, 0, 0, 0, 0, and then 1 repeated because they agree in the first 5 digits. But that sequence 0 repeated and the sequence 0, 0, 0, 0, 0, 0, 0 about 20 times, then 1 repeated, those two sequences are much closer.

That gives us an idea of what the setup of this sequence space is, what we call the topology of this sequence space. As I said, there's a way to measure those distances exactly. One other thing, notice that the itinerary of x_0 is the sequence $s_0, s_1, s_2, \ldots$ What is the itinerary of $f(x_0)$? If $S(x_0)$, the itinerary of x_0 is $s_0, s_1, s_2, \ldots$, that means that x_0 lives in I_{s_0}, I_0 or I_1 depending on what s_0 is. $f(x_0)$ lives in I_{s_1}, so that's either I_0 or I_1 depending on what that digit s_1 is. The next digit is s_2, that means that $f^2(x_0)$ lives in I_{s_2}, etcetera.

What's the itinerary of $f(x_0)$? The tail of the itinerary of x_0 tells us exactly that. The itinerary of $f(x_0)$ is to throw away that first digit and keep the rest. The itinerary of $f(x_0)$ is s_1, that's where $f(x_0)$ lives from before; s_2, that's where $f(f(x_0))$ lives from before; s_3, s_4, etcetera. You drop the first digit.

Let me define a function on this sequence space, I'll call it little sigma (σ). It's called the shift map. Sorry, too many sigmas, but this is traditional; σ takes a point in the sequence space to another point in the sequence space. What does the shift map do to the sequence $s_0, s_1, s_2, \ldots$? It drops the first digit. That is, $\sigma(s_0, s_1, s_2, \ldots) = (s_1, s_2, s_3, \ldots)$. that's a function that takes sequences to sequences. That's a function on our sequence space, Σ, to itself.

For example, let's take the sequence 0 1 repeated. What does the shift map do to that? Takes 01, drops the 0, gives you 1, then 0 1 01 0 1. ... Iterate this shift map again. Then it drops the 1 and gives you 0 1 repeated. The sequence 0 1 repeated lies on a 2 cycle for the shift map. What about the sequence 0 1 0 repeated? That lies on a 3 cycle because if you iterated 3 times, you'll drop the first 3 digits, you'll drop the 0 1 0, and what's left is 0 1 0 repeated, the same thing. Similarly, 0 1 0 0 repeated, that's a 4 cycle for the shift map. Think about it.

How can you find cycles of period n? You take any string of bits, any string of 0s and 1s of length n, s_0 all the way up to s_{n-1}, that lies on an n-cycle for the shift map. That means we have exactly 2^n cycles of period n for our shift map. Think about it. Just like that, we've found all the cycles for this function. Suppose I asked you to go back to the quadratic function or the logistic function and find a cycle of period 103? No way, but you could do that. In fact, you can find all of them for the shift map. It's a very interesting function that we understand completely.

The fact is that the dynamics of our function, f, a logistic function on our space, x, is exactly the same as the shift map on Σ. The dynamics, the behavior of the orbits, the cycles, all the other stuff that's happening for f is exactly the same as the shift map, σ, on Σ. This is a little hard to prove in the remaining time, but I do this in my sophomore level or junior level chaos course. It's relatively easy to prove.

Let me explain the chaos. There are actually three properties of a chaotic function. Here's the first. The first is that the set of periodic points, the set of cycles, is dense in our space. What do I mean by a dense set? A dense set is a set whose points come arbitrarily close to every point in the big space. We're looking at the subset of our sequence space, which is the periodic points, that's a subset, and I'm saying it's dense. That means that if you give me any point whatsoever in Σ, any infinite sequence, there's a periodic point arbitrarily close to it. Periodic points are everywhere.

Do you see why that is? Take a point in Σ, say $(s_0, s_1, s_2, \ldots)$, can you find a periodic point close to that point in Σ? Sure you can, take your sequence, whatever it starts with $(s_0, s_1, \ldots s_n)$, that's the beginning of your given sequence, and now throw away the rest and put in s_0 through s_n repeated. The sequence ($\overline{s_0 \ldots s_n}$), that's first a periodic point for the shift map and secondly agrees with your given sequence all the way up to digit n, so it's close to it.

Do you want a period point closer? Go out further; take your given sequence $(s_0, s_1, s_2, \ldots)$, go all the way out to say $n + k$, take those digits, s_0 up to s_{n+k}, put a line over them, overbar, repeat it, and you've got another sequence that's A, periodic under the shift map, and B, even closer to your given point.

If you can find a set that has the property that's arbitrarily close to any point in the big set, you've got a point there, then that's a dense set. Maybe thinking about dense sets would be easier if we did something on the real line that's dense. Take the real line and show some dense subsets. What about the set of all rational numbers. That's a subset of the real line, and that's dense, arbitrarily close to any real number, you've got infinitely many rational numbers. The set of rational numbers is a dense subset of the real line just as the set of periodic points is a dense subset of Σ.

What about the irrationals? Yes, the irrational numbers on the real line are also dense, arbitrarily close to any fraction, and you've got infinitely many irrationals. There's another dense subset of the real line just like the set of periodic points are dense in the sequence space. They're everywhere.

That was the first property of a chaotic function. Here's the second. The second property is there is a dense orbit. There's a point in the sequence space whose orbit comes arbitrarily close to any other point. There's a point that runs around forever, and you give me any point in the sequence space, that orbit gets closer and closer and closer to it as well as to every other point. Can you see a point that has that property?

Here's a point in the sequence space, let me call it s*, the sequence corresponding to s* starts out 0, 1, then 00, 01, 10, 11. What I've done is I've taken all blocks of length 1 for my sequence, that's 0 and 1, then all possible blocks of length 2, that's 00, 01, 10, 11, and listed them, and now I'll list all possible 3 blocks, 000, 001, 010, 011, etcetera. Those are the next entries in s*. Then I'll list all 4 blocks, all 5 blocks, all n blocks, and whatever. That's a point in the sequence space.

What happens to the orbit of this point when it's shifted? It comes arbitrarily close to every sequence in the sequence space. Think about it. Give me a sequence in the sequence space, your favorite one, $s_0, s_1, \ldots$ up to s_n. does the orbit of s* come close to that point? Yes, look out in that string of digits for s*. Look out to where we had all those n blocks or $n + 1$ blocks. You'll see a block that's called s_0 up to s_n, way out there to the right.

What happens when we shift s* a huge number of times? Get rid of the 0, the 1, all the 2 blocks, all the 3 blocks .... Eventually $s_0 \ldots s_n$ comes to the front. At some point on the orbit of s*, you're very close to your given sequence. Do you want to get closer? Yes, look way out further to the right, you'll see a block of much longer length called s_0, s_1, s_n, s_{n+1} all the way up to s_{n+k} agrees with your sequence to $n + k$ digits. Start shifting, taking the orbit of s* under the shift map. Shift, shift, shift, a huge number of times, but boom, eventually your first digits are s_0 up to s_{n+k}, the exact digits that you gave me.

The orbit of s* comes arbitrarily close to any point in the sequence space. That's the second property of a chaotic function. Then the third is as we saw with the Lorenz equation and several others, you've got sensitive dependence. Arbitrarily close to any point in the sequence space, there has to be another point whose orbit goes far away. That's sensitive dependence. Nearby points have orbits that go far away. Do you see sensitive dependence for the shift map? Think about it.

Give me a sequence, a point, $s_0 \ldots s_n, s_{n+1}$, that's a sequence. We know what its orbit is under the shift map. Can you find a nearby sequence very close to that one whose orbit goes very far away? Yes, change all the last digits there. For example, if you have s_{n+1} is 0, make that 1. If you have s_{n+2} is 0, make that 1, s_{n+3} is 1, make that 0. Change the tail of that sequence, so there's another point in your sequence space. It starts out s_0 up to s_n, agrees with your sequence, but then thereafter, it disagrees. If you shift that a bunch of times, eventually you'll get to a sequence. Your sequence will be s_{n+1}, s_{n+2}, ..., but the nearby sequence will be the exact opposite. You'll have completely different digits, far away. That shows that in fact you've got sensitive dependence arbitrarily close to any point in your sequence space, there's another one that goes very far away.

To summarize, for a chaotic function, you need to show three properties: that cycles are dense, sort of nice behavior is everywhere; that there's a dense orbit, there's an orbit that just goes crazy, goes running around, visits every area, and everywhere you see there's this dense orbit; and then finally, you have as we saw in Lorenz, sensitive dependence, nearby initial points that have orbits that go far away.

There is no way to prove this directly for our logistic function, but it's easy to do so for the shift map. As I said, what we can do is show that a shift map on Σ is exactly the same as the logistic function on that set, X. That means that that set F, on the Cantor-like set, is chaotic, kind of an interesting journey away from the real line. We wanted to understand the dynamics on the real line. We now understand it by moving to the sequence space.

We can't find exactly the x values of points that lie in that set X, but we know the totality of behaviors of orbits. That's how we've come to understand, how we've come to comprehend chaotic behavior.

This technique of symbolic dynamics has allowed us to understand the chaotic behavior in many situations. For example, in certain cases involving the Lorenz system, and in certain cases involving pendulums, we go from the realm of three dimensional space or two dimensional space to a very complicated sequence space where we understand everything.

Next time we're going to take a slightly different sojourn. Here we move from the real line to the sequence space to understand the chaotic behavior of a logistic function. Next time, I'm going to move from the real line to the complex plane to understand our bifurcation diagram or our orbit diagram, and remember what that picture is. That's that very complicated sequence of events that led into the chaotic regime. What we'll do is introduce the concepts of the Julia set and the Mandelbrot set in the complex plane to comprehend what's going on here.

Conquering Chaos—Mandelbrot and Julia Sets
Lecture 24

The goal of this final lecture is to give a brief glimpse at how mathematicians are coming to grips with the compete picture of systems with chaotic behavior, and in particular, how these systems evolve as parameters vary.

The simplest example of an iterated map that exhibits a wealth of chaotic behavior is the logistic family $F(x) = kx(1 - x)$. The big question is whether we understand everything that is happening as the parameter k varies. For example, how and when do the various periodic windows arise? How do the chaotic regimes arise and change? These questions were only finally answered in the mid-1990s, and the way we found the answer to this question was to pass to the complex plane and iterate complex rather than real functions.

For historical as well as technical reasons, we will consider the quadratic function $F(z) = z^2 + c$ rather than the complex logistic maps. Here z and c are both complex numbers. The iteration of $z^2 + c$ arose in modern times thanks to the foresight of Benoit Mandelbrot. Mandelbrot used the computer to plot what is now called the Mandelbrot set (as well as the Julia sets) for these maps. The images he produced around 1980 have had a major impact on mathematics. First of all, these fractal images are amazingly intricate and beautiful, and as a consequence, they led to quite a bit of curiosity in the scientific world. More importantly, these images allowed mathematicians to use tools from complex analysis to investigate real dynamics.

The simplest example of a complex iteration occurs when $c = 0$ (i.e., iteration of the complex function $F(z) = z^2$). It is easily seen that if $|z| < 1$, then the orbit of z tends to 0, which is an attracting fixed point. If, on the other hand, $|z| > 1$, then the orbit of z tends to ∞. Finally, if $|z| = 1$, then the orbit of z stays on the unit circle in the plane. Indeed, the behavior of F on this circle is quite chaotic. We certainly see sensitive dependence in any neighborhood of a point on the circle since, arbitrarily close by, there are points whose orbits go far away, either to 0 or to ∞.

For the complex quadratic function $F(z) = z^2 + c$, the set of seeds z_0 for which the orbit does not tend to infinity is called the **filled Julia set**. This set is named for the French mathematician Gaston Julia, who pioneered the study of complex iteration back in the 1920s. It turns out that there are only 2 different types of filled Julia sets for $z^2 + c$: Either the filled Julia set is a connected set (just one piece), or else it is a Cantor set (infinitely many point components, sometimes called fractal dust).

As for the logistic map, the function $F(z) = z^2 + c$ has a single critical point, this time at 0. Amazingly, it is the orbit of 0 that tells us everything about the filled Julia set. For if the orbit of 0 goes to ∞, the filled Julia set is a Cantor set, but if the orbit of 0 does not tend to ∞, the filled Julia set is a connected set. This is the result that prompted Mandelbrot to plot the set of c-values in the complex plane for which the orbit of 0 does not escape (and so the filled Julia set is connected). This intricate and beautiful image is what is known as the Mandelbrot set.

Figure 24.1

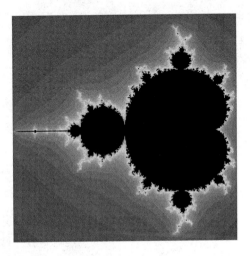

It was the geometry of the Mandelbrot set that finally allowed mathematicians to understand the real dynamical behavior of $x^2 + c$, and so also the logistic family. There are 2 reasons for this. First, the area of mathematics known as complex analysis offers many more mathematical tools to study iteration of functions like polynomials. And second, looking at objects in the plane gives us many more geometric tools to work with.

Rather than showing how the Mandelbrot set allows us to understand the behavior of $x^2 + c$ for real c-values, let's look instead at another pattern that appears. Attached to the main cardioid are infinitely many smaller bulbs. Each of these bulbs contains c-values for which the corresponding orbit of the critical point 0 tends to an attracting cycle of some period. This period is the same for all c-values drawn from a given bulb; this is the period of the bulb. So how are these periods arranged as we move around the main cardioid?

We first see that the number of spokes in the antenna attached to the bulb gives us the period of the bulb.

Figure 24.2

period-3 bulb   period-5 bulb

We next attach a fraction to each bulb. The denominator is the period of the bulb. The numerator can be defined in 3 different ways. (1) The location of the smallest spoke in the antenna relative to the principal spoke (the spoke that connects to the bulb itself) going in the counterclockwise direction gives us the numerator. Above are the 1/3 and 2/5 bulbs. (2) If we plot a filled Julia set corresponding to a *c*-value from the bulb, the smallest ear hanging off the central portion of the set, again counted in the counterclockwise direction, also determines the numerator. Below are filled Julia sets drawn from the 1/3 and 2/5 bulbs. (3) The rotation number of the attracting cycle also gives the numerator.

Figure 24.3

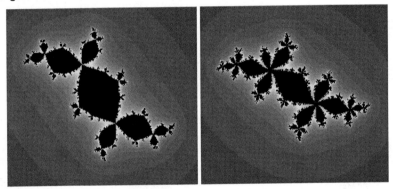

It turns out that the bulbs are arranged in the exact order of the rational numbers. So we see a lot of connections between the geometry of the Mandelbrot set and Julia sets and the dynamical behavior. Proving these facts is not that easy, but this nonetheless shows how mathematicians use tools from diverse areas of mathematics to understand the behavior of complicated systems of differential equations.

Important Term

filled Julia set: The set of all possible seeds whose orbits do not go to infinity under iteration of a complex function.

Suggested Reading

Devaney, *A First Course in Chaotic Dynamical Systems*, chap. 17.

———, *The Mandelbrot and Julia Sets*.

Mandelbrot, *Fractals and Chaos*, chap. 1.

Peitgen, Jurgens, and Saupe, *Chaos and Fractals*, chaps. 13–14.

Relevant Software

The Quadratic Map, http://math.bu.edu/DYSYS/applets/Quadr.html

The Mandelbrot Set Iterator, http://math.bu.edu/DYSYS/applets/M-setIteration.html

Orbit Diagram for $x^2 + c$, http://math.bu.edu/DYSYS/applets/bif-dgm/Quadratic.html

Problems

1. Compute the orbit of i under $F(z) = z^2$ and describe its fate.

2. Compute the orbit of $2i$ under $F(z) = z^2$ and describe its fate.

3. Compute the orbit of $i/2$ under $F(z) = z^2$ and describe its fate.

4. Which orbits of $F(z) = z^2$ tend to infinity?

5. Which orbits of $F(z) = z^2$ tend to the origin?

6. For the quadratic function $x^2 + c$, we saw that there were no fixed points on the real line when $c > 1/4$. What happens in the complex plane?

7. What is the fate of the orbit of 0 under $z^2 + i$? Is the Julia set connected?

8. Use the computer to investigate what happens to the filled Julia sets for c-values along the path $c = -3/4 + iA$ (where A is a parameter).

9. Use the computer to look at the portion of the Mandelbrot set that lies along the real axis. What object in the Mandelbrot set corresponds to the windows in the orbit diagram for $x^2 + c$?

10. Discuss the bifurcation that occurs at $c = 1/4$ along the real axis, but now from the complex point of view.

Exploration

Consider the bulbs hanging off the period-2 and period-3 bulbs attached to the main cardioid of the Mandelbrot set. How are the periods of these bulbs arranged? You may use The Mandelbrot Set Iterator software located at http://math.bu.edu/DYSYS/applets/M-setIteration.html to find these periods.

Conquering Chaos—Mandelbrot and Julia Sets
Lecture 24—Transcript

Welcome to this final lecture. In the previous lecture, I gave you a little glimpse of how mathematicians come to understand the chaotic behavior of a specific system. In this lecture, I'd like to give you a glimpse of how mathematicians come to see the big picture of all this chaotic behavior, how these systems evolve as parameters vary. How all these bifurcations accumulate, etcetera.

I could consider the logistic map, a constant times x times $(1 - x)$ and ask what's the big picture there. We've seen that big picture; let me remind you. Here is the logistic orbit diagram. We saw lots of interesting behavior, lots of bifurcations, period doubling bifurcations, and another period doubling bifurcation, etcetera. The period 3 window, remember, Sharkovsky's theorem, lots of chaotic regimes.

Instead of looking at the logistic family, I'm instead going to look at the quadratic family $x^2 + c$. This is the orbit diagram for $x^2 + c$. It looks exactly the same, except it's reversed, period doubling bifurcations, chaotic regime, period 3 window, etcetera. We'd like to understand the complete behavior of $x^2 + c$.

The reason I'm going to $x^2 + c$ is mainly for historical as well as some technical reasons. We're going to pass the complex plane. Back in the 1920s, the French mathematician Gaston Julia and Pierre Fatou were looking at $z^2 + c$, and they could see there were lots of interesting geometric and chaotic behaviors, although the word "chaos" hadn't even been invented back then. They were very interested; unfortunately, they had no computer pictures, so the study of iterating things in the complex plane disappeared only to be resurrected in 1980 by Benoit Mandelbrot who used the computer to plot what's now known as the Mandelbrot set and the Julia sets. Incredibly beautiful images, but more importantly, these images brought in all sort of techniques, from geometric to complex analytic, into play. Once we started iterating complex functions, then we could finally understand what was happening when we iterated real functions.

Let me begin quickly with an example. We're going to look at $z^2 + c$ where z is a complex number. Let me choose the simplest case, $c = 0$. We've seen this before. If z is a complex number that's larger than 1, then the orbit of z goes to infinity. Take $2i$: $2i^2 = -4$; -4^2 is 16, 16^2 is big, big^2 is bigger, and it goes off to infinity. If z is smaller than 1 in magnitude, then the orbit of z goes to 0, an attracting fixed point. For example, $i/2$ gets sent to $-1/4$ to $+1/16$, something small and even smaller, that orbit goes to 0. But on the unit circle, you have a function that turns out to be chaotic. That's the angle doubling function.

What we call the filled Julia set is the set of all points in the complex plane whose orbits do not go off to infinity. When c was 0, the filled Julia set was all those seeds' initial points that lie on or inside the unit circle. The boundary of the filled Julia set is what's known as the Julia set, and that's the place where chaos occurs.

It turns out that for quadratic functions, there are only two types of filled Julia sets, either connected sets, that means they're one piece, or what we call totally disconnected sets, that's a scatter of infinitely many points, in fact, something much like the Cantor set that we saw in the last lecture. Sometimes that's called fractal dust.

As we've seen before, the critical points to these functions know it all. Amazingly, for $z^2 + c$, the orbit of the critical point, namely 0, that's where the derivative of 0 determines what types of filled Julia sets you have. If the orbit of 0 goes to infinity, the filled Julia set is fractal dust, a scatter of points, or a Cantor set. If not, the filled Julia set is just one piece.

Mandelbrot decided to draw the picture of those c-values for which the orbit of 0 does not escape, and now it's named for him, the Mandelbrot set. There are a couple of interesting things here. The reason he got into that was his uncle, Szolem Mandelbrot, was actually a mathematician back in the 1920s who knew Julia and Fatou, and he came to Mandelbrot much later in his life and said, you know, they had some interesting things to look at. You should look at them. At that time, Mandelbrot was at the IBM Watson Lab, and he had access to some of the world's largest computers and went ahead and founded this field.

Let me show you the Mandelbrot set and some of the corresponding filled Julia sets. Here is the Mandelbrot set on the right. The Mandelbrot set is a picture of the c-values for which the orbit of 0 does not escape. All these black points are c-values for which the orbit of 0 does not escape; it stays bounded. Colored points are c-values for which the orbit of 0 goes off to infinity. How I'm coloring these points is red points consist of c-values for which the orbit of 0 escapes very quickly, followed by orange, yellow, green, blue, violet, etcetera. The color tells me how quickly the orbit of 0 escapes—red: fastest; green, blue, and violet: much more slowly.

What I've done is I'll choose a c-value from the Mandelbrot set. In fact, I chose this c-value right here that's c-value 0, and we just saw that the filled Julia set was the unit disk. That's what I'm plotting over here; again, black points do not escape, colored points do escape. If I go into this region here, let me choose a c-value there, we get a very different looking filled Julia set. In fact, this is a fractal. What you see is a big black disk in the middle. If I zoom in on the top piece right above it, you see another black disk on top of it. If I zoom in on that, you see again the same structure over and over again. The more you zoom, the more you see the same structure. That's the hallmark of these objects called fractals.

If I plot the orbit of 0, what you see is the orbit of 0 eventually cycles with period 2. If I actually go in there and put in 0 and 0 and view the orbit, you'll see that over in the filled Julia set, it actually cycles with period 2. We call this the period 2 bulb. Were I to go up here and choose a c-value, again we're in the Mandelbrot set, so the Julia set is one connected piece. This is what we call the fractal rabbit, and again, the orbit of 0 does not escape. It eventually cycles with period 3.

That's what happens in the Mandelbrot set, but if I leave the Mandelbrot set, I choose a c-value slightly outside, then the filled Julia set explodes. It's not one piece anymore, it's actually infinitely many pieces. You don't see any black over here, but as your eye goes from red to orange to yellow, you can see some green and maybe some blue. There's actually a scatter of points here. The Julia set has imploded. It has become like our Cantor set.

Go up into this region, and again, the Julia set disappears. It consists of infinitely many pieces. Let me show you that in a more animated fashion. Let me take you on a tour out of the Mandelbrot set exiting due right. We're starting here at $c = 0$ where the filled Julia set is the disk, and now as c moves to the right you see suddenly the filled Julia set implodes, becomes one of these fractal dust objects. You go from having one component to your filled Julia set to having infinitely many components.

Another example of that, let me take you on another tour around the Mandelbrot set. Let me start again inside this big, black region, go to the left, and then I'll take a right hand turn and exit. When I do, again you see the Julia set suddenly falls apart. It goes from one connected piece to leaving the Mandelbrot set. It disappears into infinitely many little pieces.

What does this Mandelbrot set do to tell us about the iteration of real functions? It turned out it was the geometry of the Mandelbrot set together with the field of mathematics known as complex analysis that finally allowed us to understand real quadratic functions in the mid 1990s. Think about it. We only understood x^2 plus a constant in 1990.

The way we're going to do this is that rather than looking on the real line, we'll look instead around what's happening in the Mandelbrot set, on these little bulbs in the Mandelbrot set. When we understood the real line, we were actually going right down the real line, and the bulbs were a little bit different. The same procedure allows us to understand what's happening for real dynamics as I'm going to show you now.

Let's look at these different bulbs in the Mandelbrot set. It turns out that each of these bulbs contain parameters, c-values, for which you have an attracting cycle of some period. If you look at the geometry of those bulbs, you can understand the dynamics. You can understand the behavior of that attracting cycle and the geometry of the filled Julia set.

Let me show you some of these. We saw up here that if I chose a c-value in this region, we got the rabbit, and if I look at the behavior of 0, eventually it cycles with period 3. If I had chosen the c-value right in the center of that bulb, you'd just see three points. All orbits that are colored black here eventually start to cycle with period 3, and look at this bulb. Let me magnify it. There's a magnification of this bulb. Look at the bulb; it has got an antenna with three spokes attached, and the period of that bulb was 3. You see a junction point right here and three spokes attached. The period of that bulb we say is 3 because all of the black points go to a cycle of period 3.

Let me look at this bulb over here. If I magnify that bulb and choose a c-value right in the middle, you see a different-looking filled Julia set. This is sort of a degenerate rabbit. We have a main body, 0 is right in the middle, and now we have three ears attached. If I look at the orbit of 0, you see that it cycles with period 4.

Back to this bulb in the Mandelbrot set, you see an antenna with a junction point from which 4 spokes emanate, and that bulb consists of c-values that have attracting cycles of period 4. One more, if I magnify this bulb right here and compute a c-value, the filled Julia set of a c-value inside, you get a different looking filled Julia set, but now the fate of the orbit of 0 is it cycles with period 5, and look at the geometry of this bulb. It has an antenna that has a junction point with 5 spokes attached, so there's an interesting connection between the geometry of this object and the behavior of orbits.

The geometry says you've got a period 5 cycle in there, and then the geometry of the filled Julia set tells us what's happening on the boundary, the chaotic regime. The question is how do we understand the arrangement of these bulbs. The real question is, how do we understand the arrangement of those windows in the orbit diagram. For that, we would take different bulbs. For that we would take bulbs lying along the real axis. That's that sort of central region in the Mandelbrot set.

That's a little harder to explain. A lot more geometry and algebra goes into it, so let me just constrain myself to what's happening to the bulbs that lie around that main, black cardioid-like region. What I'm going to do is assign a fraction to each bulb hanging off the main cardioid. What is that fraction going to be? It's going to be p/q, and the denominator will be the period of the bulb.

For example, what is the denominator corresponding to this bulb in the northern section of the Mandelbrot set? Look at it. It's got a junction point with three spokes emanating, so this will be the something over 3 bulb. What over three? Look at this bulb. It has a junction point. Attached to it is what we call the principal spoke leading down to that bulb, to the period 3 bulb, and now there are two other spokes attached. Where is the smallest spoke? That's located, roughly speaking, 1/3 of a turn in the counterclockwise direction from the principal spoke. You start at the principal spoke and rotate 1/3 of a turn around, you'll end up at the smallest spoke sticking out of that junction point, not including the principal spoke.

For that reason, this is the 1/3 bulb. The period is 3, and the antenna says 1/3 bulb. That's kind of a visual way of determining what the p/q is, what the fraction is. There are two other ways of determining what p/q is. What we do is let's take a c-value from that period 3 bulb, and plot the filled Julia set. There it is. There's the fractal rabbit. Right in the middle, that main bulb contains 0, so 0 is inside the filled Julia set. It doesn't escape, of course. The filled Julia set is a connected set, but this is the fractal rabbit. Attached to the main body are a pair of ears and it's fractal. You see there are ears attached to ears, ears attached to ears, etcetera. In, say, the upward direction, where is the smallest ear? If you start in the main body and take 1/3 of a turn in the counterclockwise direction, you'll get to the smallest ear. The smallest ear is, again, located 1/3 of a turn in a counterclockwise direction from the main body, another reason why this is the 1/3 bulb.

Then a third reason, let's look and see what happens to the orbit of 0. The orbit of 0 just bounces around this filled Julia set cycling by 1/3 of a turn each time you iterate. All orbits that are counted black tend to that cycle of period 3 that rotates 1/3 of a turn each time you iterate in the counterclockwise direction. That's another reason why that bulb is the 1/3 bulb.

Another question, here's another bulb located on the northeast side of the Mandelbrot set. What bulb is this? Look at the antennas. We see 4 spokes emanating from that antenna, the principal spoke and three others, so this is the period 4 bulb. The denominator of our fraction is 4.

What's the numerator? Where is the smallest antenna sticking out from that junction point that is not the principal spoke? It's a little hard to tell here, and that's why this part of the reasoning is a little iffy. It is certainly not the spoke that's pointing directly up. Was it the first or the second spoke in the counterclockwise direction? It

turns out that it's the first spoke, and there are other ways, mathematically, to see it, so the smallest spoke sticking out of the junction point is located 1/4 of a turn in the counterclockwise direction.

This is the 1/4 bulb. As we said, there are two other ways of determining which bulb this is. First, take a c-value from inside that period 4 bulb, and plot its filled Julia set. As we saw earlier, it's the degenerate rabbit. We have a main body with now three ears attached, and if you zoom in, you see 3 ears and 3 ears and 3 ears, again, a fractal object. Where is the smallest ear attached to this filled Julia set? The main body is right in the middle. If you look, you see the smallest ear is located 1/4 of a turn in the counterclockwise direction, another reason why this is the 1/4 bulb.

Finally, let's plot the fate of the orbit of 0. The orbit, the cycle of period 4 to which 0 tends, and what you see is, this cycle, as we iterate, runs around these ears jumping a quarter of a turn at each iteration, a third reason why this is the 1/4 bulb.

Let's keep going. What bulb is this? Look at the antenna. You see a junction point with 5 spokes, so this is the period 5 bulb. Something over 5, what is the numerator? Look at the junction point. The first counterclockwise spoke is not the smallest; that's the longest now. It's the second spoke that's the shortest, so this is the 2/5 bulb.

There is an amazing connection between the geometry of this Mandelbrot set and what's happening dynamically because take a filled Julia set from inside that bulb. There it is. Plot the corresponding orbit of 0, and in both ways you look at this, you see 2/5. The orbit of 0 was cycling around this filled Julia set jumping 2/5 of a turn at each stage, the 2/5 bulb, and now this is a 4-eared rabbit. The smallest ear is located 2/5 of a turn in a counterclockwise direction from the main body, the 2/5 bulb. The geometry and the dynamics, the fate of orbits comes together somehow in the Mandelbrot set.

Here are a few more just to get everything in place, what is this bulb? It is a little hard to count here. There are more spokes coming out. If you count the spokes, and remember to count the principal spoke. You'll see 7 of them, so this is the something over 7 bulb. What's the numerator? Look closely. You'll see that the third spoke in the counterclockwise direction is the shortest. This is the 3/7 bulb.

Turn back to the dynamics, plot the filled Julia set, you see that the orbit to which 0 is attracted is jumping around by 3/7 of a turn each stage, and if you look at this multi-eared rabbit, you see that the smallest ear is located 3/7 of a turn in the counterclockwise direction. All three of these issues tell you that this is the 3/7 bulb.

Here are a couple more, what is this bulb? We saw earlier that we had a cycle of period 2 in that left-hand side bulb, so this is the something over 2 bulb. What's the numerator? There's only one junction point, so there's only one spoke, and so the smallest spoke is located halfway around the junction point. This is the 1/2 bulb, and plot the filled Julia set, the one we earlier with all the fractal bulbs on top of it, you see that it's a one-eared rabbit, the smallest ear is halfway around, and the cycle to which 0 is attracted is a cycle of period 2. It jumps halfway around each time, the 1/2 bulb.

Finally, what's this bulb? You see three spokes, and it's the something over 3 bulb. What's the numerator? Remember to go in a counterclockwise direction. Now you see you have to jump 2/3 of a turn to get to the smallest spoke. This is the 2/3 bulb. If you plot the filled Julia set, you see that the filled Julia set is a rabbit again, although this time the smallest ear is located 2/3 of a turn in the counterclockwise direction from the main body. The cycle, it was running around 1/3 of a turn in the clockwise direction, so 2/3 of a turn in a counterclockwise direction.

Look at what we have. Look at the Mandelbrot set. Look at the order of these bulbs. We went from 1/4 to 1/3 to 2/5 to 3/7 to 1/2 to 2/3. These bulbs are ordered in the exact order of the fractions. In fact, for any fraction whatsoever, there's a bulb attached to that big, black region that has exactly the right geometry. You give me 23,243/36,947, there's a bulb located exactly in that right place that has 23,243/36,947, exactly that geometry. The bulbs are ordered in the exact order of the fractions. That's an amazing ordering. Somehow the ordering of the bulbs, the geometry of the filled Julia set, and the dynamics of the cycles to which we're attracting, all come together in the Mandelbrot set and allow us to understand what's happening in certain cases for $z^2 + c$.

As I said, to understand real $x^2 + c$, we'd have to take a different tour around the Mandelbrot set that's much more difficult, but nonetheless, in the 1990s, that's what people did. They used the ideas from complex analysis together with the Mandelbrot set to understand $x^2 + c$.

Unfortunately, we still don't understand $z^2 + c$. When you stay off in a complex plane, much, much more is going on. It's much more difficult to do and still a major open problem in mathematics. What happens for all c-values for the quadratic function $z^2 + c$.

Think of it. We've really come full circle in this course. We've done a lot of stuff here. We've done a lot of analytic, qualitative and numerical approaches to differential equations. We've seen lots of different pictures that help us analyze these differential equations, analyze qualitatively the differential equations ranging from phase planes to bifurcation planes, trace determinant planes, a lot of qualitative methods have come in to help us understand differential equations.

Lots of fields of mathematics have come in. We've used linear algebra. We've used calculus. We've seen a lot of geometry. All of this allows us to understand certain of these differential equations. If you were to continue in this study, you'd see even more mathematics coming in, the area of dynamical systems that provides all this qualitative information, topology, and as we saw today, complex analysis.

What I hope you've really seen in this course is that some really interesting and new things are happening in mathematics. In the old days, when people went to school and studied mathematics, what did they see? All through

12 grades in high school they'd see 4th century B.C. geometry, 11th century algebra, and if they were really good and stuck with it, they'd see 17th century calculus. Even undergraduates just a few years ago would see nothing but historical mathematics, 17th or 18th century mathematics.

But now, notice what's happened. We're seeing some incredibly new and interesting things happening in mathematics, and they're talking about just quadratic functions. We don't understand $z^2 + c$. This is a real change in mathematics education, so I can go up to a high school student and show him or her some of the things that we've been seeing like the Julia sets or the Mandelbrot set or the bifurcation diagram or the orbit diagram. They'll say, that's not mathematics. They'll say, that's beautiful. That's interesting, and I'll say, no, that's mathematics. Things have really changed in mathematics and especially in this area of mathematics governed by differential equations.

Let me end by telling you that there's a lot more to be done in this field. Remember, we don't understand $z^2 + c$. If you figure out $z^2 + c$ and what's happening for all c-values, you have the Fields Medal, the equivalent of the Nobel Prize in mathematics. You don't want to work on something as easy as $z^2 + c$? Try cubic functions, $x^3 + bx + d$. A cubic function, that's infinitely more difficult. Why is that? Remember the critical points, $x^2 + c$ has a single critical point, but a cubic function has a pair of critical points. Two things control what's going on, infinitely more difficult, but still beautiful and interesting.

Let me end by thanking you for staying here. I've really enjoyed very much giving these lectures. I know things went pretty quickly at times. I hope you were able to stay with me.

Solutions

Lecture 1

1. **a.** $y'(t) = 3t^2 + e^t$.

    b. $y''(t) = 6t + e^t$.

    c. $f(t) = t^4/4 + e^t$.

    d. The graph is always positive and increasing. Moreover, e^t tends to 0, and t tends to $-\infty$ and to ∞ as t tends to positive infinity.

    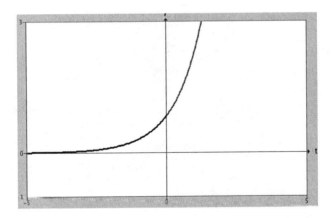

    e. $t = 0$.

2.

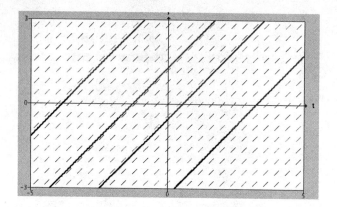

3. Solutions are of the form $y(t) = t + C$, where C is a constant.

4.

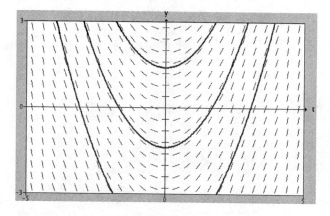

Solutions are of the form $y(t) = t^2/2 + C$.

5. All solutions tend to 0 since $y' < 0$ when $y > 0$. As before, $y(t) = e^{kt}$, but now $k < 0$.

6. The general solution here would by $y(t) = t^2/2 + $ constant.

7. Equilibria occur at $y = 1$ and $y = -1$. We have $y' > 0$ if $y > 1$ and $y < -1$, whereas $y' < 0$ if $-1 < y < 1$. So solutions tend to infinity if $y > 1$ and to -1 if $y < 1$.

8. A solution to this differential equation for any constant is $y(t) = $ constant. So solutions never move; they stay put.

9. First write $y'' = g/m$. Then we must have $y' = (g/m)t + A$ for some constant A. And then $y(t) = (1/2)(g/m)t^2 + At + B$, where B is any other constant.

Lecture 2

1. a. $t^3/3 + t^2/2$.

   b. $y = 0, 2,$ and -2 are solutions.

   c. $y(t) = 0$ when $t = 0, 1,$ and -1. $y(t) > 0$ if $t > 1$ or $-1 < t < 0$.

d. The graph of $y(t) = t^2 - 1$ is a parabola opening upward and crossing the (horizontal) t-axis at $t = 1$ and $t = -1$.

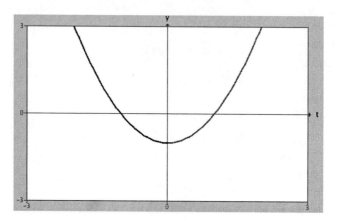

e. This graph crosses the horizontal t-axis at $t = 0$ and $t = 2$. The graph is a parabola opening downward since $y(t)$ tends to $-\infty$ as t approaches $\pm\infty$.

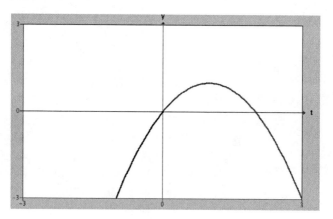

2.

3.

4. The only equilibrium point is 0.

5. There are no equilibria for this differential equation.

6. For n even, all solutions with $y > 0$ tend to ∞ while all solutions with $y < 0$ tend to 0. When n is odd, again all solutions with $y > 0$ tend to ∞ while all solutions with $y < 0$ now tend to $-\infty$. So the answer does depend on n.

7. There are equilibria at -1 and 0. If $y > 0$, then $y' > 0$, so solutions go to ∞. If $-1 < y < 0$, then $y' > 0$, so solutions increase to 0. If $y < -1$, then $y' < 0$, so solutions tend to $-\infty$.

8. The function $\sin(y)$ has equilibria at $n\pi$ for each integer n. Between 0 and π, $\sin(y)$ is positive, so solutions increase to $y = \pi$. Between π and 2π, $\sin(y)$ is negative, so solutions decrease to $y = \pi$. Similar behavior occurs in other intervals of length 2π; solutions always tend to the equilibria at $y = n\pi$ when n is odd and away from the equilibria at $y = n\pi$ when n is even.

9. What about $y' = \sin^2(\pi y)$? y' is always positive except at the integers where $y' = 0$.

10. One example would be $y' = |y(1 - y)|$ since $y' > 0$ everywhere except $y = 0$ and $y = 1$, but of course there are many others, like $y' = y^2(1 - y)^2$.

Lecture 3

1. **a.** $y'(t) = 2t$, so $y(t)$ increases when $t > 0$.

   b. $y(t)$ decreases when $t < 0$.

   c. $y(t) = (t + 1)^2$ so the only root is $t = -1$.

   d. $y' = 2t + 2$, so this function increases when $t > -1$.

   e.

   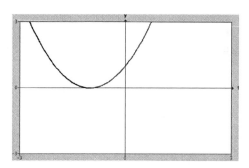

   f.

   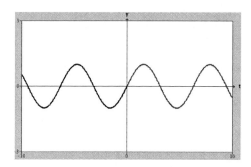

   g. $-3\sin(3t + 4)$.

2. The only equilibrium point is $y = -1$, which is a source.

3. The only equilibrium point is $y = 0$, which is a node.

4.

5. Equilibria at $y = 1$ (source) and $y = -1$ (sink).

6. The only equilibrium point is at $y = 1$, and there we have that the derivative of $y^3 - 1$ is $3y^2$. At 1 we get $y' = 3$, so 1 is a source.

7. Technically, the existence and uniqueness theorem does not apply when $y = 0$ since the function $|y|$ is not differentiable there. However, we have an equilibrium solution $y = 0$ there, and all other solutions are given by $y(t) = Ce^t$ when $C > 0$ or Ce^{-t} when $C < 0$, so we do have existence and uniqueness at $y = 0$.

8. The equilibrium points are given by $\pm A^{1/2}$ when $A > 0$ and 0 when $A = 0$. There are no equilibrium points when $A < 0$. Since the derivative of $y^2 - A$ is $2y$, we have that $+A^{1/2}$ is a source while $-A^{1/2}$ is a sink. When $A = 0$, the equilibrium point at 0 is a node.

9. The equilibria are at 0 and 1 and the derivative is $A(1 - 2y)$. So for $y = 0$, this point is a source when $A > 0$ and a sink when $A < 0$. For $y = 1$, we have a sink when $A > 0$ and a source when $A < 0$. When $A = 0$, all points are equilibrium points, so they are all nodes.

10. This equation has an equilibrium point at $y = -1$. For $y < -1$, $y' < 0$, so these solutions tend to $-\infty$. For $-1 < y < 1$, solutions now increase until they hit $y = 1$, where the slope becomes infinite, so the solutions stop there. If $y > 1$, then $y' < 0$, so solutions decrease until they hit $y = 1$, when again the slope becomes infinite and solutions stop.

Lecture 4

1. The only equilibrium point is at $y = A$, and this point is a source.

2. No.

3. There are no equilibrium points when A is nonzero and infinitely many when $A = 0$; every y-value is an equilibrium point when $A = 0$.

4. A bifurcation occurs when $A = 0$.

5. The equilibria occur at $y = 0$ and $y = -1/A$ (as long as A is nonzero).

6. When $A > 0$, there is an equilibrium point at $y = 0$ that is a source. When $A < 0$, the equilibrium point at $y = 0$ is a sink. But when $A = 0$, all points are equilibrium points, so we have a bifurcation at $A = 0$.

7. We have equilibria at $y = 0$ and $y = A$ for each A. When $A = 0$, there is a single equilibrium point at $y = 0$, a node, with all other solutions decreasing. When $A > 0$, the equilibrium point at $y = A$ is a sink while 0 is a source. The opposite occurs when $A < 0$.

8. Only at those that are nodes.

9. When $B = 0$, we have a similar situation to that in problem 7. When $B > 0$, we always have a pair of equilibria for each A, given by

$$\frac{A \pm \sqrt{A^2 + 4B}}{2}.$$

When $B < 0$, these 2 equilibria only exist if $A^2 + 4B \geq 0$ (i.e., for $B \geq -A^2/4$). When $B = -A^2/4$, there is a single equilibrium point at $A/2$. When there are 2 equilibria, the larger one is a sink and the other is a source.

10. When $B = 0$, we have a single equilibrium point at $y = 0$ when $A \leq 0$. This equilibrium point is a sink. There are 3 equilibria if $A > 0$: one at $y = 0$, a source; and 2 sinks at $y = \pm A^{1/2}$. When $B \neq 0$, the graph of $F(y) = B + Ay - y^3$ has derivative equal to 0 when $3y^2 = A$. Therefore if $A < 0$, the graph of $F(y)$ is strictly decreasing, so there is always a single equilibrium point that is a sink. If $A > 0$, there are now 2 places where $F'(y) = 0$, at $y = \pm(A/3)^{1/2}$. $F(y)$ has a local minimum at the point $y = -(A/3)^{1/2}$ and a local maximum at $y = +(A/3)^{1/2}$. For B-values for which the value of F at the local minimum is greater than 0, the graph of F shows that there is only one equilibrium point, a sink. Similarly, there is only one equilibrium point if the value of F at the local maximum is less than 0, again a sink. But if the local minimum is less than 0 and the local maximum is greater than 0, then there are 3 equilibrium points. The largest and smallest equilibria are sinks, and the other is a source. When either the local maximum or local minimum value is 0, then a sink and a source merge to become a node.

Lecture 5

1.
   a. $t^2/2 + 5t + C$.

   b. $t^3/3 + t^2 + t + C$.

   c. $-t^{-1} + C$.

   d. $-e^{-t} + C$.

   e. C.

2. $t = e^4$.

3. $t = \ln(4)$.

4. $y(t) = Ce^t$.

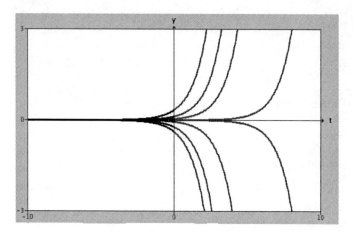

5. $y(t) = Ce^t - 1$.

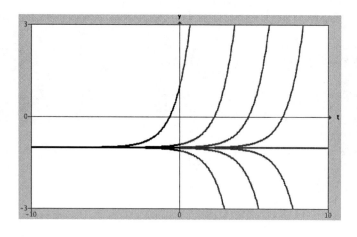

6. We have the supposed solution

$$y(t) = \frac{De^t}{1+De^t}.$$

First use the quotient rule to compute that

$$y'(t) = \frac{De^t}{\left(1+De^t\right)^2}.$$

Next compute

$$y(1-y) = \frac{De^t}{1+De^t}\left(1-\frac{De^t}{1+De^t}\right) = \frac{De^t}{\left(1+De^t\right)^2}$$

so we see that this is indeed a solution.

7. Separating and integrating, we find

$$\frac{-1}{y} = \int \frac{dy}{y^2} = \int dt = t+c$$

so that $y(t) = -1/(t+c)$. This is not the general solution, since when $y(0) = 0$, we have $0 = y(0) = -1/c$, or multiplying through by c, we find that $0 = -1$. Is that true? Obviously not. What's wrong here? The answer is that the solution to the initial value problem $y(0) = 0$ is just the constant function $y(t) = 0$ (i.e., an equilibrium solution). So the solutions $-1/(t+c)$ is not quite the general solution; we must also add in the solution $y(t) = 0$ to solve all initial value problems.

8. The differential equation here is $y' = k(y - 80)$. This equation is separable, so we have

$$\ln|y-80| = \int \frac{dy}{y-80} = \int k\,dt = kt + c.$$

Assuming $y(t) > 80$, we find by exponentiating both sides that $y(t) = 80 + e^{kt+c} = 80 + De^{kt}$. Since $y(0) = 200$, we also have $200 = y(0) = 80 + De^0$ so that $D = 120$. So our solution so far is $y(t) = 80 + 120e^{kt}$. But we also have $y(1) = 180$, so that $180 = 80 + 120e^k$. Therefore $5/6 = e^k$ or $k = \ln(5/6)$. Thus the full solution is $y(t) = 80 + 120e^{(\ln(5/6))t} = 80 + 120(5/6)^t$.

9. First off, the solution satisfying the initial condition $y(0) = 1$ is clearly the constant solution $y(t) = 1$ (i.e., one of the 2 equilibrium solutions). For the other solutions we can again separate and integrate

$$\int \frac{dy}{1-y^2} = \int dt = t + c.$$

To integrate $1/(1-y^2)$, note that we can break up this fraction into

$$\frac{1}{1-y^2} = \frac{1/2}{1+y} + \frac{1/2}{1-y}.$$

First assume $y(0) = 0$. Then our solution lies below the equilibrium solution $y = 1$, so we may integrate to find

$$\int \left(\frac{1/2}{1+y} + \frac{1/2}{1-y}\right) dy = \tfrac{1}{2}\ln(1+y) - \tfrac{1}{2}\ln(1-y) = t + c.$$

Exponentiating both sides then yields

$$\sqrt{\frac{1+y}{1-y}} = ke^t$$

or

$$\left(\frac{1+y}{1-y}\right) = De^{2t}.$$

Solving for y then yields

$$y(t) = \frac{De^{2t}-1}{1+De^{2t}},$$

which satisfies $y(0) = 0$ when $D = 1$. Similar calculations show that this expression also solves the initial value problem $y(0) = 2$ when $D = -3$.

10. Notice that our solution above includes the equilibrium solution $y(t) = -1$ when $D = 0$. Also, solving the initial value problem $y(0) = A$ shows that $A = (D-1)/(1+D)$ so that $D = -(A+1)/(A-1)$, which is fine as long as A is not equal to 1. But we know the solution that solve the initial value problem $y(0) = 1$; it is our equilibrium solution $y(t) = 1$, so we must add this solution to the other solution to get the full general solution.

Lecture 6

1. **a.** 3/2.

   b. $(y_1 - y_0)/(t_1 - t_0)$.

   c. $y = -t + 1$.

   d. $y = 0$.

   e. $t = 1$.

2. $y' = 2t$, so the slope at $t = 1$ is 2. So our equation so far is $y = 2t + B$. To determine B, we know that the point $(1, 1)$ lies on this straight line. So we have $1 = 2 + B$, so $B = -1$. Thus the equation is $y = 2t - 1$.

3. $t_1 = 0.1$, and $y_1 = 1.1$.

4. $t_2 = 0.2$, $y_2 = 1.21$, $t_3 = 0.3$, $y_1 = 1.331$.

5.

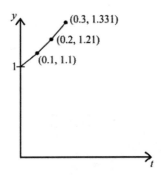

6. a. Clearly, this solution is just e^t.

   b. Using a spreadsheet, I calculate that $y(1) = 2.593743$ when the step size is 0.1.

   c. With step size 0.05, I find $y(1) = 2.653298$ and with step size .01 I find $y(1) = 2.704814$ (your approximations may differ slightly depending on the software you use). Clearly, we are getting better approximations of the actual solution.

   d. Using the value of $y(1) = 2.718281$, the error when the step size is 0.1 is 0.124538; when the step size is 0.05, the error is 0.064983; when the step size is 0.01, the error is 0.013467. So when we cut the step size in half by going from step size 0.1 to 0.05, the error decreases by approximately one half. And when we decrease the step size by 1/5 when we go from step size 0.05 to 0.01, the error also decreases by approximately 1/5.

7. Any solution that starts above the equilibrium point at $y = -1$ tends toward the value $y = 1$ where the slope field has infinite slope. Here the numerical method "goes crazy," and we again see chaotic behavior.

Lecture 7

1. Only the origin is an equilibrium point.

2. All points along the line $y = -x$ are equilibrium points.

3. **a.**

    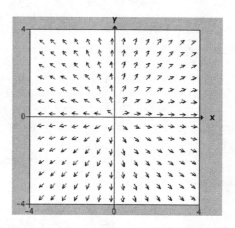

   b. It appears that all solutions (except the equilibrium point at the origin) move away from the origin along a straight line.

   c. Since this system decouples, solutions are of the form

   $x(t) = k_1 e^t$

   $y(t) = k_2 e^t$.

4. $y' = v$
    $v' = y$

5. **a.** The direction field is always tangent to the circles that are centered at the origin and point in the clockwise direction.

    b. One solution is $x(t) = \cos(t)$ and $y(t) = -\sin(t)$. Another is $x(t) = \sin(t)$ and $y(t) = \cos(t)$. Any constant times each of these is also a solution.

6. **a.** The vector field is horizontal along the x-axis (pointing to the right if $0 < x < 1$ and to the left if $x < 0$ or $x > 1$). The vector field is vertical along the y-axis and always points toward the origin. At any point off the axes, the x and y directions of the vector field are the same as the corresponding directions on the axes.

    b. The equilibria are $(0, 0)$ and $(1, 0)$.

    c. On the x-axis, when $x > 0$, all solutions tend to 1, and when $x < 0$, all solutions tend to $-\infty$. Meanwhile, on the y-axis all solutions tend to 0. So if $x > 0$, the solution tends to $(1, 0)$ and if $x < 0$, the solution tends off to ∞ to the left.

Lecture 8

1. $y'' + 3y' + 2y = 0$.

2. $y' = v$
   $v' = -2y - 3v$.

3. The only equilibrium point lies at the origin.

4. $-4\sin(2t)$ and $-4\cos(2t)$.

5.

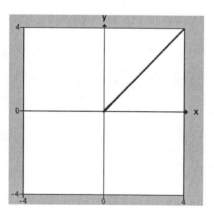

6. We need to solve $k_1 + k_2 = A$, $-k_1 - 2k_2 = B$ for any given values of A and B. Adding these equations yields $-k_2 = A + B$, so $k_2 = -A - B$. Then the first equation implies that $k_1 = 2A + B$.

7. a. The characteristic equation is $(s + 3)(s + 2)$, and the general solution is $k_1 e^{-3t} + k_2 e^{-2t}$.

   b. We must solve $k_1 + k_2 = 0$, $-3k_1 - 2k_2 = 1$, which yields $k_1 = -1$ and $k_2 = 1$.

   c. The graph of $y(t)$ increases at first, then reaches a maximum, and then slowly decreases to 0.

   d. Initially the mass moves upward, but then it turns around and glides directly back to its rest position.

Lecture 9

1. $2e^{2t}\cos(3t) - 3e^{2t}\sin(3t)$.

2. $e^{2t}(\cos(3t) + i\sin(3t))$.

3. $\dfrac{-b \pm \sqrt{b^2 - 4k}}{2}$.

4. In the clockwise direction.

5. The characteristic equation is $s^2 + 5s + 6 = (s+3)(s+2)$ with roots -2 and -3, so this system is overdamped.

6. L'Hopital's rule says that the limit as $t \to \infty$ of t/e^t is the same as the limit of the quotient of the derivatives (i.e., $1/e^t$). This quotient tends to 0 as $t \to \infty$.

7. The roots of the characteristic equation are $-1 \pm I$, so the general solution is $k_1 e^{-t}\cos(t) + k_2 e^{-t}\sin(t)$.

8. The solution that satisfies $y(0) = 0$ and $y'(0) = 1$ is $y(t) = e^{-t}\sin(t)$. The derivative is $y'(t) = -e^{-t}\sin(t) + e^{-t}\cos(t)$. This derivative vanishes when $\sin(t) = \cos(t)$, so the first positive t-value where $y'(t) = 0$ is when $t = \pi/4$. Then we have $y(\pi/4) = e^{-\pi/4}\sin(\pi/4) = e^{(-\pi/4)/\sqrt{2}}$.

9. The characteristic equation is $s^2 + bs + 1$, whose roots are

$$\frac{-b \pm \sqrt{b^2 - 4}}{2}.$$

So the system is undamped when $b = 0$, underdamped when $0 < b < 2$, critically damped when $b = 2$, and overdamped when $b > 2$.

10. Clearly $y(t) = 1$ is a constant solution to the nonhomogeneous equation. So the general solution, as in the first-order case, is $k_1\cos(t) + k_2\sin(t) + 1$. That is, the mass just oscillates about a point 1 unit removed from the natural equilibrium position.

Lecture 10

1. a. $y(t) = ke^{-t} + (1/2)\sin(t) - (1/2)\cos(t)$.

  b. All solutions tend to the periodic solution $(1/2)\sin(t) - (1/2)\cos(t)$.

  c.

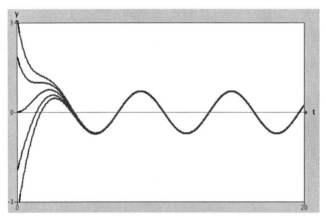

  d. The solutions are now $y(t) = ke^{t} - (1/2)\sin(t) - (1/2)\cos(t)$, so solutions no longer tend to the periodic solution given by $-(1/2)\sin(t) - (1/2)\cos(t)$ as t increases.

  e. As time tends to $-\infty$, solutions now tend to the periodic solution.

477

2. Make the guess of $A\cos(t) + B\sin(t)$ to find the solution with $A = -1/2$ and $B = 1/2$.

3. The general solution is $k_1\cos(t) + k_2\sin(t) + (1/2)e^{-t}$.

4. When $w = \sqrt{3}$, the system is in resonance.

5. In order for $\cos(wt) + \cos(t)$ to be periodic, we must find a constant A for which

$$\cos(w(t+A)) + \cos(t+A) = \cos(wt) + \cos(t)$$

for all t-values. In particular, this must be true when $t = 0$. But then we have $\cos(wA) + \cos(A) = 2$. This implies that we must have $\cos(wA) = 1 = \cos(A)$. Therefore we need both wA and A to be integer multiples of 2π. So we must have $wA = 2n\pi$ and $A = 2m\pi$ for some integers n and m, so $w = n/m$. That is, w must be a rational number.

6. When w is a rational number, solutions are then periodic in t.

Lecture 11

1. $\begin{pmatrix} 8 \\ 6 \end{pmatrix}$.

2. a. $Y' = \begin{pmatrix} 0 & 1 \\ -1 & 0 \end{pmatrix} Y$.

   b. Only the origin.

3. No.

4. a. Yes.

   b. No.

5. a. $Y' = \begin{pmatrix} 1 & 0 \\ 1 & 2 \end{pmatrix} Y$.

   b. The first solution is $x(t) = k_1 e^t$. Then we must solve $y' = 2y + k_1 e^t$. The equation $y' = 2y$ has general solution $y(t) = k_2 e^{2t}$. Therefore for the nonhomogeneous equation, we guess Ce^t so that $C = -k_1$. So our solutions are $x(t) = k_1 e^t$ and $y(t) = k_2 e^{2t} - k_1 e^t$.

   c. $Y(t) = \begin{pmatrix} k_1 \exp(t) \\ k_2 \exp(2t) - k_1 \exp(t) \end{pmatrix} = k_1 e^t \begin{pmatrix} 1 \\ -1 \end{pmatrix} + k_2 e^{2t} \begin{pmatrix} 0 \\ 1 \end{pmatrix}$.

d. We must be able to solve $x(0) = A$ and $y(0) = B$ for any A and B. The first equation gives $A = k_1$, and the second then says $k_2 - A = B$ or $k_2 = A + B$, so this is indeed the general solution.

e. When $k_1 = 0$, we have a straight line solution along the y-axis moving away from the origin. When $k_2 = 0$, we find a straight line solution along the line $y = -x$ again moving away from the origin. All other solutions also tend away from the origin.

Lecture 12

1. The determinant is -5.

2. Since the determinant is nonzero, the only equilibrium point is at the origin.

3. The trace is 4, and the determinant is 3.

4. a.

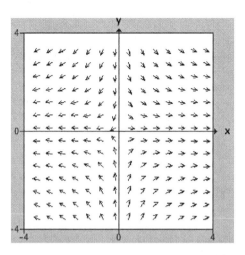

b. It appears there are straight line solutions along the x- and y-axes.

5. 2 and -1.

6. The eigenvalues are just a and b, since the characteristic equation is $(a - \lambda)(b - \lambda) = 0$.

7. The eigenvalues are both 0, since this is an upper triangular matrix. However, every nonzero vector is an eigenvector since multiplying this vector by the matrix yields (0, 0): that is, 0 times the given vector.

8. The characteristic equation here is $\lambda^2 - (1 + 3\sqrt{2})\lambda$. So the eigenvalues are 0 and $1 + 3\sqrt{2}$. The eigenvector corresponding to the eigenvalue 0 is given by any nonzero solution of the equation

481

$x + 3y = 0$, so for instance, the vector $(1, -1/3)$ is one such eigenvector. The eigenvector corresponding to the eigenvalue $1 + 3\sqrt{2}$ is given by solving the equation $\sqrt{2}\,x - y = 0$. So one eigenvector is $(1, \sqrt{2}\,)$.

9. The characteristic equation here is $\lambda^2 - 6\lambda + 8 = 0$, so the eigenvalues are 4 and 2. An eigenvector corresponding to the eigenvalue 4 is any nonzero vector of the form $y = x$, so $(1, 1)$ is an eigenvector for this eigenvalue. An eigenvector corresponding to the eigenvalue 2 is any nonzero vector satisfying $y = -x$, so $(1, -1)$ is one such eigenvector. Thus the general solution is

$$k_1 e^{4t} \begin{pmatrix} 1 \\ 1 \end{pmatrix} + k_2 e^{2t} \begin{pmatrix} 1 \\ -1 \end{pmatrix}.$$

The solutions with k_1 or k_2 equal to 0 are straight line solutions moving away from the origin and lying along the lines $y = x$ and $y = -x$. All other solutions also move away from the origin tangentially to the straight line solutions along $y = -x$.

10. The solution satisfying this initial condition is found by solving the system of equations

$k_1 + k_2 = 1$

$k_1 - k_2 = 0$

so $k_1 = k_2 = 1/2$ yield this solution.

Lecture 13

1. **a.** The characteristic equation is $\lambda^2 + 1 = 0$.

    b. The eigenvalues are $\pm i$.

    c. One eigenvector associated to the eigenvalue i is
    $$\begin{pmatrix} 5 \\ i-2 \end{pmatrix},$$
    and an eigenvector associated to $-i$ is
    $$\begin{pmatrix} 5 \\ -i-2 \end{pmatrix}.$$

2. **a.** The eigenvalues are 0 and -1.

    b. One eigenvector corresponding to 0 is $(1, 0)$ and corresponding to -1 is $(-1, 1)$.

3. As a system, $y'' + by' + ky = 0$ may be written
    $$y' = v,\ v' = -ky - bv \text{ or}$$
    $$Y' = \begin{pmatrix} 0 & 1 \\ -k & -b \end{pmatrix} Y.$$

The eigenvalues are roots of $\lambda^2 + b\lambda + k = 0$ (which is the same characteristic equation that we saw for the second order equation), and so are given by

$$\frac{-b \pm \sqrt{b^2 - 4k}}{2}.$$

4. a. The characteristic equation is $\lambda^2 - 2a\lambda + a^2 + b^2 = 0$, which has roots given by $a + ib$ and $a - ib$. The eigenvector for $a + ib$ is found by solving

$$-ibx + by = 0$$
$$-bx - iby = 0,$$

so $y = ix$. One complex eigenvector is therefore $(1, i)$. For the eigenvalue $a - ib$, the equations are

$$ibx + by = 0$$
$$-bx + iby = 0,$$

so $y = -ix$. One eigenvector in this case is $(1, -i)$.

b. For the eigenvalues $a \pm ib$, we have a spiral source if $a > 0$, a spiral sink if $a < 0$, and a center if $a = 0$. If $a = b = 0$, we have real and repeated 0 eigenvalues.

5. a. The characteristic equation is

$$\lambda^2 - (1+3\sqrt{2})\lambda = 0,$$

so the eigenvalues are 0 and $1 + 3\sqrt{2}$. The eigenvector corresponding to 0 is given by $x + 3y = 0$ or $(3, -1)$. The eigenvector for $1 + 3\sqrt{2}$ is given by solving $\sqrt{2}x - y = 0$ or $(1, \sqrt{2})$. In the phase plane, we have a straight line of equilibrium points along the line $x + 3y = 0$. All other solutions lie on straight lines with slope $\sqrt{2}$, and these solutions tend directly away from the single equilibrium point on this line as time goes forward.

b. Given the general solution

$$k_1 \begin{pmatrix} 3 \\ -1 \end{pmatrix} + k_2 e^{(1+3\sqrt{2})t} \begin{pmatrix} 1 \\ \sqrt{2} \end{pmatrix},$$

we must solve

$$3k_1 + k_2 = 1$$

$$-k_1 + \sqrt{2}\, k_2 = 0$$

so we have $\sqrt{2}\, k_2 = k_1$. Then equation 1 implies $3\sqrt{2}\, k_2 + k_2 = 1$ so that $k_2 = 1/(1 + 3\sqrt{2})$ and so $k_1 = \sqrt{2}/(1 + 3\sqrt{2})$. So the solution is given by

$$\frac{\sqrt{2}}{1+3\sqrt{2}} \begin{pmatrix} 3 \\ -1 \end{pmatrix} + \frac{e^{(1+3\sqrt{2})t}}{1+3\sqrt{2}} \begin{pmatrix} 1 \\ \sqrt{2} \end{pmatrix}.$$

Lecture 14

1. a.

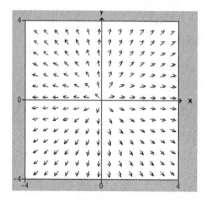

b. The eigenvalues are 2 repeated.

c. Every nonzero vector is an eigenvector.

d. One of many possible forms for the general solution is

$$k_1 e^{2t}\begin{pmatrix}1\\0\end{pmatrix}+k_2 e^{2t}\begin{pmatrix}0\\1\end{pmatrix}.$$

e. All nonzero solutions move away from the origin along straight lines.

2. We have repeated eigenvalues given by -1. An eigenvector corresponding to -1 is given by solving $x + 0y = 0$, so one eigenvector is $(0, 1)$. Next we solve the equations $0x + 0y = 0$ and $x + 0y = 1$ to find the special vector $(1, 0)$. Then the general solution is

$$k_1 e^{-t}\begin{pmatrix}0\\1\end{pmatrix}+k_2 te^{-t}\begin{pmatrix}0\\1\end{pmatrix}+k_2 e^{-t}\begin{pmatrix}1\\0\end{pmatrix}.$$

3. We could solve this using the eigenvalue/eigenvector method as in the previous question, but it is much easier to proceed as follows. Our equations read $x' = 0$ and $y' = x$. So we have $x(t) = k_1$, and then integration yields $y(t) = k_1 t + k_2$.

4. a. We have $T = a$ and $D = -a$, so the path lies along the line $D = -T$ in the TD-plane. When $a = -4$ and $a = 0$, this line meets the repeated eigenvalue parabola $T^2 - 4D = 0$. When $a < -4$, we are in the real sink region. For $-4 < a < 0$, we have a spiral sink. And when $a > 0$, we have a saddle point at the origin.

   b. So we have bifurcations at $a = -4$ (change from real to spiral sink) and $a = 0$ (change from spiral sink to saddle).

5. Here we have $T = a$ and $D = 4$, so our path is along the horizontal line $D = 4$. We cross the repeated eigenvalue parabola $T^2 - 4D = 0$ when $a^2 = 16$, so at $a = \pm 4$. When $a < -4$, we are in the real sink region. Then we cross into the spiral sink region. We cross the D-axis when $a = T = 0$ and then enter the spiral source region, and then finally enter the real source region when $a = 4$. So bifurcations occur when $a = \pm 4$ and $a = 0$.

Lecture 15

1. **a.** The *x*-nullcline is the *x*-axis, while the *y*-nullcline is the *y*-axis.

   b.

   c. The only equilibrium point is the origin.

   d.

   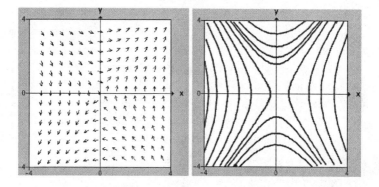

   e. Since this system is linear and has real eigenvalues $\pm\sqrt{2}$, there is only one straight line through the origin containing solutions that tend to the origin.

2. The *x*-nullclines are given by the *y*-axis and the line $y = -x/3 + 50$. The *y*-nullclines are given by the *x*-axis and the line $y = -2x + 100$. Most solutions that start with *x* and *y* nonzero will tend to either the equilibrium point at (150, 0) or the one at (0, 100).

3. The x-nullcline is given by the y-axis and the line $y = 50 - x/2$. The y-nullclines are given by the x-axis and the line $y = 25 - x/6$. Most solutions will tend to the equilibrium point at (75, 12.5).

4. The x-nullclines are the lines $x = 0$ and $x = 1$, and the y-nullcline is the parabola $x = y^2$ opening to the right. So, there are equilibrium points at the origin, (1, 1), and (1, −1). We have $x' < 0$ when $x < 0$, so all solutions in the left half of the plane tend off to infinity. From the directions of the vector field, it appears that (1, 1) is a sink and (1, −1) is a saddle.

5. a. The x-nullclines are the y-axis and the line $y = 1/A - x/A$, and the y-nullclines are the x-axis and the line $y = 1 + x$. There are equilibrium points (1, 0), (0, 1), and (0, 0). When $A < 1$, solutions that begin with x and y nonzero tend to (1, 0). When $A > 1$, solutions tend to (0, 1). So a bifurcation occurs when $A = 1$. When $A = 1$, there is a straight line of equilibria along $y = 1 - x$.

   b. Now the x-nullclines are the y-axis and the line $y = 1 - x$, and the y-nullclines are the x-axis and $y = 1 + Bx$. Again the equilibrium points are at (1, 0), (0, 1), and (0, 0). Now if $B > -1$, all solutions tend to (0, 1) when the initial conditions are not on the x-axis, whereas if $B < -1$, these solutions tend to (1, 0) (unless the solution is on the y-axis).

Lecture 16

1. The partial derivative with respect to x is $2xy + 3x^2$ and with respect to y is $x^2 + 3y^2$.

2. The Jacobian matrix is just the coefficient matrix

   $$\begin{pmatrix} a & b \\ c & d \end{pmatrix}.$$

3. **a.** The only equilibrium point is at the origin.

   b. The Jacobian matrix is

   $$\begin{pmatrix} 2x & 1 \\ 1 & 0 \end{pmatrix}.$$

   At the origin, this matrix becomes

   $$\begin{pmatrix} 0 & 1 \\ 1 & 0 \end{pmatrix}.$$

   c. The eigenvalues of the Jacobian matrix are ± 1, so the origin is a saddle.

4. The only equilibrium point is at the origin, and when linearized, the system becomes

$$Y' = \begin{pmatrix} 0 & 1 \\ -1 & 1 \end{pmatrix} Y.$$

The characteristic equation is $\lambda^2 - \lambda + 1$. The eigenvalues are

$$\frac{1 \pm \sqrt{-3}}{2},$$

so the origin is a spiral source. Using a computer, you can see that all other solutions spiral toward a periodic solution that surrounds the origin.

5. One of many possibilities is $x' = x^2$, $y' = y^2$, which has a single equilibrium point at the origin. The Jacobian matrix is the zero matrix, so both eigenvalues are 0.

6. The equilibrium points are given by $y = 1$ and $x = \pm 1$. The Jacobian matrix is

$$\begin{pmatrix} 0 & 1 \\ -2x & 1 \end{pmatrix}.$$

At $(1, 1)$ the eigenvalues are $(1 \pm \sqrt{-7})/2$, so we have a spiral source. At $(-1, 1)$ the eigenvalues are 2 and -1, so we have a saddle.

7. The equilibria are given by $(n\pi, \pi/2 + m\pi)$, where n and m are integers. Linearization shows that the equilibrium point is a saddle if n and m are both even (or both odd), a sink if n is odd and m is even, and a source if n is even and m is odd. The lines $x = n\pi$ are the x-nullclines, so the vector field is tangent to these lines, and solutions

remain on them. Similarly, the *y*-nullclines are the lines $y = \pi/2 + m\pi$, and the vector field is again tangent to these lines. These vertical and horizontal lines therefore bound squares whose corners are a pair of saddles, one sink and one source. In each square, all solutions not on the bounding lines tend to the equilibrium that is the sink.

8. The equilibrium points are at $y = 0$ and $x = \pm\sqrt{A}$, so we clearly have a bifurcation at $A = 0$. Linearization yields the eigenvalues 1 and $-2\sqrt{A}$ at the equilibrium point ($\sqrt{A}$, 0), so this point is a saddle when $A > 0$. The eigenvalues at the other equilibrium point are 1 and $2\sqrt{A}$, so this point is a source when $A > 0$.

Lecture 17

1. **a.** The equilibria are (0, 0) and (1, 1).

   b. The general Jacobian matrix is

   $$\begin{pmatrix} 1-y & -x \\ -y & 1-x \end{pmatrix}.$$

   At (0, 0) and (1, 1) this matrix becomes

   $$\begin{pmatrix} 1 & 0 \\ 0 & 1 \end{pmatrix} \text{ and } \begin{pmatrix} 0 & -1 \\ -1 & 0 \end{pmatrix}.$$

   c. At (0, 0) we have a source, and at (1, 1) we have a saddle.

d. The x-nullclines are $x = 0$ and $y = 1$. The y-nullclines are $y = 0$ and $x = 1$. The different regions are:

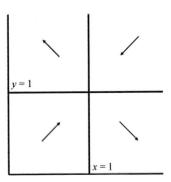

e.

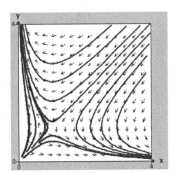

2. a. The coexistence equilibrium point when $a = b = 1/2$ is given by $x = y = 800/3$. The Jacobian matrix at this point is

$$\begin{pmatrix} -2/3 & -1/3 \\ -1/3 & -2/3 \end{pmatrix},$$

so the eigenvalues are the roots of $\lambda^2 + (4/3)\lambda + 1/3 = 0$. These eigenvalues are then -1 and $-1/3$, both of which are negative, so this equilibrium point is a sink.

b. The Jacobian matrix is given by

$$\begin{pmatrix} 1-x/200-ay/400 & -ax/400 \\ -by/400 & 1-y/200-bx/400 \end{pmatrix}.$$

At (0, 400), this matrix is

$$\begin{pmatrix} 1-a & 0 \\ -b & -1 \end{pmatrix},$$

so the eigenvalues are $1 - a$ and -1. Therefore, we have a sink if $a > 1$ and a saddle if $a < 1$. At the point (400, 0) we find eigenvalues -1 and $1 - b$, so this point is a sink if $b > 1$ and a saddle if $b < 1$.

3. a. The equilibria are (1, 0), and $x = y = 1/2$. (Technically, the equation for y' is not defined if $x = 0$).

b. The Jacobian matrix is

$$\begin{pmatrix} 1-2x-y & -x \\ y^2/x^2 & 1-(2y/x) \end{pmatrix}.$$

At (1, 0) the eigenvalues are -1 and 1, so we have a saddle. And at the other equilibrium point, (1/2, 1/2), we have eigenvalues that are complex with negative real part, so this equilibrium is a spiral sink.

c. The x-nullclines are given by $x = 0$ and $1 - x = y$, and the y-nullclines are given by $y = 0$ and $y = x$. So the nullclines meet at a single point that is not on the axes. In the regions between the nullclines, we see that the vector field indicates that solutions spiral around the equilibrium point. But that does not tell us the complete story, as we could have periodic solutions in this region. But the computer shows otherwise.

Lecture 18

1. a. There are no limit cycles since for any point (except the origin), the vector field points directly northwest.

   b. All solutions lie along straight lines with slope equal to 1. Since the origin is the only equilibrium point, all solutions tend to infinity except for those on the line $y = x$, where x and y are less than 0. These solutions tend to the equilibrium point at the origin.

2. On the circle $x^2 + y^2 = 1$, the system reduces to

   $x' = -y$

   $y' = x,$

   which is a vector field that is everywhere tangent to the unit circle.

3. On the unit circle, the vector field is now given by

   $x' = 0$

   $y' = x - y,$

   so the vector field points vertically on this circle, which means that the unit circle is not a periodic solution.

4. The second equation says that $x = xy/(1 + x^2)$. Substituting this into the first equation yields $x - 10 = -4x$, or $x = 2$. Then the second equation gives $2 = 2y/5$ so that $y = 5$.

5. **a.** We have $r' = 0$ when $r = 1$, so there is a periodic solution along the unit circle. (At the origin, we have an equilibrium point.) Since $\theta' = 1$, this solution moves counterclockwise around the circle. When $0 < r < 1$, we have $r' > 0$, and when $r > 1$, $r' < 0$, so all nonzero solutions spiral in to this limit cycle at $r = 1$, which is therefore stable.

    b. We now have periodic solutions when $r^3 - 3r^2 + 2r = 0$ and when $r(r-2)(r-1) = 0$. That is, $r = 1$ and 2 are periodic solutions as in the previous question. Choosing one point in each of the intervals between these circles, we see that $r' > 0$ when $0 < r < 1$; $r' < 0$ when $1 < r < 2$; and $r' > 0$ when $r > 2$. So $r = 1$ is a stable limit cycle while $r = 2$ is unstable.

    c. In similar fashion, we have a limit cycle at $r = n\pi$, where n is a positive integer. Odd integers yield stable limit cycles; even integers yield unstable limit cycles. The rotation is now in the clockwise direction.

6. The roots of $ar - r^2 + r^3 = r(r^2 - r + a)$ are 0 and

    $$r_\pm = \frac{1 \pm \sqrt{1-4a}}{2}.$$

    Besides the equilibrium point at the origin, there are two limit cycles when $a < 1/4$, a single limit cycle when $a = 1/4$, and no limit cycles when $a > 1/4$. The limit cycle $r = r_-$ is stable; $r = r_+$ is unstable. When $a > 1/4$, $r' > 0$, so all nonzero solutions spiral out to infinity.

Lecture 19

1. When the pendulum is in the downward position ($\theta = 0$).

2. When the pendulum is in the upward position ($\theta = \pi$).

3. The Jacobian matrix is

    $\begin{pmatrix} 0 & 1 \\ -g\cos(\theta) & 0 \end{pmatrix}$, so when $\theta = \pi$, we have the matrix $\begin{pmatrix} 0 & 1 \\ g & 0 \end{pmatrix}$.

    The eigenvalues for this matrix are $\pm\sqrt{g}$, so this equilibrium point is a saddle.

4. Now the eigenvalues are $\pm i\sqrt{g}$, so linearization does not give any information. But we know that the system is Hamiltonian, and the level curves surrounding this equilibrium point are ellipses, so this equilibrium point is a center.

5. Since $\theta = 2n\pi$, the Jacobian matrix is

$$\begin{pmatrix} 0 & 1 \\ -g & -b \end{pmatrix}$$

with characteristic equation $\lambda^2 + b\lambda + g = 0$ and roots

$$\frac{-b \pm \sqrt{b^2 - 4g}}{2}.$$

If $b^2 > 4g$, the roots are both real and negative, so we have a real sink. If $0 < b^2 < 4g$, the roots are complex with negative real part, so we have a spiral sink.

6. The Jacobian matrix now has a $+g$ in the lower left entry, so the determinant is now $-g < 0$. This means the equilibrium point is a saddle.

7. We have $\partial F/\partial x = 2x = -\partial G/\partial y$, so this system is Hamiltonian.

8. Let $F(x, y) = ax + by$ and $G(x, y) = cx + dy$. This linear system is Hamiltonian if $\partial F/\partial x = a = -d = -\partial G/\partial y$. The trace of the corresponding matrix is then equal to 0, so we can only have a center or a saddle equilibrium point (or repeated 0 eigenvalues).

9. The Hamiltonian function is given by $H(x, y) = by^2/2 + axy - cx^2/2$ (plus possibly a constant).

Lecture 20

1. **a.** $(0, 0)$, $(1, 0)$, and $(-1, 0)$.

   b. Computing the respective partial derivatives

   $$\frac{\partial H}{\partial v} = v, -\frac{\partial H}{\partial y} = y - y^3$$

   shows that the system is Hamiltonian.

   c.

   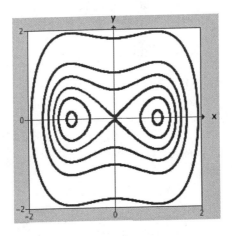

   d. If the beam starts to the right or left of the center of the two magnets with small velocity, then it just oscillates back and forth in either the left or right region, depending on its starting position. But if the beam starts with large velocity, then it will move periodically back and forth to the left and right regions.

2. The function $x(t) = 0$, $y(t) = 0$, and $z(t) = Ce^{-8/3t}$ is such a solution.

3. We first compute that $dL/dt = -20(x^2 + y^2 - (1 + R)xy) - (160/3)z^2$. So we need to show that the term $x^2 + y^2 - (1 + R)xy$ is always positive away from the origin. This is certainly true when $x = 0$. Along any other straight line $y = Mx$, this quantity is equal to $x^2(M^2 - (1 + R)M + 1)$, but the quadratic term $M^2 - (1 + R)M + 1$ is always positive if $R < 1$.

4. All solutions must tend to $(0, 0, 0)$ when $R < 1$.

5. We compute $dV/dt = -20(Rx^2 + y^2 + (8/3)(z^2 - 2Rz)) = -20(Rx^2 + y^2 + (8/3)(z - R)^2) = (8/3)R^2$. Note that the equation $Rx^2 + y^2 + (8/3)(z - R)^2 = K$ defines an ellipsoid when $K > 0$. So if $K > (8/3)R^2$, we have $dV/dt < 0$. So we may choose a constant C large enough so that the ellipsoid $V = C$ strictly contains the ellipsoid $Rx^2 + y^2 + (8/3)(z - R)^2 = (8/3)R^2$ in its interior. Then we have $dV/dt < 0$ on the ellipsoid $V = C + \alpha$ for any constant $\alpha > 0$.

6. Far enough away on the ellipsoid $V = C + \alpha$, for example, all solutions descend toward the ellipsoid $V = C$.

Lecture 21

1. **a.** 0 and 1.

   b. If $|x| < 1$, the orbit of x tends to the fixed point at 0. If $|x| > 1$, the orbit tends to infinity. If $x = -1$, the orbit lands on 1 after 1 iteration and so is eventually fixed.

2. All orbits tend to infinity.

3. **a.** All orbits tend to the fixed point at 0.

   b. For each a, 0 is always fixed. So we consider orbits of other x-values. If $a > 1$, all these orbits tend to infinity. If $a = 1$, all these orbits are fixed. If $a = -1$, all these orbits lie on 2-cycles. If $-1 < a < 0$, all these orbits tend to 0. And if $a < -1$, all orbits tend to $\pm\infty$, alternating between the positive and the negative axis.

4. The fixed points are 0 and $(k-1)/k$ (which only exists in the unit interval if $k > 1$).

5. The points 1 and -1 lie on a 2-cycle for $-x^3$. There is no 2-cycle for x^3 since the graph of x^3 is always increasing.

6. $1/3 \to 2/3 \to 1/3$, so $1/3$ lies on a 2-cycle. $1/7 \to 2/7 \to 4/7 \to 1/7$, so $1/7$ lies on a 3-cycle. $1/15 \to 2/15 \to 4/15 \to 8/15 \to 1/15$, so $1/15$ lies on a 4-cycle.

7. Only 1/3 and 2/3 have prime period 2. The points 1/7 ... 6/7 have prime period 3, and 1/15, ... , 14/15 have prime period 4 (with the exception of 5/15 and 10/15, which have prime period 2).

8. The graph of D crosses the diagonal two times, D^2 four times, D^3 eight times, and D^n 2^n times. Points of the form $k/(2^n - 1)$, where k is an integer and $1 \leq k < 2^n - 1$, have (not necessarily prime) period n.

Lecture 22

1. The fixed points are 0 (attracting) and ± 1 (repelling).

2. **a.** This cycle is attracting.

   b.

   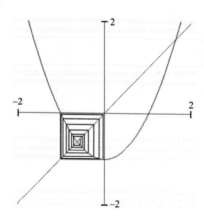

3. **a.** If $-1 < a < 1$, 0 is attracting; if $a > 1$ or $a < -1$, 0 is repelling. If $a = 1$, all other orbits are fixed, or if $a = -1$, all other orbits lie on two-cycles. So, in both of these cases, 0 is neutral.

   b. Bifurcations occur at $a = 1$ and $a = -1$.

4. The derivative of the logistic function is $k - 2kx$. So the fixed point at 0 is attracting for $0 < k \leq 1$ and repelling for $k > 1$. The other fixed point is attracting if $1 < k \leq 3$ and repelling if $k > 3$.

5. If $F(x) = -x^3$, then $F^2(x) = x^9$, and we know 1 and -1 lie on a 2-cycle. But the derivative of $F^2(x)$ is $9x^8$, so this cycle is repelling.

6. We have $D' = 2$ at all points, so the derivative of D^n is 2^n everywhere, so all cycles are repelling.

7. The derivative of the logistic function at the fixed point $(k-1)/k$ is equal to -1 when $k = 3$, so this should be the place where the bifurcation occurs. The graph of F^2 then shows the emergence of two new fixed points as k increases through 3.

8. This function always has a fixed point at $x = 0$, and the derivative at this point is equal to a. When $a > 1$, all other points have orbits that tend to infinity. When $a < 1$, there are two more fixed points at $\pm\sqrt{1-a}$. When $a = -1$, the derivative at 0 becomes -1. For $a < -1$, there is a 2-cycle at a pair of symmetrically located points given by $\pm\sqrt{-(a+1)}$.

Lecture 23

1. **a.** $(1/27, 2/27)$, $(7/27, 8/27)$, $(19/27, 20/27)$, and $(25/27, 26/27)$.

   b. The removed lengths are $1/3$, $2/9$, and $4/27$, which add up to $19/27$.

   c. At stage 4, the removed intervals have total length $8/81 = 2^3/3^4$.

   d. At stage n, the removed intervals have length $2^{n-1}/3^n$.

   e. The infinite series is

   $$\frac{1}{3}+\frac{2}{9}+\ldots+\frac{2^{n-1}}{2^n}+\ldots = \frac{1}{3}\sum_{n=0}^{\infty}\left(\frac{2}{3}\right)^{n-1} = \frac{1}{3}\frac{1}{1-\tfrac{2}{3}} = 1,$$

   so the length of the remaining Cantor set is 0.

2. Any sequence that ends in all zeros eventually lands on $(0, 0, 0, \ldots)$ under the shift map, so any sequence of the form $(s_0, s_1, s_2, \ldots, s_n\, 0000 \ldots)$ has this property.

3. The sequences above correspond to the points that lie at the endpoints of the intervals we called A_n.

4. If you just interchange all the blocks of length n in the original sequence, you get a new point in the sequence space whose orbit is dense. Or you could just take the original sequence and throw in a 0 before each block. There are countless different ways to find such interesting orbits.

5. For example, (01001000100001000001...).

6. There are 3^n points that are fixed under the nth iterate of the shift map in this case; they correspond to all possible blocks of length n, which are then repeated infinitely often in the sequence space to get the periodic point. As in the case of 2 symbols, if we use the sequence that consists of all possible blocks of length 1 in order (i.e., 0, 1, and 2), then all possible blocks of length 2 (00, 01, 02, 10, 11, etc.) have a dense orbit.

Lecture 24

1. $i \to -1 \to 1 \to 1 \ldots$, so this orbit is eventually fixed.

2. $2i \to -4 \to 16 \to 256 \to \ldots$, so this orbit tends to infinity.

3. $i/2 \to -1/4 \to 1/256 \to \ldots$, so this orbits goes to zero.

4. Any point $z = x + iy$ where $x^2 + y^2 > 1$ (i.e., z lies outside the unit circle) has an orbit that tends to infinity.

5. Any point z inside the unit circle has an orbit that tends to the fixed point at the origin.

6. The fixed points always exist in the complex domain and are given by the roots of $z^2 - z + c = 0$, or

$$\frac{1 \pm \sqrt{1-4c}}{2}.$$

Note that these fixed points are complex when $c > 1/4$.

7. The orbit of 0 eventually lands on the 2-cycle given by $-1 + i$ and $-i$, so the filled Julia set is connected.

8. The filled Julia sets are always a scatter of points when A is non-zero. When $A = 0$, this is the only place where the filled Julia set is connected. When $|A| \neq 0$ is small, it may appear that the filled Julia set consists of a single piece, but changing the number of iterations to be a much larger number shows that this is not the case.

9. Each window in the orbit diagram corresponds to a baby Mandelbrot set.

10. As in question 6, the fixed points still exist when c moves above $1/4$. But now the filled Julia set immediately shatters into a totally disconnected set as soon as c increases above $1/4$.

Types of Differential Equations Cited

FIRST ORDER

Linear: can be homogenous or nonhomogeneous, autonomous or nonautonomous.

	Autonomous	Nonautonomous
Homogeneous	$y' + ky = 0$	$y' + G(t)y = 0$
Nonhomogeneous	$y' + ky = 2$	$y' + G(t)y = 2t$

Nonlinear:

Autonomous: $y' = y(1-y)$

Nonautonomous: $y' = t^2 + y^2$

SECOND ORDER

Linear: can be homogenous or non-homogeneous, autonomous or nonautonomous.

	Autonomous	Nonautonomous
Homogeneous	$y'' + by' + ky = 0$	$y' + G(t)y = 0$
Nonhomogeneous	$y'' + by' + ky = 2$	$y' + G(t)y = 2$

Nonlinear:

Autonomous: $y'' = y^2$

Nonautonomous: $y'' = t^2$

SYSTEMS

Linear: $Y' = AY = \begin{matrix} x' = ax + by \\ y' = cx + dy \end{matrix}$

Nonlinear: $x' = y$
$y' = \sin(x)$

	Autonomous	Nonautonomous
Linear	$Y' = AY$ where A is a constant matrix	$Y' = AY + (\sin(t), \cos(t))$
Nonlinear	$Y' = (x^2, \sin(y))$	$Y' = F(Y) = (x^2 + t, \sin(y) + \cos(t))$

Both linear and nonlinear systems can be autonomous or nonautonomous; however, we did not deal with the nonautonomous cases in this course.

Using a Spreadsheet to Solve Differential Equations

Here are some tips for creating spreadsheets as introduced in Lecture 6: "How Computers Solve Differential Equations." As of this writing, most steps were virtually the same for both Mac and PC users, and for Microsoft Excel and OpenOffice; however, do consult your software documentation if you have difficulties.

To enter a formula in a cell:

First type "=" and then enter the corresponding formula, clicking on the cell row/column to indicate the variables. Use * for multiplication.

	A	B	C
1	10	5	3
2	=A1*C1		
3			

For a constant reference to a cell, say cell B1, type "B1."

	A	B	C
1	10	5	3
2			
3	=B1		

After hitting Enter:

	A	B	C
1	10	5	3
2	**30**		
3	**5**		

This works on a PC with Excel 2007 as well as the spreadsheet program available in OpenOffice.

To fill down a formula:

Whether in Excel or OpenOffice, grab the lower right corner of the cell or cells you want to fill down, then drag it down as far as you wish. (Be careful not to <u>move</u> the data from the first cell into the next cell.) In this process, cell references will automatically increase, e.g., G3 will change to G4 in the next row, to G5 in the second row, and so forth. But constant references such as "F3" will not change; that is, just insert a "$" before the letter, and another "$" before the number, of any cell that you want to remain unchanged across calculations to determine more than one cell.

To insert a chart:

As demonstrated in Lecture 6, first highlight the data you wish to plot. In that lecture, I wanted to plot the t and y values generated by Euler's method, so I highlighted all the entries containing these two values. Then click on Insert and choose Chart. Many chart options will be shown. I chose the scatter plot for the Euler's method spreadsheet. In that case, we then had a choice of how to connect the dots; select the appropriate type of plot. (Also, though not demonstrated in the lecture, you can then modify the size and color of the lines as well as the scale of the axes by clicking on the object in the plot that you wish to modify.)

Another way to do this in Excel 2007 for PC is to first insert a blank chart and then select the data you wish to use. To insert a blank chart, go to the Insert tab and select the type of Chart you want. Once you've inserted the chart, you will be given Design options in the Chart Tools toolbar, and you can select the option to Select Data.

To insert a scrollbar:

This can be a little complicated, but many people who have seen the demonstration become curious about how to do this. In Excel, select View → Toolbars → Forms. For OpenOffice, select View → Toolbars → Form Controls. Then choose the scrollbar from this menu. To insert the scrollbar into your spreadsheet, highlight the area where you wish to place the scrollbar. This could be a horizontal or vertical rectangle.

For Excel 2007 for PC, you must in addition have the Developer tab displayed. To display this tab, click on the circular Microsoft Office button in the top left corner of the spreadsheet, which opens a list of options and your recent documents, and then click on the Excel Options button at the bottom of the window. Once you are in the Excel Options window, select Popular from the menu on the left, and check the box for "Show Developer tab in the Ribbon." It should be listed among the "Top options for working with Excel." Click OK.

Once the Developer tab is displayed in Excel, go to it and click on Insert and then select the scrollbar option under Form Controls. To find the scrollbar, hover your mouse over the various options and read the descriptions that pop up for each icon. Click on the scrollbar icon and then click in the spreadsheet where you would like the scrollbar to appear.

To resize the scrollbar:

- For Excel, click on the scrollbar to reveal the points around the outline. Place your mouse over the point in the bottom right corner of the scrollbar so that an arrow appears. Click on that point and drag the edge of the scrollbar in or out until it is the appropriate size. The default orientation is vertical, but you can change it to a horizontal orientation. To change the orientation from vertical to horizontal, follow the same steps as for resizing but drag the bottom right edge of the scrollbar up and to the right at the same time until its orientation shifts.

- For OpenOffice, first make sure the Design Mode On/Off button in the Form Controls toolbar is in the On mode (the button will appear highlighted), and then click on the scrollbar button. Then, using your cursor, highlight the cells where you wish to place the scrollbar, and it should appear. Following the same steps as above for Excel, you can drag it by the corners to resize it.

To arrange for the appropriate output of the scrollbar:

For example, in our Euler's method spreadsheet, I wanted the output to be the value of delta t, which moved from .01 to 1.01 in steps of size .01. That is, I wanted to have 100 different values of delta t as I manipulated the scrollbar. To accomplish this, you need to "format" the scrollbar. To do this, while the scrollbar you inserted is highlighted, select Format → Control.

- For Excel, right-click on the scrollbar and select Format Control.

- For OpenOffice, right click on the scrollbar and select Control, which opens the Scrollbar Properties window. Again, the Design Mode On/Off button in the Form Controls toolbar must be in the On mode (highlighted).

One immediate problem is that a scrollbar only puts out nonnegative integers. So you will need to use a little algebra to change the types of outputs.

- The first thing to do is to select a cell link where the output will be placed. Since the output is an integer and we want something else, I would choose an output cell link that is off the given spreadsheet page, such as cell Z25. To assign the output cell, enter that cell (say it's Z25) in the Cell Link field of the Format Control window.

- For OpenOffice, select the Data tab in the Scrollbar Properties window and enter Z25 in the Linked Cell field.

To choose a minimum and maximum output for the spreadsheet:

- In our case, we wanted the steps to go from .01 to 1.01 in 100 different steps. So our minimum value would be 0 and the maximum would be 100. If this is not the default in the Minimum value and Maximum value fields, enter "0" and "100," respectively. Once you have entered these values, click OK, and when you move the scrollbar, you will see the entries in cell Z25 move from 0 to 100.

- For OpenOffice, enter these values in the Scroll value min and Scroll value max fields in the Data tab of the Scrollbar Properties window. For OpenOffice, there is no OK button; simply close the Scrollbar Properties window. NOTE: To test the scrollbar functionality, make sure the Design Mode On/Off button in the Form Controls toolbar is in the Off mode.

Since we want the entries to change from .01 to 1.01, we then enter in some cell that's visible, say cell F2, the formula =.01*Z25 + .01 and hit Enter.

	A	B	C	D	E	F	G
1							
2						=0.01*Z25 + 0.01	
3							

After hitting Enter:

	A	B	C	D	E	F	G
1							
2						0.01	
3							

Then, as you click down the scrollbar (whether in Excel or OpenOffice), you see the entries in cell F2 change from .01 to 1.01 as desired.

After clicking down once on the scrollbar:

	A	B	C	D	E	F	G
1							
2						0.02	
3							

After dragging the scrollbar all the way to the bottom:

	A	B	C	D	E	F	G
1							
2						1.01	
3							

Timeline

1693	Isaac Newton's "fluxions" and Gottfried Leibniz's "calculus of differences" offer the first published solutions to differential equations.
1700s	Followers of Leibniz—including Daniel Bernoulli, Joseph Lagrange, Pierre Laplace, and Leonhard Euler—continue the development of "differential calculus," while Newton's "calculus of fluxions" remains more influential in England.
1768	Euler develops a method for approximating the solution to a differential equation.
1835	William Rowan Hamilton defines the special system of differential equations in which there is a function that is constant along all solutions of the given differential equation.
1841	Carl Jacobi advances the study of determinants.
1858	Arthur Cayley initiates the use of matrices to reduce systems of differential equations to

1866 .... Jacobi refines Hamiltonian functions and pioneers the Jacobian matrix for evaluating systems of partial derivatives.

1880s .... A geometric approach to nonlinear differential equations by Henri Poincaré initiates the qualitative theory of differential equations.

1890s .... Aleksandr Lyapunov defines a function that is nonincreasing along all solutions of a system of differential equations.

1901 .... Poincaré's conjecture for showing the existence of a limit cycle is proven by Ivar Bendixson.

1895–1905 .... Alternative methods for iterative and numerical solutions of differential equations are offered by Carl Runge and Martin Kutta.

1918 .... Gaston Julia and Pierre Fatou pioneer iteration theory and lay a foundation for work on fractals.

1931 .... The differential analyzer, an electrically-powered mechanical device for solving differential equations, is invented by Vannevar Bush and others at M.I.T.

Timeline

1942	An M.I.T. differential analyzer with 2000 vacuum tubes begins to make important contributions during WWII.
1950	The first digital difference analyzer is created by Northrop Corporation, a leading U.S. aircraft manufacturer.
1950s	Boris Belousov discovers that certain chemical reactions could oscillate rather than go to equilibrium.
1963	While studying an extremely simplified problem in meteorology, Edward Lorenz offers the first system of differential equations shown to possess chaotic behavior.
1960s	Russian mathematician Anatol Zhabotinsky revives Belousov's work on oscillating chemical reactions, with their findings beginning to diffuse to other countries in 1968.
1960s	Sharkovsky's Theorem, describing the ordering of cycle periods for any iterated function, is published in Russian.
1975	A paper by Tien-Yien Li and James Yorke proves that the presence of a periodic point of period 3 implies the presence of periodic points of all other periods, and therefore chaos.

1976	An influential paper published in *Nature* by biologist Robert May encourages mathematicians to turn from an exclusive emphasis on higher-dimensional differential equations and give more attention to the complicated dynamics in low-dimensional difference equations such as the logistic difference equation.
1970s	Benoit Mandelbrot's work on fractals and use of computers extends the work of Julia and Fatou on iterated functions, drawing worldwide attention to visual solutions of real-number problems using the complex plane.
1980s	Eminent mathematicians such as Dennis Sullivan, John Milnor, and Curtis McMullen make major advances in the field of complex dynamics, motivated by Mandelbrot's introduction of the set that bears his name.
1998	Swedish mathematician Warwick Tucker proves the existence of chaotic attractor in the Lorenz equations.

Glossary

beating modes: The type of solutions of periodically forced and undamped mass-spring systems that periodically have small oscillations followed by large oscillations.

bifurcation: A major change in the behavior of a solution of a differential equation caused by a small change in the equation itself or in the parameters that control the equation. Just tweaking the system a little bit causes a major change in what occurs. *See also* **Hopf bifurcation, pitchfork bifurcation**, and **saddle-node bifurcation**.

bifurcation diagram (bifurcation plane): A picture that contains all possible phase lines for a first-order differential equation, one for each possible value of the parameter on which the differential equation depends. The bifurcation diagram, which plots a changing parameter horizontally and the y value vertically, is similar to a parameter plane, except that a bifurcation diagram includes the dynamical behavior (the phase lines), while a parameter plane does not.

carrying capacity: In the limited population growth population model, this is the population for which any larger population will necessarily decrease, while any smaller population will necessarily increase. It is the ideal population.

center: An equilibrium point for a system of differential equations for which all nearby solutions are periodic.

characteristic equation: A polynomial equation (linear, quadratic, cubic, etc.) whose roots specify the kinds of exponential solutions (all of the form $e^{\lambda t}$) that arise for a given linear differential equation. For a second-order linear differential equation (which can be written as a 2-dimensional linear system of differential equations), the characteristic equation is a quadratic equation of the form $\lambda^2 - T\lambda + D$, where T is the trace of the matrix and D is the determinant of that matrix.

critically damped mass-spring: A mass-spring system for which the damping force is just at the point where the mass returns to its rest position without oscillation. However, any less of a damping force will allow the mass to oscillate.

damping constant: A parameter that measures the air resistance (or fluid resistance) that affects the behavior of the mass-spring system. Contrasts with the **spring constant**.

determinant: The determinant of a 2-by-2 matrix is given by $ad - bc$, where a and d are the terms on the diagonal of the matrix, while b and c are the off-diagonal terms. The determinant of a matrix A (det$A$, for short) tells us when the product of matrix A with vector Y to give zero ($A\ Y = 0$) has only one solution (namely the 0 vector, which occurs when the determinant is non-zero) or infinitely many solutions (which occurs when the determinant equals 0).

difference equation: An equation of the form $y_{n+1} = F(y_n)$. That is, given the value y_n, we determine the next value y_{n+1} by simply plugging y_n into the function F. Thus, the successive values y_n are determined by iterating the expression $F(y)$.

direction field: This is the vector field each of whose vectors is scaled down to be a given (small) length. We use the direction field instead of the vector field because the vectors in the vector field are often large and overlap each other, making the corresponding solutions difficult to visualize. Those solutions and the direction field appear within the phase plane. *See also* **vector field**.

eigenvalue: A real or complex number usually represented by λ (lambda) for which a vector Y times matrix A yields non-zero vector λY. In general, an $n \times n$ matrix will have n eigenvalues. Such values (which have also been called proper values and characteristic values) are roots of the corresponding characteristic equation. In the special case of a triangular matrix, the eigenvalues can be read directly from the diagonal, but for other matrices, eigenvalues are computed by subtracting λ from values on the diagonal, setting the determinant of that resulting matrix equal to zero, and solving that equation.

eigenvector: Given a matrix A, an eigenvector is a non-zero vector that, when multiplied by A, yields a single number λ (lambda) times that vector: $AY = \lambda Y$. The number λ is the corresponding eigenvalue. So, when λ is real, AY scales the vector Y by a factor of lambda so that AY stretches or contracts vector Y (or $-Y$ if lambda is negative) without departing from the line that contains vector Y. But λ may also be complex, and in that case, the eigenvectors may also be complex vectors.

equilibrium solution: A constant solution of a differential equation.

Euler's formula: This incredible formula provides an interesting connection between exponential and trigonometric functions, namely: The exponential of the imaginary number ($i \cdot t$) is just the sum of the trigonometric functions $\cos(t)$ and $i \sin(t)$. So $e^{(it)} = \cos(t) + i \sin(t)$.

Euler's method: This is a recursive procedure to generate an approximation of a solution of a differential equation. In the first-order case, basically this method involves stepping along small pieces of the slope field to generate a "piecewise linear" approximation to the actual solution.

existence and uniqueness theorem: This theorem says that, if the right-hand side of the differential equation is nice (basically, it is continuously differentiable in all variables), then we know we have a solution to that equation that passes through any given initial value, and, more importantly, that solution is unique. We cannot have two solutions that pass through the same point. This theorem also holds for systems of differential equations whenever the right side for all of those equations is nice.

filled Julia set: The set of all possible seeds whose orbits do not go to infinity under iteration of a complex function.

first derivative test for equilibrium points: This test uses calculus to determine whether a given equilibrium point is a sink, source, or node. Basically, the sign of the derivative of the right-hand side of the differential equation makes this specification; if it is positive, we have a source; negative, a sink; and zero, we get no information.

fixed point: A value of y_0 for which $F(y_0) = y_0$. Such points are attracting if nearby points have orbits that tend to y_0; repelling if the orbits tend away from y_0; and neutral or indifferent if y_0 is neither attracting or repelling.

general solution: A collection of solutions of a differential equation from which one can then generate a solution to any given initial condition.

Hopf bifurcation: A kind of bifurcation for which an equilibrium changes from a sink to a source (or vice versa) and, meanwhile, a periodic solution is born.

initial value problem: A differential equation with a collection of special values for the missing function such as its initial position or initial velocity.

itinerary: An infinite string consisting of digits 0 or 1 that tells how a particular orbit journeys through a pair of intervals I_0 and I_1 under iteration of a function.

Jacobian matrix: A matrix made up of the various partial derivatives of the right-hand side of a system evaluated at an equilibrium point. The corresponding linear system given by this matrix often has solutions that resemble the solutions of the nonlinear system, at least close to the equilibrium point.

limit cycle: A periodic solution of a nonlinear system of differential equations for which no nearby solutions are also periodic. Compared with linear systems, where a system that has one periodic solution will always have infinitely many additional periodic solutions, the limit cycle of a nonlinear system is isolated.

limited population growth model with harvesting: This is the same as the limited population growth model except we now assume that a portion of the population is being harvested. This rate of harvesting can be either constant or periodic in time.

Lyapunov function: A function that is non-increasing along all solutions of a system of differential equations. Therefore, the corresponding solution must move downward through the level sets of the function (i.e., the sets where the function is constant). Such a function can be used to derive the stability properties at an equilibrium without solving the underlying equation.

mass-spring system: Hang a spring on a nail and attach a mass to it. Push or pull the spring and let it go. The mass-spring system is a differential equation whose solution specifies the motion of the mass as time goes on. The differential equation depends on two parameters, the spring constant and the damping constant. A mass-spring system is also called a harmonic oscillator.

Newton's law of cooling: This is a first-order ODE that specifies how a heated object cools down over time in an environment where the ambient temperature is constant.

node: An equilibrium solution of a first-order differential equation that has the property that it is neither a sink nor a source.

orbit: In the setting of a difference equation, an orbit is the sequence of points $x_0, x_1, x_2, \ldots$ that arise by iteration of a function F starting at the seed value x_0.

orbit diagram: A picture of the fate of orbits of the critical point for each value of a given parameter.

ordinary differential equation (ODE): A differential equation that depends on the derivatives of the missing functions. If the equation depends on the partial derivatives of the missing functions, then that is a partial differential equation (PDE).

phase line: A pictorial representation of a particle moving along a line that represents the motion of a solution of an autonomous first-order differential equation as it varies in time. Like the slope field, the phase line shows whether equilibrium solutions are sinks or sources, but it does so in a simpler way that lacks information about *how quickly* solutions are increasing or decreasing. The phase line is primarily a teaching tool to prepare students to make use of phase planes and higher-dimensional phase spaces.

phase plane: A picture in the plane of a collection of solutions of a system of two first-order differential equations: $x' = F(x, y)$ and $y' = G(x, y)$. Here each solution is a parametrized curve, $(x(t), y(t))$ or $(y(t), v(t))$. The term "phase plane" is a holdover from earlier times when the state variables were referred to as phase variables.

pitchfork bifurcation: In this bifurcation, varying a parameter causes a single equilibrium to give birth to two additional equilibrium points, while the equilibrium point itself changes from a source to a sink or from a sink to a source.

predator-prey system: This is a pair of differential equations that models the population growth and decline of a pair of species, one of whom is the predator, whose population only survives if the population of the other species, the prey, is sufficiently large.

resonance: The kind of solutions of periodically forced and undamped mass-spring systems that have larger and larger amplitudes as time goes on. This occurs when the natural frequency of the system is the same as the forcing frequency.

saddle-node bifurcation: In an ODE, this is a bifurcation at which a single equilibrium point suddenly appears and then immediately breaks into two separate equilibria. In a difference equation, a fixed or periodic point undergoes the same change. A saddle-node bifurcation is also referred to as a tangent bifurcation.

saddle point: An equilibrium point that features one curve of solutions moving away from it and one other curve of solutions tending toward it.

separation of variables: This is a method for finding explicit solutions of certain first-order differential equations, namely those for which the dependent variables (y) and the independent variables (t) may be separated from each other on different sides of the equation.

separatrix: The kind of solution that begins at a saddle point of a planar system of differential equations and tends to another such point as time goes on.

shift map: The map on a sequence space that just deletes the first digit of a given sequence.

sink: An equilibrium solution of a differential equation that has the property that all nearby solutions tend toward this solution.

solution curve (or graph): A graphical representation of a solution to the differential equation. This could be a graph of a function $y(t)$ or a parametrized curve in the plane of the form $(x(t), y(t))$.

source: An equilibrium solution of a differential equation that has the property that all nearby solutions tend away from this solution.

spiral sink: An equilibrium solution of a system of differential equations for which all nearby solutions spiral in toward it.

spiral source: An equilibrium solution of a system of differential equations for which all nearby solutions spiral away from it.

spring constant: A parameter in the mass-spring system that measures how strongly the spring pulls the mass. Contrasts with the **damping constant**.

steady-state solution: A periodic solution to which all solutions of a periodically forced and damped mass-spring system tend.

Taylor series: Method of expanding a function into an infinite series of increasingly higher-order derivatives of that function. This is used, for example, when we approximate a differential equation via the technique of linearization.

trace: The sum of the diagonal terms of a matrix from upper left to lower right.

vector field: A collection of vectors in the plane (or higher dimensions) given by the right-hand side of the system of differential equations. Any solution curve for the system has tangent vectors that are given by the vector field. These tangent vectors (and, even more so, the scaled-down vectors of a corresponding direction field) are the higher-dimensional analogue of slope lines in a slope field. *See also* **direction field**.

Bibliography

Alligood, Kathleen, Tim Sauer, and James Yorke. *Chaos: An Introduction to Dynamical Systems*. New York: Springer, 1997. This is a midlevel mathematical text featuring many examples of chaotic systems.

Beddington, J. R., and R. May. "The Harvesting of Interacting Species in a Natural Ecosystem." *Scientific American* 247 (1982): 62–69. This article plunges more deeply into the bifurcations that arise in harvesting models.

Blanchard, Paul, Robert L. Devaney, and Glen R. Hall. *DE Tools for Differential Equations*. 3rd ed. Pacific Grove, CA: Brooks/Cole, 2011. Digital. This software is included with the text below and contains most of the software tools used in the lectures.

———. *Differential Equations*. 4th ed. Pacific Grove, CA: Brooks/Cole, 2011. This course is based mainly on the topics in this text.

Devaney, Robert L. *A First Course in Chaotic Dynamical Systems*. Reading, MA: Westview Press. 1992.. The emphasis of this book is on iteration of functions; it provides more details on the topics of the last 4 lectures in the course.

———. *The Mandelbrot and Julia Sets*. Emeryville, CA: Key Curriculum Press. 2000. This book, written for high school teachers and students, introduces the concepts of the Julia and Mandelbrot sets as well as the ideas surrounding iteration of complex functions.

Edelstein-Keshet, Leah. *Mathematical Models in Biology*. New York: McGraw-Hill, 1988. An exceptional book featuring numerous mathematical models that arise in all areas of biology.

Guckenheimer, John, and Philip Holmes. *Nonlinear Oscillations, Dynamical Systems, and Bifurcations of Vector Fields*. New York: Springer-Verlag, 1983. This book is a more advanced treatment of some of the topics covered in the course, with specific emphasis on nonlinear systems.

Hirsch, Morris, Stephen Smale, and Robert L. Devaney. *Differential Equations, Dynamical Systems, and an Introduction to Chaos*. 2nd ed. San Diego, CA: Elsevier. 2004. This book is a more advanced version of the Blanchard, Devaney, and Hall *Differential Equations* book cited above; it would be a good follow-up read after this course.

Kolman, Bernard, and David R. Hill. *Elementary Linear Algebra with Applications*. 9th ed. Upper Saddle River, NJ: Pearson/Prentice Hall, 2008. A slightly different take on linear algebra is included in this book, with a variety of different applications in different areas.

Lay, David C. *Linear Algebra and Its Applications*. 3rd ed. Boston: Pearson/Addison Wesley, 2006. This book is an excellent introduction to linear algebra, without any major connections to linear systems of differential equations.

Li, T.-Y., and James Yorke. "Period Three Implies Chaos." *American Mathematical Monthly* 82 (1975): 985–992. This fundamental article showed the very first portion of Sharkovsky's Theorem, which at the time of its publication was unknown in the West. More importantly, it was the first use of the word "chaos" in the scientific literature and opened up this field for many scientists and mathematicians.

Mandelbrot, Benoit. *Fractals and Chaos*. New York: Springer, 2004. Mandelbrot is the father of fractals and the discoverer of the Mandelbrot set. This book summarizes a lot of the geometric aspects covered in latter portions of this course.

Peitgen, Heinz-Otto, Hartmut Jurgens, and Dietmar Saupe. *Chaos and Fractals: New Frontiers of Science.* New York: Springer-Verlag, 1992. This is one of the first books written at an undergraduate level that exposes students to many different forms of chaos.

Roberts, Charles. *Ordinary Differential Equations: Applications, Models, and Computing.* Boca Raton, FL: Chapman and Hall, 2010. This book presents a different take on differential equations—the more analytic approach.

Schnakenberg, J. "Simple Chemical Reactions with Limit Cycle Behavior." *Journal of Theoretical Biology* 81 (1979): 389–400. This article provides more details about the oscillating chemical reactions described in this course.

Strang, Gilbert. *Linear Algebra and Its Applications.* 4th ed. Belmont, CA: Thomson, Brooks/Cole, 2006. Another classic linear algebra textbook written for undergraduates.

Strogatz, Steven. *Nonlinear Dynamics and Chaos.* Reading, MA: Addison-Wesley, 1994. This book is a more advanced treatment of the topics in this course and includes many different applications.

Winfree, A. T. "The Prehistory of the Belousov-Zhabotinsky Reaction." *Journal of Chemical Education* 61 (1984): 661. The history of the Belousov-Zhabotinsky reaction is indeed worth reading!

Notes

Notes

Notes

Notes

Notes

Notes